HANDBOOK OF THICK- AND THIN-FILM HYBRID MICROELECTRONICS

HANDBOOK OF THICK- AND THIN-FILM HYBRID MICROELECTRONICS

TAPAN K. GUPTA

WILEY-INTERSCIENCE

A JOHN WILEY & SONS, INC., PUBLICATION

Published by John Wiley & Sons, Inc., Hoboken, New Jersey.
Published simultaneously in Canada.

Library of Congress Cataloging-in-Publication Data Is Available

ISBN 0-471-27229-9

10 9 8 7 6 5 4 3 2 1

To the memory of my parents
Dr. Gopeswar Gupta (Father) and Dr. Kanak Lata Gupta (Mother)

CONTENTS

PREFACE

The study of hybrid microelectronic circuits involves both thick- and thin-film technologies, including semiconductor device technology. It covers a wide range of disciplines, including material science, ceramic technology, and electronic circuit engineering. This handbook is meant to be an introduction to the design, layout, and fabrication of hybrid microelectronic circuits for electrical engineering students as well as a reference book for practitioners in this discipline. As such, the book will be suitable for an advanced undergraduate course as well as for a graduate-level course for beginners.

The text has grown out of notes prepared for a one-semester graduate-level course in the Electrical Engineering Department of Tufts University. This course was offered every semester from the fall of 1987 to the fall of 1994. The book is dedicated to those readers with no previous knowledge of hybrid circuit design and fabrication who need a working understanding of the subject. It will also provide a broadened perspective for experienced hybrid engineers, applications in industrial training programs, as an interface for groups using or planning to use hybrid circuit technologies, and as a resource for a basic understanding of the subject. It may also serve as a reference book in hybrid circuit technology.

The author is aware of the rapid advances being made in hybrid microelectronic engineering. New designs and materials are under continual development. As a result, the increasing device count is accompanied by a shrinking minimum size. It is hoped that by concentrating on fundamental processes occurring in hybrid technology and discussing contemporary devices as specific examples of these processes, this book will meet present and future requirements on this subject.

The brief introduction at the beginning of each chapter and short summary at the end of the chapter are intended to assist readers in determining the contents of each chapter and/or to describe the main topics covered in the chapter. Moreover, each chapter contains references for further reading.

Chapter 1 provides an overview of hybrid technologies, with a short history of the development of this emerging technology and includes several typical examples of applications of hybrid microcircuit technology. In Chapter 2, the mathematical equations used most often in hybrid circuit technology are given together with design and layout rules for hybrid circuits.

In Chapter 3, computer-aided industrial design is discussed, including artwork, pattern generation techniques, selective etching, and photo processing. This chapter will prepare readers to understand pattern generation and artwork transfer using lithographic tools, resists, and micrometer-level etching methods.

Chapter 4 deals with the fundamental properties of the materials used in thick-film hybrid circuits, and Chapter 5 deals with deposition techniques used in the fabrication of thick films. In Chapter 4, the physical, chemical, and engineering aspects of the materials are discussed.

Chapter 6 deals with the material aspects of thin film, including their properties and uses in the fabrication of passive components. Thin-film technology is another fast-growing branch of hybrid technology. Chapter 7 describes various deposition techniques for thin-film circuits.

In Chapter 8, fundamental technologies involved in assembling various discrete circuit elements are described. Fundamentals and the applicability of various assembly systems together with interconnecting systems are presented in this chapter.

Chapter 9 covers the technologies behind fine adjustments of the passive elements used in hybrid circuits. Figures, tables, and calculations are provided for a better understanding of the technology.

Chapter 10 presents packaging. Materials and the thermal and electrical properties of various packaging units, such as DIP, pin grid array, and some other common surface-mounting chip carriers, are described. Recent developments in silicon-on-silicon packaging and polyimide/ceramic multichip carriers are included.

Chapter 11 deals first with some fundamentals of multichip modules then with some fundamental materials for high-frequency hybrid circuits, microwave engineering, and applications of microwave devices. It is not possible to cover all of high-frequency hybrid technology in a single book. However, as a reference and to compare circuit elements in the low- and high-frequency ranges, this chapter will be beneficial to users of hybrid circuits.

ACKNOWLEDGMENTS

A book of such diversity would not have been possible without the indirect and direct assistance of many other research workers, engineers, and teachers. Virtually all of the information presented is based on my lectures to the graduate students at Tufts University. The notes were prepared from the research work of countless scientists and engineers engaged in this field. Their contributions are recognized to a small degree by citing some of their work in the references at the ends of chapters. I also wish to acknowledge the work of people who are not cited directly but who have contributed indirectly to the development of this book. My sincerest thanks go to them.

The entire book was reviewed by Dr. Harry K. Charles, Jr., assistant department head for the Engineering Technical Services Department, Johns Hopkins University, and past president of the International Society for Hybrid Microelectronics (ISHM). The author is very grateful for his help.

I also want to thank the members of ISHM for allowing me to borrow research material from the ISHM journal. I have taken many illustrations from a variety of magazines and journals. I want to express my sincerest thanks to all staff members and authors for their permission to reproduce these.

I am thankful to Frederick C. Nelson, dean, and Edward J. Maskalenko, associate dean, School of Engineering, and chairman of the Electrical Engineering Department, and the rest of the faculty of Tufts University for their encouragement and support. I am also grateful to the staff members of the Electrical Engineering Department of Tufts University who helped me with computer and technical services, and to the secretaries for their tireless typing. The comments and criticisms of anonymous reviewers have helped in shaping the book. The encouragement received from George Telecki, Associate Publisher, and Andrew Smith of Wiley-Interscience is very much appreciated.

Most of all, I want to thank my wife, Arundhati, and my daughter, Atreyee, for the love, understanding, patience, and impatience that made preparation of this book possible.

TAPAN K. GUPTA

HANDBOOK OF THICK- AND THIN-FILM HYBRID MICROELECTRONICS

1

INTRODUCTION

Hybrid is a generic term that defines the offspring of genetically dissimilar parents or stock. In a similar way, *hybrid microelectronics* can be defined as the packaging and interconnection of dissimilar microelectronics devices from thick- and thin-film technologies, combined with semiconductor device technology and discrete devices. The packaging design is intermediate between conventional design, in which discrete components are mounted on a circuit board, and monolithic integrated circuit design in a single unit or package. The interconnections may include deposited resistors, capacitors, and inductors, originating from thick- or thin-film technology or both, or these passive components may be mounted in chip form on the surface of an insulated substrate. The principal benefit of hybrid microelectronics is the ability to choose and mix different technologies to best meet the requirements of the twenty-first century. The microelectronic industry is divided primarily into two classes of technologies: monolithic integrated circuit (IC) technology and the hybrid microcircuit technology. These two technologies differ primarily in the degree of integration and sophistication involved in producing products that might appear in seemingly identical final packages. However, in combination, hybrid microcircuit technology represents numerous advanced packaging concepts that might encompass the precision integration of ICs with various discrete components on suitable substrates. The size of the original transistor is comparable to that of the hybrid microcircuit: about $4.5 \times 5.0\ \text{cm}^2$. In truth, hybrid microcircuit technology is a bridge between a monolithic IC and a printed circuit board product. Moreover, hybrid microcircuit system modules are used in

Handbook of Thick- and Thin-Film Hybrid Microelectronics, By Tapan K. Gupta
ISBN 0-471-27229-9

all facets of today's electronics industry, starting with their initial application in the computer and defense industries. As the state of the art grew, composition suppliers broadened their lines, developing new high-performance materials to meet challenging and sophisticated circuitry. In many applications a hybrid circuit is used because there is no alternative technology, especially when the application is in the high-frequency microwave region.

Advanced hybrids generally comprise more ICs, with discrete devices mounted on an insulated substrate together with screen-printed thick-film and/or vacuum-deposited thin-film components. The hybrid segment of the microelectronic industry is in transition as new technology [e.g., the multichip module (MCM)] has emerged in the market. One definition of MCM is that it is a combination of two or more semiconductor devices, possibly with passive elements, on a common substrate to produce an electronic circuit function. The new technology allows for rapid, inexpensive development of MCM with small to medium-sized volumes.

In 1964, Fairchild's Gordon Moore estimated that the density of microcircuit devices would double each year, and necessitating a high level of miniaturization in the microelectronics industry. In 1975, Intel assembled a complete computer on a single board with the help of a hybrid structure, combining monolithic devices with thick- and thin-film circuitry. Thus a new concept in hybrid microelectronic circuitry was born and industries found a use for hybrid technology as an alternative microelectronics module. If there is indeed to be a new revolution in society and/or the economy, hybrid microelectronics has already established footholds in all sectors, and it is infiltrating, although quietly as yet, the entire fabric of society.

1.1 HYBRID MICROCIRCUIT FAMILY

The elements of the hybrid microcircuit family are shown in Figure 1.1.

1.1.1 Printed Circuit Board

A printed circuit board (PCB) [1] is a predetermined printed electrical circuit produced by printing or forming an electrical conducting pattern on a base

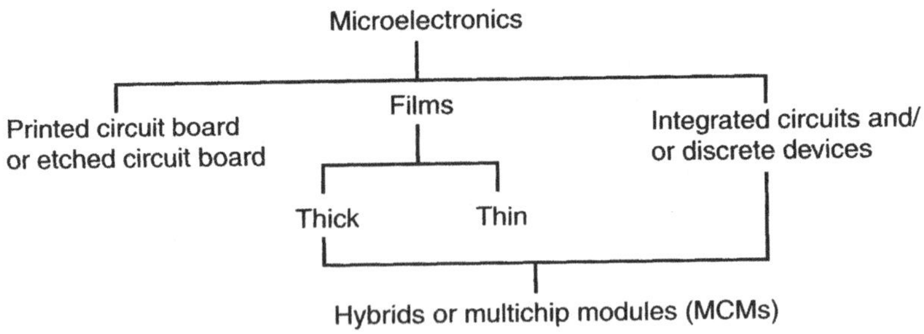

Figure 1.1 Hybrid microcircuit family.

substrate. The plastic copper-laminated substrate may be rigid or flexible and coated with an insulating material whose surface will be printed on. Discrete components such as resistors, capacitors, inductors, or printed conducting lines may be part of a PCB. Unlike common electrical circuits that employ wires from point to point, the printed circuit is a two-dimensional array, usually produced as a film on a flat surface [2].

1.1.2 Thick Film

The term *thick film* [3] is derived from the fact that the fired film is fairly thick, varying from 0.2 mil (5.08 μm) to 2 mils (50.8 μm). The film is deposited through a screen (either silk or stainless steel) by selective deposition of a paste or slurry, which is often called *ink*. The substrate material is generally a ceramic or refractory material.

1.1.3 Thin Film

Thin and thick films are defined not only by their thickness but also by the way they are deposited. Thin film [4] is deposited on an insulating substrate generally by vacuum deposition, and finer structures are made either by an additive method through masks or by a subtractive method by means of photolithography. The thickness of a thin film can vary from perhaps 50 angstroms to several thousands of angstroms ($1\ \text{Å} = 10^{-10}$ m $= 4.0 \times 10^{-6}$ mil).

1.1.4 Integrated Circuit

An IC is also called a *monolithic IC* (*monolithic* is derived from the Greek *monos*, meaning single, and *lithos*, meaning stone) [5]. It consists of a single chip, generally of silicon [gallium arsenide (GaAs) is also used], typically 50 by 50 mils (1.27 by 1.27 mm) containing both active elements (transistors, diodes, etc.) and passive elements (resistors and capacitors) and their interconnections. Thus a monolithic IC is built on a single crystal (single stone) substrate [6]. Almost all the circuit elements are fabricated within the substrate by such processes as epitaxial growth, masked impurity diffusion, oxide growth, and oxide etching, using photolithography for pattern definition. Finally, internal contacts are made with an alloy of aluminum with 1 to 2% silicon and 2 to 4% copper [7–9]. However, two much Cu (more than 2%) creates problems in a dry etching system. The overall dimension of a typical transistor in an IC is 140 by 95 μm [10]. The feature size is shrinking at an appropriate annual rate of 11% and is thus expected to reach a minimum feature size of 10^2 nm (0.1 nm = 0.004 mil) by the year 2006 [11]. In fact, a device with a feature size as small as 0.005 mil (0.13 μm) has been reported [12]. As scientists have reduced feature size (from 1 μm in 1990 to 0.13 μm in 2002), faster transistors, and packing them together with the existing conductor (aluminum alloy), have become a problem. Aluminum alloy (Al, Cu, and Si) has long been the conductor of choice, but pushing electrons through smaller dimensions with aluminum alloy lines just is not fast enough. Thus copper as the conducting material and low-*K* dielectric materials (organic and inorganic) for faster devices and reduced *RC* delays have been introduced and are in production. When the feature size is small and the trenches and vias have high aspect ratios, copper damascene technology has proved to be the answer. A physical vapor deposition seed layer with electroplated copper with spin-on or vacuum-deposited low-*K* dielectrics (between 1 and 3, compared to about 4 for SiO_2) has gained support rapidly. However, the basic unwritten rule for low-*K* materials is that for a given application, they should integrate no differently than silicon oxides (i.e., they should show similar or better chemical and mechanical integrity). Dry etching of copper is difficult, and the metal has a high diffusion rate through silicon and its oxides. These two problems have been addressed very efficiently by chemical mechanical planarization and providing an additional barrier layer of the nitrides of transition metals between copper and silicon dioxide on silicon substrate. As the feature size is shrinking and new materials for conductor and dielectrics are introduced, deep ultraviolet (DUV) photolithography seems to be mandatory. Since DUV resist materials are transparent for low-*K* materials, an additional antireflection coating has become necessary. In the field of thin-film deposition, atomic layer chemical vapor deposition technology is competing for atomic-scale (for 100-nm devices) metal deposition, intermetal layer deposition, and multilayer circuits. The advancement in very large scale integrated (VLSI) semiconductor technology is largely attributed to advances in materials, in particular advances in polymeric materials [13,14]. Recently, all-polymer ICs have been fabricated at Philips Research Laboratories in Eindhoven, The Netherlands, on flexible substrates that can sustain sharp bends while maintaining functionality [15].

1.1.5 Modules

Modules are defined as separable units, components, or circuits in which each unit has identical dimensions, characteristics, or forms. Modules are standard units or building blocks that can be assembled quickly or economically. The module concept is the result of a research program in the 1950s. The project, known as *Tinkertoy*, offered the idea of interchangeability of parts that could be obtained and mass-produced quickly by changing from one type of module to another [16]

1.2 NEED FOR HYBRID MICROCIRCUITS

Hybrid microcircuit technology has a long history, dating back to the 1940s. At that time microelectronics circuits consisted of resistors and electrodes printed on titanate ceramic substrates. That was the state of the art until 1947, when a small group of researchers and technical managers invented the transistor at Bell Labs (Figure 1.2). Figure 1.2(*a*) shows the newly born transistor at Bell

(*a*)

(*b*)

Figure 1.2 (*a*) The first transistor, produced at Bell Labs, December 1947 (courtesy of Andrew Wylie); (*b*) a typical thick- and thin-film hybrid circuit (courtesy of SST International, Downey, California). The two devices are comparable in size.

laboratories in December 1947, and Figure 1.2(*b*) is a high-density advanced hybrid microcircuit packaging unit, a product of the twenty-first century. With this invention, the need for miniaturization of semiconductor devices became the major issue in industry and in research laboratories. In the late 1950s and early 1960s, Radio Corporation of America (RCA) came out with some new ideas for miniaturization when they were working on a micromodule program funded by the Department of Defense (DoD). RCA developed small ceramic substrates for microcircuit fabrication with excellent structural strength and low thermal and electrical conduction. This DoD program provided the momentum for a new phase of industrial development. A coordinated attack by industry and research organizations gave birth to the IC [17]. Thus a completely workable microelectronics system evolved, and a new technology called *ceramic integrated circuits* emerged on the market. This new ceramic technology was soon joined with silicon technology to give birth to what is now termed *hybrid microcircuit technology* [18,19].

The needs of the DoD have long been a driving force in the evolution of sophisticated hybrid microelectronic systems, which package more functions into smaller volumes at lower cost. The DoD recognized very early the need for hybrid microelectronics and their significance for defense systems. Substantial sums thus became available to industry, and during the decade of the 1960s this hybrid technology matured under the impetus of DoD programs. Its explosive growth continued during the 1990s at an even faster pace. As hybrid microcircuits became increasingly powerful, much of the innovation was stimulated by consumer electronics, such as calculators, watches, television sets, and automotive applications such as voltage regulators [20]. As a matter of fact, the history of hybrids has been fast moving, dynamic, frenetic, exciting, at times frustrating, and fun.

1.2.1 Multilayer Circuits

Higher system performance, lower-cost material systems, and cost-effective manufacturing systems represent some of the major factors shaping advancements in microelectronic packaging [21]. However, the driving force is the need for increased circuit density. Multilayer technology [22] allows one to increase circuit density in the most cost-effective way while keeping the size and weight of the module as low as possible. Thick- and thin-film ceramic hybrid technology is the only technology, that can address this. Figure 1.3 shows a three-conductor-layer thick-film hybrid circuit that is used for sensing circuitry in a prototype biomedical system. Because of the low firing temperatures and low cost, many microcircuit manufacturers favor, thick-film processes over thin-film processes [23]. Moreover, thick-film technology is reliable and flexible, and the availability of compatible high-performance, low-cost resistors accounts for heavy industry penetration [24]. The process has become cost-effective because there are multiple layers in one module. This type of circuitry is possible only in a hybrid circuit system.

1.2.2 Military Applications

In the military market segment, new programs require the use of very high speed ICs (VHSICs) and microwave/millimeter ICs (MMICs) [25,26]. VHSICs

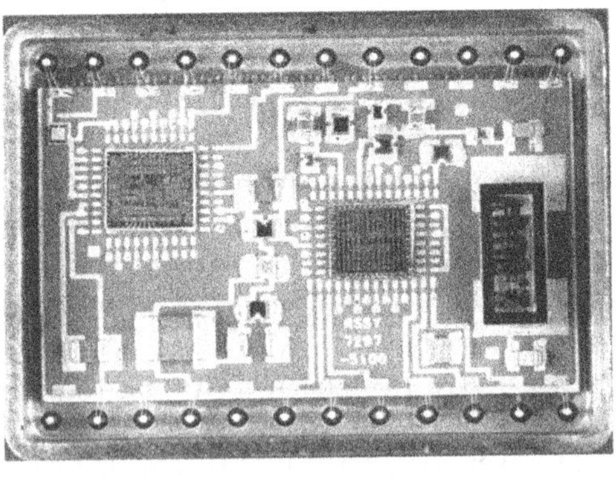

Figure 1.3 Multilayer thick-film hybrid circuit [20].

are complex in nature and require small leads and compact packaging. A unit may have 300 to 600 leads, and such high-density packaging requires a good heat dissipation capacity. A multiple-layer hybrid circuit approach, coupled with fine lines and spaces and small vias within a single package, has proven to be very successful. This multilayer design has cut down on the size and manufacturing cost of a circuit. This is possible because designers of hybrid circuits have been able to incorporate VHSICs into multilayer hybrid microcircuits [22]. Ceramic substrate is generally used in hybrid circuits, as it offers higher thermal conductivity than that of silicon or gallium arsenide [27]. In some cases, organic film-forming materials are used in a hybrid circuit to provide good high-temperature characteristics and insulation between the substrate and the circuit elements. This is accompanied by a thin-film copper conductor, which provides a low-temperature environment for deposition on the ceramic substrate. The undesirable inductance levels at high frequencies due to the skin effect are avoided by rectangular leads provided by tape-automated bonding (TAB). Furthermore, the preassembly electrical test capability offered by TAB allows all devices to be tested electrically prior to package assembly [28]. This is possible only with hybrid microcircuit technology.

1.2.3 Data Processing

The predominant trend in the data processing market segment is increased processing speed [29]. To cope with this, the data processing engineer has introduced a different version of hybrid microelectronic circuitry in a double-sided hybrid circuit board (HCB). For the ICs to talk to one another, the signals must travel through long distances along the board, and crosstalk and interference are common events [30]. To eliminate these problems, ICs with more functions are used and combined with other ICs mounted directly to a multilayer ceramic substrate (generally, ceramics have high mechanical strength, excellent thermal properties, and low cost compared to silicon). Recently, the MCM concept [31] is introduced in the hybrid family to eliminate the need for traditional first- and second-level packaging. The basic implications of such a scheme are higher density through an increased number of functions with a shorter connection length. This has reduced input/output (I/O) spacing, incorporated higher speeds, improved thermal performance, and lowered the cost.

1.2.4 Telecommunications

For microelectronics, trends in the telecommunication market segment [32,33] are similar to those of data processing since the two industries are beginning to cross and merge. The circuits are characterized by the need for high density with fine lines and spaces, faster switching speeds, reduced size, and high-conductivity interconnection. The Integrated Service Digital Network (ISDN) is an example of where the telecommunication industry is headed in the near

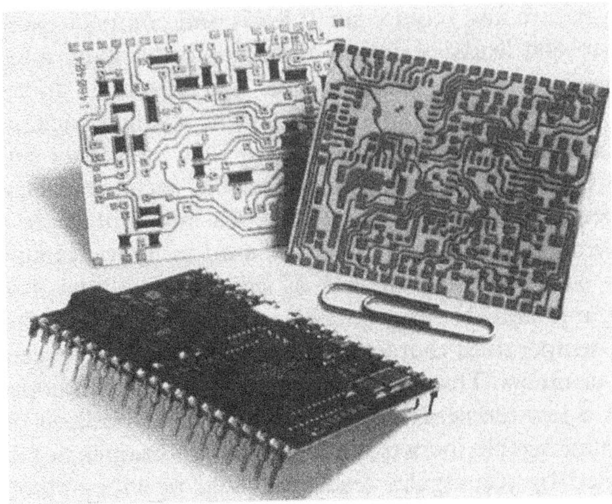

Figure 1.4 GTD-5 digital switch used in a line hybrid telecommunication system [35]. (Courtesy of DuPont, Wilmington, Delaware.)

future [34]. ISDN is described as the ultimate convergence of voice, data, facsimile, and video and is expected to have the fastest growth in the market segment. It is expected that ISDN will cover a broad range of field in equipment and device segments (e.g., intelligent equipment, optoelectronic devices, digital loop carrier gear, and special service equipment).

Thick-film hybrid technology has a long history of meeting demanding, yet cost-sensitive applications in the telecommunications segment. A recent example is the line hybrid manufactured by AG Communications System (AGCS) of Genoa, Illinois [35]. Figure 1.4 shows the GTD-5 digital switch designed and manufactured by AGCS. With the growth in the volume of voice and data services, including emergency services and the Internet, system reliability is critical. Although the public-switched telephone network in the United States is regarded as among the best in the world, it still experiences service interruptions. These interruptions, however infrequent, are well documented in the press and serve as unpleasant reminders of our increasing dependence on the public telephone network. However, during 1998–1999, the GTD switch designed and manufactured by AGCS, with more than 17 million line hybrids, was installed in the United States. The system is the most reliable in the competitive and fast-growing telecommunication market.

1.2.5 Automotive Industry

Another important market segment for hybrid microelectronics is the automotive industry. It may become the largest growth area for hybrid electronics in the next decade [36]. It is expected that by the year 2010, this industry will grow fourfold. Figure 1.5 shows an automotive pressure sensor that monitors atmo-

Figure 1.5 Automotive pressure sensor fabricated following thick-film hybrid circuit technology [20].

spheric and engine manifold pressure. The device can operate nicely in a harsh environment. Some other electronic devices in which hybrid microelectronics is an essential part of devices used in automobiles are antilock breaking systems, direct fire ignition, electronic engine management, and multiplex linking of electronic components.

All the electronic circuits used in the automotive devices cited above come from hybrid microelectronics. There is no alternative to the hybrid circuit when double-sided substrates with holes or crossovers are needed. It is expected that the automobile industry, also, will need multilayer circuits to maximize space and cost. There is no alternative technology to meet the demands of the fast-growing automobile industry other than hybrid microelectronics technology [37].

1.2.6 Medical Science

Advances in medical care devices have generated an important and growing market for hybrid microelectronics [38]. The need for high standards of reliability and the importance of an encapsulating package of high integrity both demonstrate a very happy fit between the requirements imposed by the application and the features that hybrid technology has to offer [39]. Whether it is a cardiac pacemaker to provide electric stimuli to the heart muscle or implantation of a cardiac defibrillator to stop the chaotic activity of the muscle forming the wall of the two major chambers of the heart, or myoelectric control of a hand prosthesis, highly reliable compact packaging of multicomponent assemblies of the hybrid microelectronic circuitry has no alternative. Figure 1.6 shows a pacemaker fabricated on ceramic tape 175 μm (ca. 7 mil) thick that has six conductor layers and approximately 400 filled vias [20]. As a matter of fact, the electronic complexity, which is needed to analyze the electrocardiogram reliably enough to detect arrhythmia and initiate defibrillation, is considerable and could not be considered for an implant without full use being made of the latest hybrid microelectronics. At the same time, the very advanced signal processing technology [40]—the magical wand that commands control of prehension of an artificial arm, rotation of the wrist, and rate-variable elbow flex-

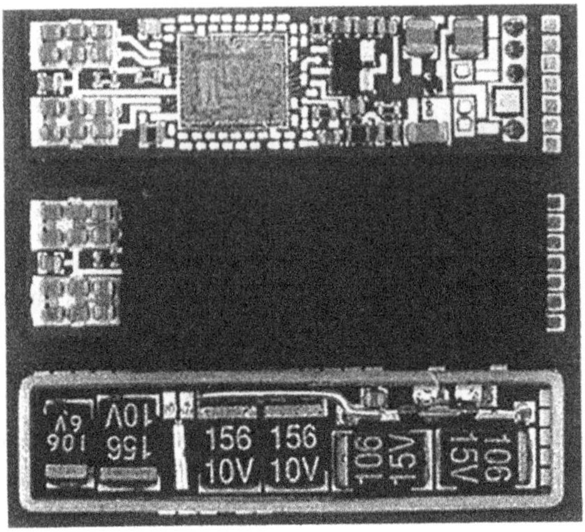

Figure 1.6 Heart pacemaker circuit fabricated on a cofired substrate [20].

ion from electromyograms of the biceps and triceps—is realized by thick-film hybrid circuit technology.

1.2.7 Aerospace Systems

An aircraft engine control system requires a full-wave rectifier designed to operate at 15 V with a 20-A current, together with a high-speed switching transistor with drivers and a dc-to-dc converter [30]. The power dissipation factor of the system is 5.5 W. Hybridization of aircraft engine control systems has been implemented successfully through integral substrate packaging (ISP), which requires very high conductivity, low inductance (beam leads are part of the design), and an electrically isolated base [41]. The use of ISP devices has resulted in a surface-area reduction of 6:1 and an overall volume savings of 12:1. A small unit with such a high power-handling capacity makes possible the development of a hybrid microcircuit that uses ceramic substrate. A 50-W triple-output dc-to-dc converter fabricated by Delco is shown in Figure 1.7.

1.2.8 High-Frequency Circuits [42]

Invariably, as operating frequency is increased, inductors need to be included in circuits. In PCB technology, inductors were included as a separate discrete component, resulting in a bulky, space-consuming circuit. However, monolithic integrated circuits cannot accommodate an inductor as a circuit element. Flat spiral inductors, both circular and rectangular [43,44], can be printed with thick-film materials on the alumina substrate of a hybrid microcircuit. The process of printing an inductor in a hybrid circuit together with other screen-

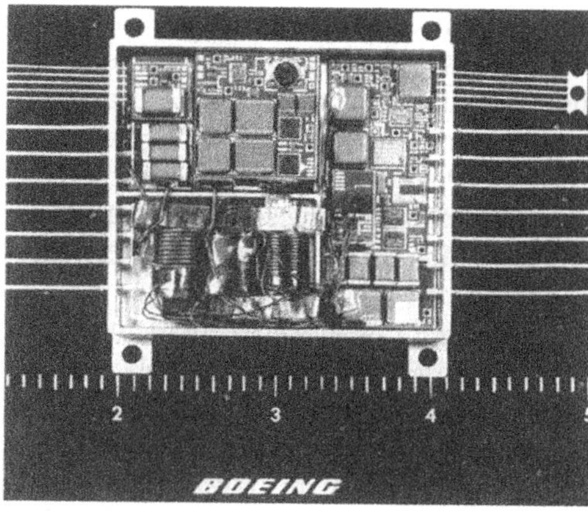

Figure 1.7 Thick-film 50-W triple-output dc-to-dc converter [20].

printable components is cost-effective and the space occupied is small compared that required for an external inductor, which was generally used in earlier PCB circuits. Thus a hybrid circuit consumes little space and the circuit is not unnecessarily bulky. Only hybrid microelectronics circuit technology can offer such a small, compact, and sophisticated circuit with the packaging of functions into a smaller volume. Hybrid microcircuitry is also unique for use in space programs.

As digital processing frequencies increase, more onboard filtering will be needed to keep line noise below the logic noise margin and within Federal Communications Commission emission limits. For example, Apple Computer is using thin-film resistor/capacitor networks in Quarter Size Outline Packages (QSOPs) as filter terminators on bus lines [45]. Similarly, unique packaging of a 1-megabit high-speed static stacked RAM module is achievable with the assistance of a multilayer hybrid RAM module that is impossible to fabricate with discrete chip carriers [46].

1.3 WHY HYBRID MICROCIRCUITS?

In many applications, hybrid circuitry is an essential part of a microelectronic module because there is no alternative technology that can fulfill all the requirements of modern sophisticated electronic circuitry [20]. This is particularly true in high-frequency and microwave circuits, where small size, line definition, reproducibility, and cost are the major criteria of the program. At lower frequencies hybrid resistor networks can be used very efficiently and in a cost-effective way compared to their discrete counterparts. In recent years, increasing limitations on space and weight in military and aerospace systems have increased demands for circuits that will achieve a higher degree of integration and maintain overall reliability and performance. Development of complex monolithic integrated circuits by increasing component density did not solve this problem because simultaneously, systems became more sophisticated. Today's high-performance military equipment is literally packed with electronic systems designed to provide the most effective vehicle utilization. These systems require new packaging techniques aimed at a higher density of circuit elements and high power dissipation. Hybridization has become a major factor in the acceleration of this trend [47].

Power hybrids [33] are more cost-effective than monolithic hybrids in low-production-volume situations. For high-power applications, power hybrids are the only solution because monolithic circuits are inadequate. Power hybrid circuits are finding more use in power amplifiers, high-fidelity, and motor control, as power switches for inverters, and in automotive ignition, regulators, and solenoid control circuits [20]. The power transistor for these circuits requires low-value (0.5 to 2.0 Ω) resistors, depending primarily on current level. Screen-printed thick-film resistors on alumina substrate are able to handle more power and are able to dissipate more heat than are their counterpart ICs. With laser trimming, resistor tolerances can be controlled within $\pm 1\%$. Power transistors in all these hybrid circuits are mounted on a copper pedestal, which spreads heat flux substantially [48].

Special-function hybrids are finding a strong niche market in medical implant devices [49]. These include pacemakers, defibrillators, and ear and other implantation. The complexity of these implants has increased and the demand for highly reliable small and compact packaging systems has increased. It goes without saying that hybrid microelectronics circuitry has no alternative in this field. Artificial arms that provide simultaneous control of prehension, wrist rotation, and rate-variable elbow flexion from electromyograms of the biceps and triceps are the result of research and development on special thick-film hybrid microcircuitry [39,50].

The increasing need for equipment providing interlocking and signaling systems under varying environmental conditions has led to more complicated devices and circuitry, especially those used in transport machines [51]. This is due to the large temperature ranges outdoors, with simultaneous day–night temperature changes, and relative humidity as high as 100%. In cars and other road vehicles, electronic equipment may be exposed to rapid temperature changes of 5°C/min. On trains, electronic equipment is exposed to vibrations that at low frequencies of 1 to 5 Hz reach a mean value of acceleration as high as $3g$. In harsh environments, hybrid microcircuits with preselected special elements have proved very useful for interlocking and signaling devices [52].

Another important area in which hybrid microcircuits demonstrate superiority over ICs and PCBs is in their capability to accommodate more circuits within a small area with good heat dissipation capability. Multilayer circuits reduce the space required for wiring, reduce weight, and allow more contacts and components within a specific area and space. For example, engineers can design a 12-layer board 7.87 by 7.87 in. (20 by 20 cm) which can accommodate more than 100,000 interconnections. Since techniques of manufacturing multilayer circuits and selection of materials are more carefully controlled, the reliability and quality of these hybrid microcircuits are great. Neither the monolithic IC chip nor the PCB is able to handle such multilayer circuits in a single module; only hybrid technology can do that [53,54].

1.4 APPLICATIONS OF HYBRID MICROCIRCUITS

Hybrid microcircuitry is an integral part of many hybrid microcircuit devices: portable wireless devices (e.g., cellular phones, pagers, PDAs, WLANs, and PC interfaces) [55–57]; a credit-card-size camera (Panasonic Cool Shot II), measuring 3.54 × 2.34 by 1.23 in. (90 by 60 by 31.2 mm^3) and weighing 142 g (5 oz.); a radio-frequency power amplifier for cellular phones (Miyoshi Electronics), chip-scale packaging (Hydelco, Inc.) under a harsh environment (85% relative humidity at 85°C for 100 h and −40 to 100°C air-to-air thermal shock testing for 1000 cycles) [35]; a power integrated module and an intelligent power module (Siemens); multilayer ceramic circuits and a ROM-AID circuit for automotive applications [58]; a hybrid power package for the military; and a high-frequency microcircuit (500 MHz to 20 GHz), thermal print head, D/A and A/D converters, sensors, and implanted cardiac defibrillators [59,60] for use in the medical industry. To quote Whitesides [61], "Almost everything from shoes to windows to children's toys will have microelectronics in a few years." Indeed, hybrid microelectronics technology is the gateway to the twenty-first century.

1.4.1 Automotive Industry

Automobile Engines The use of cost-effective ways to fabricate thick-film circuitry has been demonstrated successfully in the microcontroller of an electronic automobile engine [62,63]. The hybrid circuit contains six dies, an 8-bit microcontroller, a flash EPROM, an ASIC, a dual-quad-input NOR, a latch, and a quad-dual-input NAND [64]. Included in the hybrid are three capacitors, two resistors, and an oscillator crystal. The dimensions of the substrate are 1.6 by 1.6 in. (4.1 by 4.1 cm), and it contains three metal layers on the top and a ground plane on the back. The conductor layers are printed with DuPont 6160 silver metallization and the dielectric paste is DuPont's QM 42. DuPont's QM 32 silver via fill material is also used. The pattern has 9.75-mil (0.25 mm or 250 μm) lines and spaces and 14.62-mil (0.375 mm or 375 μm) vias. The two 10-kΩ resistors are printed on the top layer with DuPont 1739 ink. The passive components are attached by solder reflow of 62% Sn/36% Ag. solder paste, and the semiconductor dies are attached with Ablebond 84-ILMI silver epoxy.

The second hybrid microcircuit application in the automotive industry is a ROM-AID circuit used for engine controller soft wire development and in vehicle calibration [58]. The module contains a 16-bit microcontroller on the top, two 128k × 8 RAM dies, two chip resistors, and a transistor mounted on the back. The completed hybrid has the same footprint as that of the single-chip microcontroller: a plastic quad flatpack (132-lead fine pitch 0.627 mm = 24.7 mils) [65]. A single DuPont 6160 silver conductor layer with 4.87-mil (0.122-mm or 122-μm) lines and spaces and an encapsulated layer of DuPont QM 42 dielectric are printed on the top. The back contains two metal layers—a dielectric layer and an encapsulation layer.

High-Voltage Regulators for Automobiles A high-voltage regulator is required in automobiles to meet all environmental requirements [e.g., vibration (0 to 2000 Hz, $<20g$); mechanical shock ($50g$); thermal shock (-35 to 125°C)], and it has to dissipate heat because the circuit requires a high-power transistor and low-frequency ICs. Special metal insulated substrates are used and are fabricated using hybrid microcircuit technology [53]. Experimental evidence shows that a hybrid microcircuit is the only choice for such a harsh environment. A thick-film hybrid voltage regulator utilizes high-current power elements, monolithic chips, and thick-film networks using conductive, resistive, and thermistor compositions. The regulator is constructed in two independent and separate sections, input and output, so that each section can be fabricated in the most convenient and compatible manner [66].

1.4.2 Commercial Products

Laser Power Detectors In recent years, laser trimming [67], soldering [68], engraving, and micromachining [69] have been used widely in microelectronics. The power of the laser beam is a crucial parameter in these technologies. It has become necessary to provide a detector to measure different parameters of the equipment used. Various laser power detectors have already been developed, but most detectors are expensive or inaccurate [70]. Conventional laser power detectors for low-power measurements are made of semiconductors and pyroelectric thin-film technologies. These detectors are of high reliability and accuracy; however, they do not have the wide dynamic ranges required to measure the magnitude of lasers used in michromachining, due to their easily damaged surfaces [71]. Detectors made with thick-film technology are relatively less expensive and can be used for high power levels. These detectors are highly sensitive to measuring conditions. Several different types of detectors can be produced with thick-film technology and are suitable to measure laser power [72]. Thermopile- and thermister-based detectors [73] are both from the thick-film hybrid family and both show linear output characteristics with no or little hysteresis.

High-Frequency Amplifiers [74] Superior performance and smaller package size have been achieved through hybrid technology in a microwave amplifier that are not possible through monolithic silicon technology. The amplifier provides noise figures of less than 2.6 dB and gain above 29 dB over the 4- to 8-

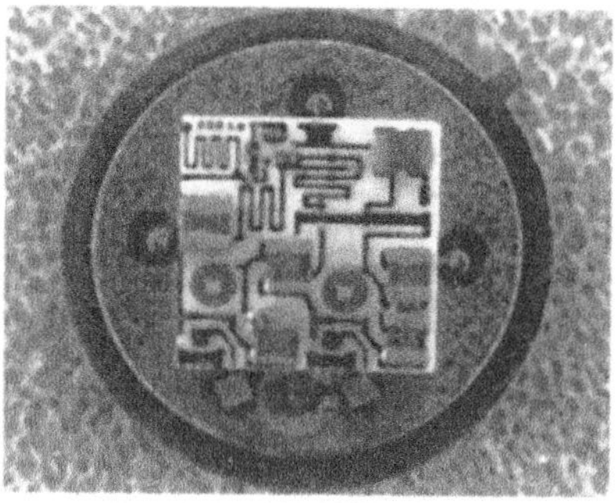

Figure 1.8 High-frequency (10 to 200 MHz) cascadable hybrid amplifier on ceramic substrate, showing the chip capacitors and inductors [74]. (Courtesy of WJ Communications, San Jose, California.)

GHz band [75,76]. The microwave circuit consists of three balanced amplifiers utilizing chip and wire techniques to interconnect the GaAs FET die, ceramic chip capacitors, and thin-film alumina-based microwave integrated circuitry (MIC) [77]. This hybrid circuit enjoys quick design turnaround and higher yields than those for monolithic microwave integrated circuits (MMICs) [25]. Advanced technologies in materials, interconnections, substrates, and packaging continue to keep the hybrid circuit industry building its base. There are many applications, environments, and functions for which hybrid circuit performance cannot be surpassed. The hybrid circuit world will increasingly become one of strong niche markets and applications [78]. Figure 1.8 shows a high-frequency thin-film hybrid circuit built on a ceramic substrate. The passive components are the chips and deposited thin films. Additive and subtractive processes following vacuum evaporation and photolithography produce fine structures for the inductors.

Isolation Amplifiers The isolation amplifier [79], which passes signals accurately across ground systems separated by thousands of volts, has been commercially achievable through a hybrid circuit system. It is unlikely that circuits of this type can be fabricated only with monolithic IC chip technology [80,81]. The function of the isolation amplifier depends on a means of withstanding very common mode voltages that is easily manufactured in hybrid form. It isolates analog signals accurately from extreme values of common-mode voltage or noise. Ideally, the input stage is galvanically separated from the output stage. The physical means of separating the stages is called the *barrier*. The potential difference across the barrier can be as high as 3500 V rms at 60 Hz, and the amplifier should still appear transparent to the signal [82]. Figure 1.9 shows an isolation amplifier, ISO 106 from Burr-Brown, which has a 70-kHz bandwidth and is another example of hybrid circuit technology used to fabricate the device.

1.4.3 Medical Science

Implanted Cardiac Defibrillators There are approximately 1.5 million heart attacks each year in the United States alone, resulting in about 600,000 deaths

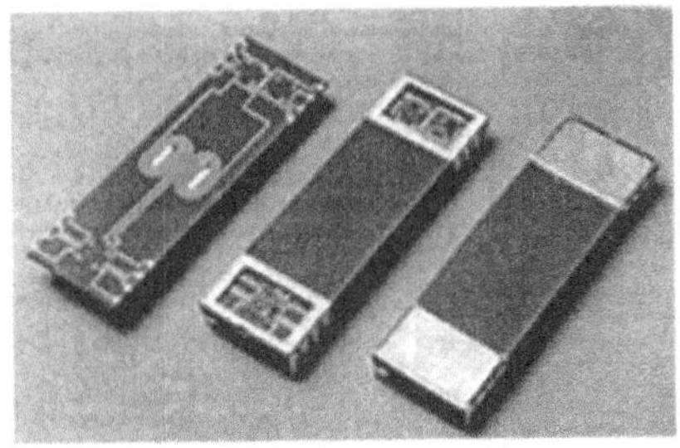

Figure 1.9 Burr-Brown (now Texas Instruments) isolation amplifier [79]. (Courtesy of Texas Instruments, Dallas, Texas.)

(three per 1000 population per year). Many of these attacks strike virtually without warning, so therapy has only a short time available between the cessation of circulation and the onset of irreversible brain damage, leading finally to death. Although there may be underlying (although perhaps undiagnosed) pathology that participates an attack, ventricular fibrillation (an uncoordinated, chaotic activity of the muscle forming the wall of the two major chambers of the heart) is a prominent feature of sudden cardiac death. An expedient that gives the shortest possible time to defibrillation [83] is the implanted automatic defibrillator, which monitors the electrogram continuously, detects dangerous disorder rhythms, and delivers a low-energy countershock within 20 seconds or so [84].

It is evident how the complexity of these electronic implants has increased and how compact packaging and the high reliability of multicomponent assemblies of hybrid microelectronic circuitry are being utilized. The electronic complexity needed to analyze an electrocardiogram reliably enough to detect arrhythmia and initiate defibrillation is considerable and would not be possible in an implant without full use of the latest hybrid microelectronics [85].

Myoelectric Control of Hand Prostheses Attempts to provide an artificial substitute for a lost limb represent one of the oldest examples of engineering applied to medicine. Considerable success with myoelectric prostheses [86] fitted to children and adults has extended control from simple control of prehension (grasp) to wrist rotation and control of the elbow mechanism [87]. An artificial arm can provide simultaneous control of prehension, wrist rotation, and rate-variable elbow flexion from the electromyograms of the biceps and triceps [88,89]. The very advanced signal processing system for this limb, probably the most successful in terms of function, has been developed and realized by use of thick-film hybrid circuit technology.

Cardiac Pacemakers The chambers of a normal healthy heart contract and relax rhythmically, with contractions of the atria preceding those of the ventricles by about 0.15 s. Some of the cells making up the muscle of the heart have the capacity for self-excitation; the natural pacemaking function of the heart results from this activity. Sometimes as a consequence of congenital abnormality or of disease, the conduction of electrical impulse and the coordinated contraction of heart muscle are impaired. The effective pumping action may cease intermittently; in that event, consciousness will rapidly be lost. Convulsions will occur if the heart function does not restart within 10 s or so, and if the arrest is prolonged, death will result. Contraction of the heart muscle and a return to normally coordinated activity can be obtained by implanting a device called a *pacemaker* [90], which applies electrical stimuli. The most technically refined pacemakers today provide for sequential stimulus to both the atrium (A) and ventricle (V) of the heart. The devices may be reprogrammed from outside the patient using a transmitter coil brought unto the chest wall. Pulse amplitude, pulse separation (between A and V stimuli), frequency, and demand threshold may be set and, if necessary, adjusted in accordance with the clinical response of the patient. In another pacemaker design, the (natural) atrial contraction is detected and used to initiate the ventricular stimulus [91]. Fifty-two years have passed since the first pacemaker was implanted in a human patient. Today, the need for smaller, faster, and cheaper implanted devices is more complex in nature, and much of the success of complex circuitry is due to multilayer assemblies in hybrid microelectronic systems [92,93]. Figure 1.10 shows the anatomy of an implanted brain pacemaker used for patients suffering from everything from Parkinson's disease to obsessive-compulsive disorder [94,95]. A part of this pacemaker circuitry is made possible by modern hybrid circuit technology.

1.4.4 Telecommunications [32,37,96]

Telephone Networks For economic reasons, a two-wire circuit is used to connect a subscriber to his or her central office, and central offices are connected

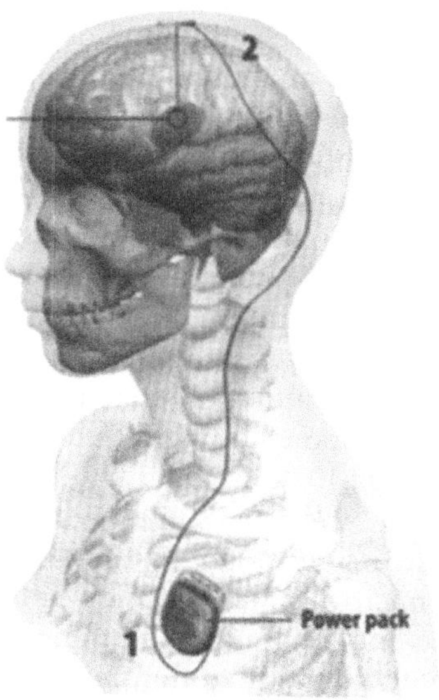

Figure 1.10 Anatomy of a pacemaker: 1, power pack; 2, implanted electrode [94]. (Courtesy of John MacNeill.)

Figure 1.11 Assembled SCF circuit [102]. (Reproduced with the express permission of EXAR Corporation, Fremont, California.)

using four-wire circuits to the outside world. For trunk calls over long distances, [3000 km (1860 miles) or more] a four-wire circuit is preferred because it uses two separate unidirectional paths for signal transmission in both directions. To provide a perfect impedance match between a two-wire system and a four-wire system, a hybrid microcircuit is placed in between. The hybrid circuit provides a perfect impedance match to the two-wire circuit by balancing impedance so that the signal coming into the four-wire system is passed directly to the two-wire circuit via a hybrid circuit, with no portion appearing in the four-wire transmission path [97–100]. In addition to the hybrid microcircuit, a typical echo reduction of about 40 dB is provided to eliminate the problem generated when both subscribers are talking.

Switched Capacitor Filters Switched capacitor filters (SCFs) are penetrating modern markets far beyond the telephone networks for which they were developed. They are now found in audio equipment, speech analyzers and recognizers, test instruments, and musical instruments. Their precision, compactness, and low cost make them preferable to discrete and hybrid filters as well as to digital filters [101]. Figure 1.11 shows an SCF circuit, a new workhorse of the hybrid family. SCFs consist of operational amplifiers, capacitors, switching transistors, and passive analog filters. Passive analog filters—resistor–capacitor networks—are implemented as discrete components or as thin or thick films on ceramic substrates. They also do well where the application tolerates up to 1% imprecision in cutoff frequencies and passband amplitudes. If precision is important, the designer must use either discrete components with precisely known values or thick-film laser-trimmed resistors [102,103].

Directional Couplers Directional couplers provide a simple, low-cost method of sampling and measuring power flow in an RF/microwave transmission path [104,105]. The device is typically utilized as a three-port device; the fourth port, termed the *local port*, must be connected to a matched resistive transmission.

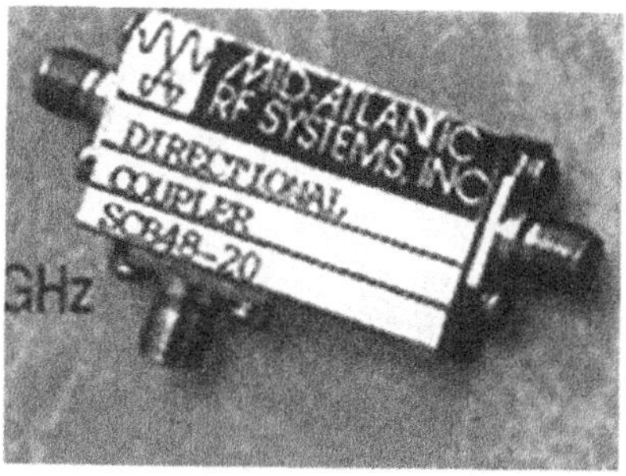

Figure 1.12 Directional coupler fabricated and assembled by Mid-Atlantic RF Systems. (Courtesy of Mid-Atlantic RF Systems, Forest Hill, Maryland.)

The power limitations of this load are the determining factor in the power-handling ability of the coupler. Thick-film fabrication enjoys the advantages of low-cost construction and automated assembly procedures. The advantages have caused thick-film circuit construction to become well established for low-frequency and dc applications. The film technology, offering improved microstrip [106] line definition, has been used extensively for fabricating circuits at microwave frequencies. Thick-film hybrid microcircuits are also being used increasingly at higher frequencies (e.g., in 3- and 10-GHz directional couplers) [107]. Figure 1.12 shows a directional coupler based on hybrid microelectronic thick- and thin-film technology.

1.4.5 Consumer Electronics

Handheld Calculators Hewlett-Packard's handheld calculator (HP 18C and HP 28C) is a multichip PCB hybrid [108]. No other packaging technology except hybrid multilayer packaging circuits could meet the combined requirements of high pin count, low package profile, environmental stability, and low cost [109] for such a small handheld calculator. The calculator is designed, developed, and fabricated by Hewlett-Packard's Northwest Integrated Circuit Division in Oregon. This complex handheld calculator (Figure 1.13) has double-sided PCB hybrids [110,111] with a total of 18 components. This hybrid circuitry with the latest surface-mounting technology has provided a unique opportunity to make this device as tiny as possible. The top of the board carries

Figure 1.13 Double-sided thick-film substrate with surface-mounting component [20].

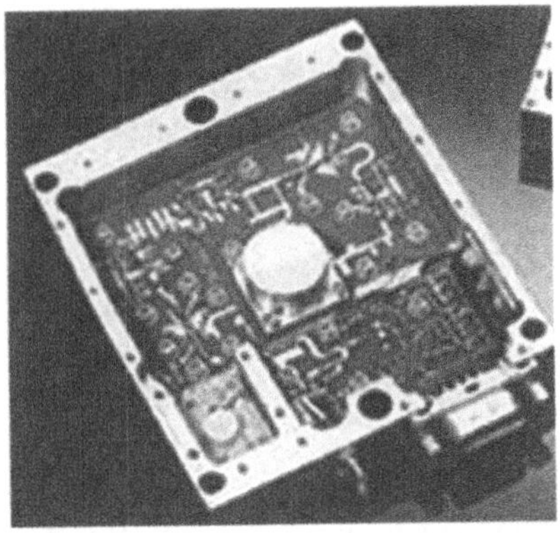

Figure 1.14 Portable transceiver circuit fabricated on a low-temperature ceramic substrate. (Courtesy of ST Microwave Corporation, Chandler, Arizona.)

three ICs, which are epoxy die attached and connected by a total of 263 gold wire bonds to gold-plated pads.

Portable Transceivers A wide range of portable transceiver products currently operating from 10 MHz to 6 GHz, such as cellular phones, global positioning satellite receivers, chordless phones [112–114], pagers, and wireless devices, owe their existence to thick- and thin-film hybrid microcircuitry. As a matter of fact, thick-film ceramic hybrid circuits have long been providing a high level of integration, thereby reducing package size and improving reliability.

Figure 1.14 shows the circuit in a portable transceiver, fabricated on a low-temperature ceramic substrate following thick- and thin-film hybrid circuit technology.

1.4.6 Military Applications

Hybrid Power Packages Hybridization of high-power semiconductor chips [115,116] has become necessity in many military applications because of reliability and hermeticity requirements. It is an integral substrate package which includes a hybrid full-wave rectifier designed to operate at 15 V with 20 A of current flow and a power dissipation of 5.6 W. The boost provides high voltage to a dc-to-dc converter from a 28-V input [117,118]. It also provides all boost functions, including peak current control, startup-less power supply, high-speed switching transistors with drivers, and an output rectifier diode. The operating frequency is 80 Hz, and the maximum power output is 250 W. The hybridization of this package has resulted in surface-area savings of 6 : 1 and overall volume savings of 12 : 1. This was made possible by the new hybrid circuit packaging concept [119].

High-Power, High-Frequency Amplifiers The AN/TPS-59 radar system [120, 121] developed by the General Electric System Division is a 100-W MIC amplifier. It is built on an alumina substrate 3.5 in. (8.89 cm) long, 2 in.

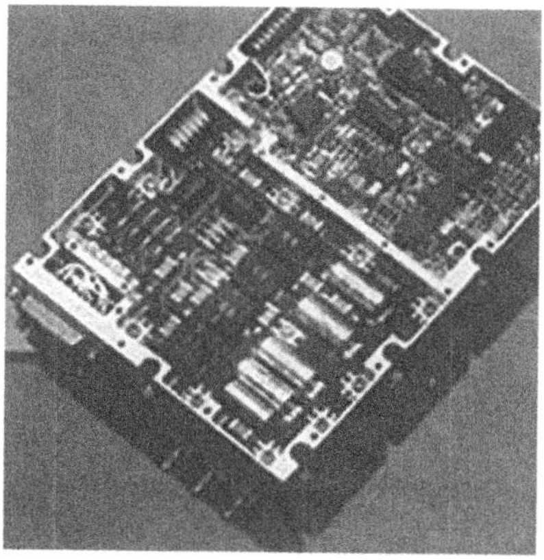

Figure 1.15 Circuit diagram of a hybrid microcircuit on low-temperature ceramic substrate and used in airborne radar. (Courtesy of ST Microwave Corporation, Chandler, Arizona.)

(5.08 cm) wide, and 0.025 in. (0.06 cm) thick with sixteen 0.060-in. (0.15-cm)-diameter holes that are used for low-impedance ground connection of the thick-film copper strip line. There are three large square holes for accommodating beryllia-mounted microwave power transistor with their tuning networks [122].

The hybrid MIC amplifier has minimum size and weight, is rugged, and provides very low thermal resistance for dissipating the considerable heat emanating from microwave transistors. Furthermore, maintenance is easy, cost is low, and the amplifier [123] meets the hermeticity requirements of MIL-STD-883, providing reliable operation for maximum system availability [124]. Figure 1.15 shows the picture of an X-band airborne radar, fabricated on a LTCC substrate following thick and thin film hybrid microcircuit technology.

1.5 TYPICAL MICROELECTRONIC PRODUCTS

1.5.1 Consumer Electronics

Cameras Panasonic's Cool Shot II camera, developed and produced by Matsusita-Kotobuki Electronics Industries in Japan, is a light, credit-card-sized camera [35]. The substrate is a low-temperature cofired ceramic (LTCC) substrate [125] with 10 dielectric and 11 conductive layers [126,127]. The conductive layers and chip resistors all belong to the thick-film hybrid family and the materials are all from DuPont. This digital camera provides improved picture quality and has been on the market since 1997. Figure 1.16 shows the Cool Shot II camera following an LTCC multilayer interconnect [22].

Cellular Phones The RF amplifier circuit [128] for use in cellular phones developed by Miyoshi Electronics in Japan is another hybrid microelectronics application area. The size of this power amplifier for handheld wireless telephones is less than 0.000488 in^3 (0.008 cm^3). The substrate is an LTCC mate-

Figure 1.16 Panasonic Cool Shot II digital camera [35]. (Courtesy of DuPont, Wilmington, Delaware.)

rial with conventional thick-film circuitry [129,130]. Figure 1.17 is the circuit of a small cellular phone fabricated on an LTCC substrate. Hybrid microcircuit technology has been used successfully to make this cellular phone small (palm size), lightweight, and versatile.

The RF amplifier shown in Figure 1.18, a 2.1-GHz varactor-controlled oscillator (VCO) with a 22-nH RF choke, a minimum Q value of 3, and a ladder network consisting of inductors, capacitors, and resistors for matching the impedance of the circuit used in such portable transceiver circuits, is fabricated using thick-film hybrid technology. Laser-trimmed thick-film resistors and capacitors are cost-effective for high-volume, tight-tolerance applications. The circuit complexity needed for miniaturization of these portable circuits has been possible for hybrid microelectronic circuits. The additional circuit chips are added to the circuit following surface-mounting technology [131–133]. To improve thermal conductivity, silver-based thermal via composition and DuPont 6344 paste (ink) screen-printed to fill through–holes in conventional alumina substrates are used to form high-thermal-conductivity paths to dissipate heat. Fine lines of silver–platinum conductors form 5.9-mil (150-μm)

Figure 1.17 Inside circuit diagram of a small cellular phone [35]. (Courtesy of DuPont, Wilmington, Delaware.)

Figure 1.18 Thick-film power amplifier circuit on alumina substrate for cellular phones [35]. (Courtesy of DuPont, Wilmington, Delaware.)

signal line with an 11.8-mil (300-μm) pitch, and DuPont's QS87 series inks are used for thick-film resistors [35]. Figure 1.19 shows the circuit of a VCO fabricated using MMIC technology. The substrate can be LTCC or GaAs single crystal.

Schoolchildren's Knapsacks An Italian company has developed an innovative product for mothers-to-be and schoolchildren. The company's project team has designed a roomy knapsack made of waterproofed nylon with a special rubberized pocket to house a flashing electroluminescent lamp (Figure 1.20). The lamp is overlaid with tinted electroluminescent inks from DuPont screen-printed on a 4.92-mil (125-μm)-thick polyester film. The knapsack has been well received in the European and Mexican markets [35].

1.5.2 Industrial Applications

Supercomputers The performance of computer systems ranging from personal computers to the highest-performance supercomputers [134] has been improving at explosive rates year after year [135–137]. Supercomputer technology owes much to hybrid technology; special polyimide ceramic [138,139] substrates and special conductor materials such as molybdenum, whose thermal expansion coefficient closely matches that of silicon chips, are being used. New packaging materials enhance the fatigue life of the associated chip-to-substrate

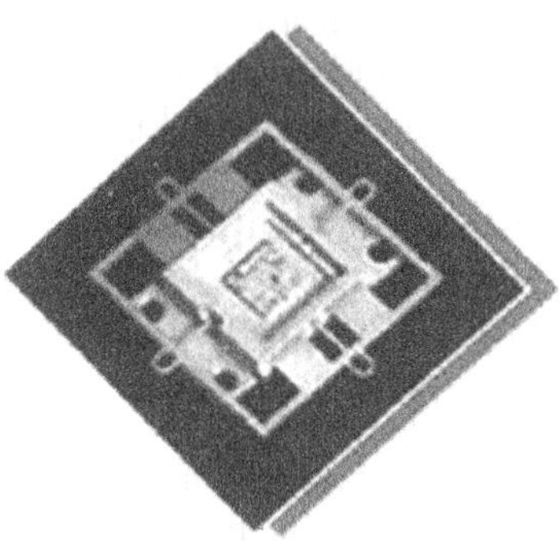

Figure 1.19 Circuit diagram of a VCO. (Courtesy of ST Microwave Corporation, Chandler, Arizona.)

Figure 1.20 Child's knapsack built by an Italian company [35]. (Courtesy of DuPont, Wilmington, Delaware.)

connections [140] and permit unprecedented packaging density. The NEX SX supercomputer uses a multilayer ceramic substrate 3.94 by 3.94 by 0.1 in. (10 by 10 by 0.25 cm) thick and uses a tungsten layer for power distribution. It has 2177 I/O pins brazed on the bottom surface on a 0.1-in. (2.54-mm) staggered grid. On top of this alumina substrate, four insulating polyimide layers and five thin-film conductor layers are deposited. The chips are placed by the use of flipped *tape automated bonding* (TAB). Thirty-six such chips are placed in a high-density multichip package. This module achieves high speed and has a high I/O count. The densely packed thin-film interconnections are placed on low-dielectric-constant polyimide layers [141]. The alumina substrate with high dielectric constant provides a large decoupling capacitance for power distribution and supports a high pin count [142].

Thermal Printheads Thermal printing, one of several forms of noncontact printing, essentially carries a resistive heating pin head developed and fabricated by Hewlett-Packard [143]. This high-speed scanning (122 characters per second) pin head is sputter-deposited, and to optimize the resistive value of the high-speed scanning pin head, fine-line structures are created by means of photoetching [144]. Such high-speed scanning requires sufficient heat dissipation, which is achieved using a glazed mesa structure on an alumina substrate. High performance coupled with low cost and convenience in high-speed noncontact printing is made possible through thin-film hybrid technology.

Cryogenic Infrared Imaging Arrays [145] The outstanding versatility of thick-film hybrid technology is demonstrated by the development of a butted five-chip shortwave infrared focal plane with 2560 pixels $\times$ 2 spectral bands operating at 125k [146]. Thick-film screen printing technology is adopted to print insulators and conductors on silicon substrate. For added reliability, and to reduce parasitic capacitance to the silicon substrate, two layers of passivation glaze are also screen printed. To form an interconnect pattern, first gold con-

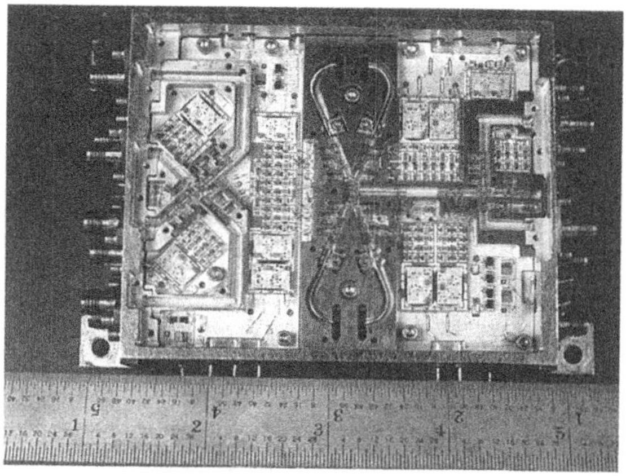

Figure 1.21 Velocity and position deception module [20].

ductor is screen-printed and fired, followed by a double print and fired multilayer dielectric, which survives repeated thermal shock immersion in liquid nitrogen. For infrared arrays, the precise mounting relationship is maintained for an assembly room ambient to cryogenic operational temperature. The focal plane is fabricated with a polysilicon substrate plane to minimize thermal expansion mismatch.

1.5.3 Military and Aerospace Applications

Velocity and Position Deception Modules The velocity and position deception module (VPDM) is a sophisticated RF input network and receiver housed in an alumina module used in military fighter aircraft that integrates a variety of RF/microwave hybrid microstrip [147] subassemblies, including seven custom multicavity circuits. As enemy radar is detected, the VPDM processes and transmits amplitude- and frequency-shifted pulses across a broad spectrum up to 18 GHz [148]. Figure 1.21 shows a VPDM housed in an alumina module.

DC-to-DC Converters [149] A 50-W triple-output dc-to-dc converter is shown in Figure 1.22. The 28-V input supplies two isolated outputs: +15.0 V, 500 mA and +5.0 V, 7.0 A. The power supply is designed for military and aerospace applications. The circuit includes a number of magnetic components and capacitors common to a power circuit [148].

Hybrid Silicon Carbide Differential Amplifiers Applications of high-temperature electronics (175 to 600°C) include instrumentation for jet aircraft engines, nuclear reactors, geothermal wells, automotive underhood electronics, and space-based power systems [150]. The hybrid differential amplifier [151] uses 6-H silicon carbide substrate using both dielectric isolation and bonded wafer technology and silicon monolithic devices. The gain ratio of this amplifier is more

(*a*)

(*b*)

Figure 1.22 (*a*) Isolated DC/DC converter; (*b*) non-isolated DC/DC converter. (Reprinted with permission from ISHM/IMAPS, Washington, DC.)

than 60 dB, the common-mode rejection ratio is greater than 55 dB, and the offset voltage varies from 139 to 159 mV over the entire range [152]. This hybrid circuit can operate at 350°C, which is not possible for integrated monolithic circuits built on a silicon wafer or PCB.

1.5.4 Automotive Industry

Automotive Fan Speed Controllers Thick-film hybrid circuits are finding high-volume circuit applications in fan-speed-control modules in automobiles, where performance and cost are extremely important [153]. The circuit uses a thick-film laser-trimmed resistor. The resistor circuit uses DuPont 6800 series resistor

Figure 1.23 Hybrid circuit module used in the speed controller system of a fan inside the engine of an automobile [35]. (Courtesy of DuPont, Wilmington, Delaware.)

compositions, a 6160 silver conductor, and a QS580 encapsulant (Figure 1.23). The ceramic substrate that carries the hybrid microcircuit is attached to a heat sink. The master module is capable of dissipating heat at a rate of about 15 W/cm^2. This hybrid microcircuit module is rugged and can withstand harsh environmental conditions: high temperatures, rapid temperature changes, repeated thermal cycles, vibration, corrosive gases, solvents, salt spray, and high humidity.

1.5.5 Microwave Engineering [154,155]

When two unshielded transmission lines are close together, power can be coupled between the lines due to interaction of the electromagnetic fields of each line. Such lines, called *coupled transmission lines*, usually consist of three conductors in close proximity. The coupling in a coupled line coupler is too loose to accommodate coupling factors of 3 or 6 dB. One way to increase coupling between edge-coupled lines is to have several lines parallel to each other, so that the fringing fields at both edges of a line contribute to the coupling. Probably the most practical implementation of this idea is the *Lange coupler* [156,157] (Figure 1.24), which is more compact than simpler two-line designs and provides a better coupling factor and good isolation. Recently, circuit processing technology has utilized a thick-film process with DuPont's Fodel 6050 dielectric ink to eliminate wire bonds in Lange couplers, to reduce inductance, improve coupling and repeatability, and lower the overall cost [35].

Figure 1.24 Circuit diagram of a Lange coupler [35]. (Courtesy of DuPont, Wilmington, Delaware.)

Figure 1.25 Amitech's MCM module [35]. (Courtesy of DuPont, Wilmington, Delaware.)

1.5.6 Multichip Modules

Multichip modules (MCMs) can be defined as high-density interconnections of multiple chips (may be discrete or deposited or both), including ICs nested on a single substrate, which may be a single crystal of silicon or gallium arsenide or polycrystalline ceramics or glass-filled epoxy laminate [158–159]. Figure 1.25 shows an MCM circuit fabricated by Amitech, based in Roros, Norway. This circuit consists of five layers of insulation 17 by 17 mm^2 housed in a ceramic substrate. The circuit is made with DuPont's QM silver multilayer system.

MCMs are very diverse; they may be small, nonhermetic commercial circuits (camcorders or cellular phones). They may be large, hermetic military circuits, such as a standard electronics module, E-size, SEM-E, for a naval system, or SAM, for a standard avionics module. MCMs belong to the hybrid microcircuit family, and the two have many things in common [160,161]. An MCM, however, uses a variety of substrates according to the family to which it belongs. According to the processes and assembly requirements, the MCM family is divided into three categories: MCM-L (multichip module laminated), MCM-D (multichip module deposited), and MCM-C (multichip module ceramic). MCMs minimize chip spacing and reduce inductive and capacitive discontinuities between chips mounted on the substrate by replacing the die–wire bond–pin–board–pin path with a much superior die–bump–interconnect–bump–die path. In addition, narrower and shorter wires on the ceramic substrate have much lower capacitance values than do PCB interconnections [162,163].

Researchers from Fraunhofer IBM in Berlin, IMEC in Leuven, Belgium, and the Universities of Bonn and Wuppertal in Germany have used thin-film techniques with copper in benzocyclobutane set on a depleted silicon wafer to create MCM substrate [56]. This technique is used to make a detector module for an experiments using the planned large hadron collider at CERN in Geneva [57].

1.6 SUMMARY

Thick- and thin-film systems are used in all facets of today's electronic industry, starting with applications in the computer and defense industries. As the state of the art has grown, composition suppliers broadened their lines, developing new high-performance materials to meet changing and sophisticated circuitry. The present hybrid microcircuit technology offers the advantages of simple processing, a fast and inexpensive tooling system, economy, tolerance-active devices, high reliability, and multilevel circuit capabilities. As a result, virtually all sectors of the electrical and electronic industries are turning to solid-state systems which utilize thick- and thin-film hybrid microcircuitry. Increasingly, thick- and thin-film hybrid microcircuits are finding a place in microwave and wideband applications because of their low cost in volume production, reduced parasitics, high power dissipation and the development of processing terms of line definition, and crossovers. In an effort to deliver highly reliable products in a period when microelectronics devices and technologies have been changing rapidly, accurate reliability prediction techniques must be established. The failure prediction model must use novel materials, manufacturing processes, testing, environmental stress screening, application criteria, parametric stresses and drafting, and thermal and environmental elements.

Although hybrid technology accommodates an appreciable amount of flexibility in design methodology, assembly techniques, and to realize a wide variety of electronic functions, it is not the least expansive circuit technology. Moreover, there is a cost penalty associated with reduced size, weight, and unique performance. However, it is vitally important that the cost impact of circuit design and fabrication design techniques be considered carefully because the ultimate cost of a hybrid circuit may play a key role in its lifetime. Although cost premiums may initially be justified, value-engineering considerations at the electronic system level should address replacement costs. Automation in the design and manufacture of hybrid microcircuits will continue to drive down overall computer-aided design and computer-automated manufacturing costs. Thus computerized testing of hybrids will ultimately reduce the costs of hybrid circuits. The most recent technology in hybrid microcircuitry is the multichip module.

REFERENCES

1. C. F. Coombs, Jr., ed., *Printed Circuit Handbook*, 2nd ed., McGraw-Hill, New York, 1988.
2. M. Pecht, ed., *Advanced Routing of Electronic Modules*, CRC Press, Boca Raton, FL, 1986.
3. M. L. Topfer, *Thick-Film Microelectronic Fabrication Design and Application*, Van Nostrand Reinhold, New York, 1971.

4. R. Glang and L. V. Gregor, Generation of patterns in thin films, in *Handbook of Thin-Film Technology*, 2nd ed., L. I. Maissel and R. Glang, eds., McGraw-Hill, New York, 1983, Chap. 7.
5. W. S. Ruska, Photolithography, in *Microelectronics Processing*, McGraw-Hill, New York, 1987, Chap. 4.
6. J. Millman and C. C. Halkais, eds., *Integrated Electronics: Analog and Digital Circuit Systems*, McGraw-Hill, New York, 1972, Chap. 7, p. 1966.
7. T. K. Gupta, Aluminum alloy as an interconnecting material in the fabrication of integrated circuits, *Micron. Reliab.*, Vol. 19, 1979, p. 337.
8. T. W. Mayer and S. S. Lau, eds., *Electronic Materials Science for Integrated Circuits in Si and GaAs*, Macmillan, New York, 1990, p. 284.
9. S. K. Ghandhi, ed., *VLSI Fabrication Principles*, Wiley, New York, 1980, p. 456.
10. P. R. Gray and R. G. Mayer, eds., *Analysis and Design of Analog Integrated Circuits*, 3rd ed., Wiley, New York, 1993, p. 111.
11. S. Wolf and R. N. Tauber, eds., *Silicon Processing: VLSI Era*, Vol. 1, Lattice Press, Sunset Beach, CA, 1986.
12. World News, *Solid State Technol.*, Dec. 1998, p. 20.
13. C. Cohen and M. T. Sih, Packaging and interconnections of integrated circuits, in *Electronic Packaging and Interconnection Handbook*, 3rd ed., C. A. Harper, ed., McGraw-Hill, New York, 2000, Chap. 7.
14. C. P. Wong, Materials for electronic packaging, in *Electronic Packaging and Interconnection Handbook*, 3rd ed., C. A. Harper, ed., McGraw-Hill, New York, 2000, p. 1.3.
15. Emerging Technology, *Semicond. Int.*, Vol. 20, No. 4, 1998, p. 42.
16. D. D. Zimmerman and J. R. Dinitto, eds., *Hybrid Microcircuit Design Guide*, Joint Publication ISHM-1402 and IPC-855, Oct. 1982.
17. T. R. Reid, *The Chip: How Two Americans Invented the Microchip and Launched a Revolution*, Simon & Schuster, New York, 1984; Jack St. Clair Kilby, The transistor ... the first 40 years, *Solid State Technol.*, Vol. 30, No. 12, Dec. 1987; T. S. Perry, The IC, *IEEE Spectrum*, Dec. 1988, p. 40.
18. B. V. Hiremath, ed., *Ceramic Thick and Thin Film*, American Ceramic Society, Columbus, OH, 1985.
19. P. Moran, ed., *Hybrid Microelectronic Technology*, Gordon & Breach, New York, 1984.
20. R. W. Johnson, Introduction to hybrid microelectronics, in *Hybrid Microcircuit Monograph*, R. W. Johnson, ed., ISHM, Reston, VA, 1991.
21. R. R. Tummala and E. J. Rymaszewski, eds., *Microelectronic Packaging Handbook*, Van Nostrand Reinhold, New York, 1989, Chap. 7.
22. K. Otsuka, ed., *Multilayer Ceramic Substrate Technology for VLSI Package/Multichip Module*, Elsevier Applied Science, New York, 1993.
23. J. W. Kratz, Surface mount technology, in *Handbook of Thick-Film Hybrid Microelectronics*, C. Harper, ed., McGraw-Hill, New York, 1982, Chap. 8.
24. M. F. Bender, F. H. Patterson, E. A. Kemp, and J. E. Gantzhorn, Jr., Low-temperature co-fired ceramic tape system, *Proc. ISHM*, Seattle, WA, 1988, p. 12.

25. R. Goyal, ed., *Monolithic Microwave Integrated Circuit Technology and Design*, Artech House, Norwood, MA, 1989.
26. H. Sobol, The technology of microwave integrated circuits, in *Advances in Microwaves*, Vol. 8, H. Sobol and M. Caulton, eds., Academic Press, San Diego, CA, 1974.
27. R. Brown, ed., *Materials and Processes for Microwave Hybrids*, ISHM, Reston, VA, 1991.
28. J. Fjelstad, Chip scale packaging and direct chip attach technologies, in *Electronic Packaging and Interconnection Handbook*, 3rd ed., C. A. Harper, ed., McGraw-Hill, New York, 2000.
29. E. F. Kuo, ed., *Protocols and Techniques for Data Communication Networks*, Prentice Hall, Upper Saddle River, NJ, 1981.
30. C. S. Walker, ed., *Capacitance, Inductance and Cross-Talk Analysis*, Artech House, Norwood, MA, 1970.
31. P. E. Garrou, ed., *Multichip Technology Handbook*, McGraw-Hill, New York, 1988, Chap. 1.
32. R. Freeman, ed., *Telecommunication System Engineering*, 2nd ed., Wiley, New York, 1989.
33. A. E. Riad and F. D. Barlow III, eds., *Thin-Film Technology Handbook*, McGraw-Hill, New York, 1998, Sec. 9-49.
34. W. Stallings, ed., *Data and Computer Communications*, 3rd ed., Macmillan, New York, 1988, Chap. 8, pp. 697–748.
35. Horizons, *DuPont Thick-Film News*, Global Edition, No. 12, Nov. 1998.
36. J. Revard, Automotive electronics: a perspective look at the USA, *Proc. Int. Automot. Electron. Conf.*, London, Oct. 1987.
37. H. Taraseiskey, ed., Market and applications, in *Power Hybrid Circuit Design and Manufacture*, Marcel Dekker, New York, 1996, p. 221.
38. Z. H. Meiksin, ed., Medical applications, in *Thick and Thin Films for Hybrid Microelectronics*, Lexington Books/D.C. Heath, Lexington, MA, 1976, Chap. 14.
39. H. T. Law, Medical applications of hybrid microelectronics, *Proc. ISHM*, Atlanta, GA, 1986, p. 172.
40. S. K. Mitra and J. F. Kaiser, eds., *Handbook of Digital Signal Processing*, Wiley, New York, 1993, Chap. 1.
41. J. F. Dicson and D. R. Asher, A versatile high-performance hybrid power package, *Proc. Int. Symp. Microelectron.*, Anaheim, CA, 1985, pp. 343–348.
42. J. Hardy, ed., *High-Frequency Circuit Design*, Reston Publishing, Reston, VA, 1979.
43. K. Green, Design curves for flat square spiral inductors, *Electron. Components*, Feb. 11, 1972, pp. 121–126.
44. H. E. Bryan, Printed inductors and capacitors, *Tel. Tech.*, Vol. 14, No. 2, 1955, pp. 68–69.
45. P. Horowitz and W. Hill, Microcomputer: bus signals and interfacing, in *The Art of Electronics*, 2nd ed., P. Horowitz and W. Hill, eds., Cambridge University Press, Cambridge, 1993, p. 711.

46. R. W. Koss, B. Space, and D. Dewar, One-megabit high-speed hybrid RAM module, *Proc. ISHM*, Atlanta, GA, 1987, p. 235.
47. H. Tarascisky, Custom power hybrids, *Solid State Technol.*, No. 1, Oct. 1985, p. 111.
48. L. Balents, R. D. Gold, A. K. Kaiser, and W. R. Patterson, Design considerations for power hybrid circuits, *Proc. Int. Symp. Hybrid Microelectron.*, Dallas, TX, Sept. 29, 1969.
49. R. Peura and J. Webster, Basic sensors and principles, in *Medical Instrumentation: Application and Design*, J. G. Webster, J. W. Clark, Jr., and M. R. Neuman, eds., Wiley, New York, 1997.
50. J. J. Carr and J. Brown, eds., *Introduction to Biomedical Equipment Technology*, Prentice Hall, Upper Saddle River, NJ, 1997.
51. A. Koslof and G. DeMay, eds., *Thermal Modeling and Optimization of Power Microcircuits*, IMAPS, Washington, DC, 1998.
52. L. Gabco and M. Kejzlar, Reliability of hybrid integrated-circuit applications, *Int. J. Microelectron. Packag.*, Vol. 7, No. 2, 1994, p. 205.
53. J. J. Licari and L. R. Enlow, *Hybrid Microcircuit Technology Handbook*, Noyes Publications, Park Ridge, NJ, 1998, Chap. 4, p. 113.
54. A. L. Eustice, S. J. Horowitz, and A. R. Trivis, Low-temperature co-fired ceramics multilayer interconnect hybrids, *Proc. IMC*, Kobe, Japan, May 28–30, 1986, pp. 44–60.
55. L. Harte, R. Kikta, and R. Levine, eds., *3G Wireless Demystified*, McGraw-Hill, New York, 2001.
56. J. Harrer, R. Petit, O. Brown, D. Nabatian, and S. Shahbazi, RF characterization of thick-film materials for portable transceivers, *Proc. Int. Symp. Hybrid Microelectron.*, 1996.
57. W. Y. Hwong, ed., *Materials and Processes for Wireless Communications*, Vol. 53, American Ceramic Society, Westerville, OH, 1995.
58. D. S. Meir, N. Borkowicz, J. L. Evans, and R. W. Johnson, Using MCM technology to enhance the development of electronic control modules, *Proc. Int. Conf. MCM*, Apr. 1993, pp. 102–106.
59. M. Mirowski, Prevention of sudden arrhythmic death with implanted automatic defibrillators, *Ann. Intern. Med.*, Vol. 97, No. 4, 1982, pp. 606–608.
60. K. A. Ellenbogen, G. N. Kay, and B. L. Wilkoff, eds., *Clinical Cardiac Pacing and Defibrillation*, 2nd ed., W.B. Saunders, Philadelphia, 2000.
61. G. M. Whitesides, Nanotechnology: art of the possible, *Technol. Rev. Innov.*, Vol. 101, No. 6, 1998, p. 85.
62. R. W. Johnson and J. L. Evans, Automotive multichip module, in *Advanced Electronic Packaging with Emphasis on MCM*, W. D. Brown, ed., IEEE Press, Piscataway, NJ, 1999, p. 588.
63. J. Aday, R. W. Johnson, J. L. Evans, and C. Romanczuk, Wire-bonded thick-film silver multilayer for underhood automotive applications, *Int. J. Microcircuit Electron. Packag.*, Vol. 17, No. 3, 1994, p. 302.
64. J. F. Wakerly, ed., *Digital Principles and Practices*, Prentice Hall, Upper Saddle River, NJ, 1990.

65. J. H. Lau, ed., *Flip Chip Technologies*, McGraw-Hill, New York, 1996, p. 317.

66. G. E. Hartland, High-voltage application for automotive applications, *Int. Symp. Hybrid Microelectron.*, Dallas, TX, 1969, pp. 361–368.

67. M. J. Beesley, ed., *Lasers and Their Applications*, Taylor & Francis, London, 1976, Chap. 11.

68. R. J. Klein, *Solder Engineering*, 2nd ed., Electrochemical Society, Pennington, NJ, 1989.

69. M. Madou, ed., *Fundamentals of Microfabrication*, CRC Press, Boca Raton, FL, 1997.

70. L. Percsi and G. Harasanyi, Low-power detector made by thick-film technology, *Int. J. Microcircuits Electron. Packag.*, Vol. 18, No. 1, 1995, p. 46.

71. S. Middlehoek, D. J. Noorlag, and J. K. Steenvoorden, Silicon and hybrid microelectronic sensors, *Proc. 3rd Hybrid Microelectron. Conf.*, Avignon, France, 1981, pp. 100–113.

72. T. Kwikkers, Two thick-film thermal sensors, *Hybrid Circuits*, No. 16, 1985, pp. 39–42.

73. Z. H. Meiksin, ed., Medical applications, *Thin and Thick Film for Hybrid Microelectronics*, Lexington Books/D.C. Heath, Lexington, MA, 1976, Chap. 14.

74. P. Horowitz and W. Hill, eds., High-frequency amplifier, in *The Art of Electronics*, 2nd ed., Cambridge University Press, Cambridge, 1998, Chap. 13, p. 874.

75. M. Uenohara, in *Advances in Microwaves*, L. Young, ed., Vol. 2, Academic Press, San Diego, CA, 1967.

76. S. Wang, ed., *Fundamentals of Semiconductor Theory and Device Physics*, Prentice Hall, Upper Saddle River, NJ, 1989.

77. C. E. Jowett, ed., *The Engineering of Microelectronic Thin and Thick Film*, Macmillan, London, 1976, Chap. 4; M. Caulton, Film technology in microwave integrated circuits, *Proc. IEEE*, Vol. 59, No. 10, Oct. 1971.

78. M. Bader, U.S. hybrid circuit markets, applications, and technology, *Solid State Electron.*, Feb. 1989, p. S1.

79. P. Horowitz and W. Hill, eds., *The Art of Electronics*, 2nd ed., Cambridge University Press, Cambridge, 1993, p. 711.

80. H. Queisser, ed., *The Conquest of the Microchip*, Harvard University Press, Cambridge, MA, 1988.

81. Y. J. Wong and W. E. Ott, eds., *Function of Circuits: Design and Applications*, McGraw-Hill, New York, 1976.

82. W. Meinel and G. Smith, Hermetic analog isolation amplifier, *Proc. ISHM*, 1987, p. 202.

83. K. A. Ellenbogen, G. N. Kay, and B. L. Wilkoff, eds., Basic concepts, in *Clinical Cardiac Pacing and Defibrillation*, 2nd ed., W.B. Saunders, Philadelphia, 2000.

84. R. E. Ideker, Sensing and arrhythmia detection by implanted devices, in *Clinical Cardiac Pacing and Defibrillation*, 2nd ed., K. A. Ellenbogen, G. N. Kay, and B. L. Wilkoff, eds., W.B. Saunders, Philadelphia, 2000.

85. H. T. Law and J. J. Hewson, An electro-myographically controlled elbow locking mechanism for an upper limb prosthesis, *Proc. Eur., Hybrid Microelectron. Conf.*, Avigon, France, 1981, pp. 261–268.

86. N. Neuman, Therapeutic and prosthetic devices, in *Medical Instrumentation and Design*, J. G. Webster, J. W. Clark, and M. R. Neuman, eds., Wiley, New York, 1997, Chap. 13.
87. H. T. Law, Engineering of upper limb prostheses, in *Management of Upper Limb Amputations, Orthopedic Clinics of North America*, R. W. Bessely, ed., Vol. 12.4, 1981, p. 929.
88. K. M. Ward, C. Lehmann, and A. M. Leiken, *Clinical Laboratory Instrumentation and Automation: Principles, Applications and Selection*, AACC Press, New York, 1994.
89. S. C. Jacobson and R. W. Mann, Control system for artificial arms, *IEEE Trans. Syst. Man. Cybern.*, Vol. 5, 1975, p. 252.
90. S. Furman, Cardiac pacing, Sec. 1, in *Clinical Cardiac Pacing and Defibrillation*, 2nd ed., K. A. Ellenbogen, G. N. Kay, and B. L. Wilkoff, eds., W.B. Saunders, Philadelphia, 2000.
91. J. J. Car and J. Brown, eds., Physiological pressure and other cardiovascular measurements and devices, in *Introduction to Biomedical Equipment Technology*, Prentice Hall, Upper Saddle River, NJ, 1997.
92. J. G. Webster, ed., *Encyclopedia of Medical Devices and Instrumentation*, Wiley, New York, 1988.
93. A. Weinberg, R. Krug, and K. Jones, High-density double-sided SMT packaging of vertically integrated chip carriers for cardiac pacemakers, *Proc. ISHM*, Seattle, WA, 1988, pp. 102–105.
94. S. S. Hall, Brain pacemakers, *Technol. Rev.*, Vol. 104, No. 7, Sept. 2001, p. 41.
95. R. W. Johnson, Introduction to hybrid microelectronics, in *Hybrid Microelectronics Handbook*, 2nd ed., J. E. Sergent and C. A. Harper, eds., McGraw-Hill, New York, 1995, Chap. 7.
96. P. Horowitz and W. Hill, eds., *The Art of Electronics*, 2nd ed., Cambridge University Press, Cambridge, 1998, p. 935.
97. D. L. Duttweiler, Bells echo killer chip, *IEEE Spectrum*, Vol. 17, 1980, pp. 34–37.
98. S. K. Mitra and J. F. Kaiser, eds., *Handbook for Digital Signal Processing*, Wiley, New York, 1993, p. 35.
99. A. V. Oppenheim, ed., *Applications of Digital Signal Processing*, Prentice Hall, Upper Saddle River, NJ, 1978, Chap. 1, pp. 1–28.
100. D. G. Messerschmitt, Echo cancellation in speech and data transmission, *IEEE J. Select. Areas Commun.*, Vol. 2, 1982, pp. 283–297.
101. P. Horowitz and W. Hill, eds., *The Art of Electronics*, 2nd ed., Cambridge University Press, Cambridge, 1998, p. 281.
102. C. W. Solomon, Switched capacitor filters: precise, compact, inexpensive, *IEEE Spectrum*, June 1988, pp. 28–32.
103. P. E. Allen and E. S. Sinencio, eds., *Switched Capacitor Circuits*, Van Nostrand Reinhold, New York, 1984.
104. A. F. Harvey, ed., *Microwave Engineering*, Academic Press, San Diego, CA, 1993.
105. R. E. Collin, ed., *Foundation of Microwave Engineering*, IEEE Press, Piscataway, NJ, 2001.

106. L. A. Trinogga, G. Kaizhou, and I. C. Hunter, eds., *Practical Microstrip Circuit Design*, Ellis Horwood, New York, 1991, Chap. 7.

107. P. Drexler, S. Bogina, M. Hainworth, P. Koerner, P. B. Beard, and K. Scoles, 3- and 10-GHz directional couplers, using thick-film technology, *Proc. Int. Symp. Microelectron.*, Minneapolis, MN, 1987, p. 66.

108. M. Pecht, *Integrated Circuit, Hybrid, and Multichip Module Package Design Guidelines*, Wiley, New York, 1994.

109. B. Hauge, R. Dunlap, C. Hoekstra, P. V. Loan, and C. N. Kwee, A multichip printed circuit board hybrid for the HP 18C and 28C handheld calculators, *Proc. Int. Symp. Microelectron.*, Minneapolis, MN, 1987, p. 268.

110. P. Eiser, ed., *The Technology of Printed Circuits*, Academic Press, San Diego, CA, 1959.

111. T. H. Stearns, ed., *Flexible Printed Circuits*, McGraw-Hill, New York, 1996.

112. R. Perez, ed., *Wireless Communication Design Handbook*, Academic Press, San Diego, CA, 1998.

113. L. Harte, R. Kikta, and R. Levine, eds., *3G Wireless Demystified*, McGraw-Hill, New York, 2001, p. 210.

114. U. L. Rohde and D. P. Newkirk, *RF/Microwave Circuit Design for Wireless Applications*, Wiley, New York, 2000.

115. L. Marton, ed., High-power electronic devices, in *Advances in Electronics and Electron Physics*, Vol. 41, Academic Press, San Diego, CA, 1976.

116. *Semiconductor Power Circuit Handbook*, Application Engineering Department, Motorola Semiconductor Products, Phoenix, AZ, 1968, pp. 5-30 to 5-33.

117. J. F. Dickson and D. R. Asher, A versatile high-performance hybrid power package, *Proc. Int. Symp. Microelectron.*, Anaheim, CA, 1985, pp. 343–348.

118. R. K. Jurgen, ed., *Automotive Electronics Handbook*, McGraw-Hill, New York, 1999.

119. J. K. Lumpp, Hybrid assembly, in *Electronic Packaging Handbook*, G. R. Blackwell, ed., CRC Press, Boca Raton, FL, 2000, Chap. 7.

120. M. I. Skolink, ed., *Introduction to Radar Systems*, McGraw-Hill, New York, 1962.

121. D. M. Pozar, ed., *Microwave Engineering*, Addison-Wesley, Reading, MA, 1990, p. 679.

122. S. Teszner and J. L. Teszner, Microwave power semiconductor devices, in *Advances in Electronics and Electron Physics*, Vol. 39, L. Marton, ed., Academic Press, San Diego, CA, 1975.

123. D. M. Pozar, ed., Microwave circuit, in *Microwave Engineering*, Addison-Wesley, Reading, MA, 1990, pp. 646–651.

124. J. C. Gioia, MIC package using thick-film and Direct-Bond Copper for Low 100w L band power amplifier, *Proc. Int. Symp. Microelectron.*, Los Angeles, 1989, pp. 214–218.

125. R. W. Johnson, R. K. F. Teng, and J. W. Blade, eds., *System Advantages Major Construction, and Multichip Modules*, Part III, *Cofired Ceramic Technology*, IEEE Press, Piscataway, NJ, 1991.

126. K. Otsuka, ed., *Multilayer Ceramic Substrate Technology for VLSI Package/MCM*, Elsevier Applied Science, New York, 1993.

127. H. B. Bakoglu, Multilayer interconnections, in *Circuits, Interconnections, and Packaging for VLSI*, Addison-Wesley, Reading, MA, 1990, p. 211.

128. P. K. Gray and R. G. Meyer, eds., *Analysis and Design of Analog Integrated Circuits*, 3rd ed., Wiley, New York, 1993, p. 519.

129. J. J. Licari and L. R. Enlow, *Hybrid Microcircuit Technology Handbook*, Noyes Publications, Park Ridge, NJ, 1998, p. 122.

130. R. W. Johnson, R. K. F. Teng, and J. W. Blade, *Cofired Ceramic Technology*, Part 3 in *Multichip Module*, IEEE Press, NY, 1991.

131. P. P. Marcoux, *Fine Pitch Surface Mount Technology*, Van Nostrand, New York, 1992.

132. J. Harrer, R. Petit, O. Brown, D. Navatian, and S. Shahbaz, RF characterization of thick film materials for a portable transceiver, *Int. J. Microcircuits Electron. Packag.*, Vol. 20, No. 2, 1996, pp. 460–469; M. Osawa, Y. Arai, Y. Ando, T. Wada, and M. A. Stein, A superminiaturized double balanced SMT mixer-modulator, *Microwave J.*, Vol. 37, No. 2, Feb. 1994, p. 90.

133. M. D. Gallagher and R. A. Snyder, *Mobile Telecommunication Networking with IS-41*, McGraw-Hill, New York, 1997; H. Basseches, Interconnection systems for solid-state components in hybrid ICs, *Proc. IEEE*, Vol. 59, No. 10, Oct. 1971, pp. 1468–1473.

134. P. Horowitz and W. Hill, *The Art of Electronics*, 2nd ed., Cambridge University Press, Cambridge, 1993, p. 211; T. Watari and H. Murano, Packaging technology for NEC SX supercomputers, *IEEE Trans. Comp. Hybrid Manuf. Technol.*, Vol. 8, Dec. 1985, pp. 462–467.

135. *Supercomputer Technology*, Computer Technology Research Corporation, Charleston South Carolina, 1989.

136. H. B. Bokoglu, Super computer, in *Circuits, Interconnections, and Packaging for VLSI*, Addison-Wesley, Reading, MA, 1990, p. 404.

137. J. R. Lineback, C-MOS gates key affordable supercomputer, *Electron. Week*, Oct. 29, 1984, pp. 17–19.

138. D. H. Sanders, Super computer systems, in *Computers Today*, McGraw-Hill, New York, 1983, pp. 6 and 255–257.

139. C. Marino, Supercomputer applications in automotive research and development, *Comput. Mech.*, Vol. 1, 1986.

140. H. Murano and T. Watari, Packaging technology for NEC SX-3 supercomputers, *IEEE Trans. Comp. Hybrid Manuf. Technol.*, Vol. 15, No. 1, 1992, pp. 411–417.

141. R. R. Tummala and S. Ahmed, Overview: packaging for IBM enterprise system/9000 based on glass-ceramic copper/thin film thermal conducting module, *IEEE Trans. Comp. Hybrid Manuf. Technol.*, Vol. 15, No. 4, 1992.

142. H. B. Bakoglu, *Circuits, Interconnections, and Packaging for VLSI*, Addison-Wesley, Reading, MA, 1990, p. 382.

143. E. K. Kanazawa, D. M. Gilbert, and K. E. Trueba, A high-speed, high-performance, scanning thin-film thermal printed head, *Proc. Int. Symp. Microelectron.*, Los Angeles, 1979, pp. 122–128.

144. R. Glang and L. V. Gregor, Photolithography, in *Handbook of Thin-Film Technology*, 2nd ed., L. I. Maissel and R. Glang, eds., McGraw-Hill, New York, 1983.

145. J. J. Licari and L. R. Enlow, Infrared imaging, in *Hybrid Microelectronic Technology Handbook,* Noyes, Publications, Park Ridge, NJ, 1998, Chap. 8.

146. S. Goldfarb, R. R. Bigler, and R. T. Strong, Thick-film hybrid application to cryogenic infrared imaging array, *Proc. Int. Symp. Microelectron.*, Anaheim, CA, 1985, pp. 162–167.

147. T. C. Edward and M. Steer, *Foundations of Interconnect and Microstrip Design*, Wiley, New York, 2000, Chaps. 4 and 5.

148. M. I. Skolink, *Introduction to Radar Systems*, McGraw-Hill, New York, 1962.

149. R. W. Johnson, ed., *Modular Series in Hybrid Microelectronics*, ISHM, Reston, VA, 1991, p. 6.

150. H. B. Bakoglu, Insulated materials substrates, in *Circuits, Interconnections, and Packaging for VLSI,* Addison-Wesley, Reading, MA, 1990, p. 100; G. Ellison and D. Patelzick, The thermal design of a forced air-cooled power supply, *Proc. IEPS Symp.*, 1986, pp. 829–838.

151. D. B. Tuckerman and F. Pease, High-performance heat sinking for VLSI, *IEEE Electron Devices Lett.*, Vol. 2, No. 5, May 1981, pp. 126–129.

152. C. G. Fonstad, Differential amplifier stages, in *Microelectronics Devices and Circuits,* McGraw-Hill, New York, 1994, Chap. 12.

153. M. Tomana, R. W. Johnson, R. C. Jaeger, and W. C. Dillard, *IEEE Trans. Comp. Manuf. Technol.*, Vol. 16, No. 5, 1993; D. Hopkins, F. Lee, and W. Spherson, Printing of thin-film conductors for power hybrid circuits, *Proc. ISHM*, 1988, pp. 95–101.

154. A. F. Harvey, *Microwave Engineering*, Academic Press, San Diego, CA, 1963.

155. H. J. Reich, P. F. Ordang, H. K. Krauss, and J. E. Skalink, *Microwave Theory and Techniques*, D. Van Nostrand, Princeton, NJ, 1953.

156. L. Lange, Integrated strip line quatrature hybrid, *IEEE Trans. Microwave Theory Tech.*, Vol. 17, Dec. 1959, pp. 1150–1151.

157. D. M. Pozar, The Lange coupler, in *Microwave Engineering*, Addison-Wesley, Reading, MA, 1990, p. 430.

158. P. A. Sandborn and H. Moreno, *Conceptual Design of a Multichip Modules System*, Kluwer Academic, Norwell, MA, 1994; G. Fehr, J. Long, and A. Tippets, New generation of high pin count packages, *IEEE Proc. 1985 (CICC)*, 1985, pp. 46–49.

159. P. E. Garrou and I. Turlik, *Multichip Module Technology Handbook*, McGraw-Hill, New York, 1998; J. M. Segelken, L. J. Wu, M. Lau, K. L. Tai, R. R. Shively, and T. G. Garau, Ultra dense: an MCM board 3-D digital signal processor, *IEEE Trans. Comp. Hybrid Manuf. Technol.*, Vol. 15, No. 4, Apr. 1992.

160. M. Sage, Multichip module, *Proc. Int. Symp. Microelectron.*, Seattle, WA, 1988, p. 1.

161. H. B. Bakoglu, Packaging technologies, in *Circuits, Interconnects, and Packaging for VLSI,* Addison-Wesley, Reading, MA, 1990, pp. 421–425.

162. J. E. Brewer, W. D. Brown, and S. S. Ang, Advanced topics and future technology in MCM technologies, in *Advanced Electronic Packaging with Emphasis on MCM*, W. D. Brown, ed., Wiley, New York, 1999.

163. L. Mark and J. A. Caterina, *Printed Circuit Assembly and Design*, McGraw-Hill, New York, 2000.

RECOMMENDED READING

Chopra, K. L., *Thin-Film Phenomena*, McGraw-Hill, New York, 1969.

Combs, C. F., Jr., *Printed Circuit Handbook*, McGraw-Hill, New York, 1988.

Computer Technology Research Corporation Staff, *Supercomputer Technology*, CTRC, Charleston, South Carolina, 1989.

Ellenbogen, K. A., ed., *Clinical Cardiac Pacing and Defibrillation*, 2nd ed., W.B. Saunders, Philadelphia, 2000.

Galagher, M. D., and R. A. Snyder, *Mobile Telecommunications Networking with IS-41*, McGraw-Hill, New York, 1997.

Garrou, P. E., and I. Turlik, *Multichip Module Technology Handbook*, McGraw-Hill, New York, 1998.

Hammer, D. W., and J. V. Biggers, *Thick-Film Hybrid Microcircuit Technology*, Wiley, New York, 1972.

Harper, C. A., ed., *Electronic Packaging and Interconnection Handbook*, 3rd ed., McGraw-Hill, New York, 2000.

Harper, C. A., and R. M. Simpson, eds., *Electronic Materials and Processes Handbook*, 2nd ed., McGraw-Hill, New York, 1993.

Harte, L., R. Kikta, and R. Levine, *3G Wireless Demystified*, McGraw-Hill, New York, 2001.

Horowitz, P., and W. Hill, *The Art of Electronics*, Cambridge University Press, Cambridge, 1998.

Johnson, R. W., ed., *ISHM Modular Series in Hybrid Microelectronics*, ISHM, Reston, VA, 1991.

Jones, R. D., *Circuit Design and Manufacture*, Marcel Dekker, New York, 1982.

King, J. A., J. Freer, and R. Woodard, eds., *Materials Handbook for Hybrid Microelectronics*, Artech House, Norwood, MA, 1988.

Lau, J., C. P. Wong, J. L. Prince, and W. Nakayama, *Electronic Packaging: Design, Materials, Process, and Reliability*, McGraw-Hill, New York, 1998.

Licari, J. J., and L. R. Enlow, *Hybrid Microcircuit Technology Handbook: Materials, Processes, Design, Testing and Production*, Noyes Publications, Park Ridge, NJ, 1988.

Maissel, L. I., and R. Glang, eds., *Handbook of Thin-Film Technology*, 2nd ed., McGraw-Hill, New York, 1983.

Marino, C., ed., *Supercomputer Applications in Automotive Research and Engineering Development*, Computational Mechanics Publishing, New York, 1986.

Meiksin, Z. H., *Thin and Thick Films for Hybrid Microelectronics*, Lexington Books/ D.C. Heath, Lexington, MA, 1976.

Millman, J., and C. C. Halkias, *Integrated Electronics: Analog and Digital Circuits and Systems*, McGraw-Hill, New York, 1972.

Mitra, S. K., and J. F. Kaiser, eds., *Handbook for Digital Signal Processing*, Wiley, New York, 1993.

Moran, P., *Hybrid Microelectronic Technology*, Gordon & Breach, New York, 1984.

Murarka, S. P., and M. C. Peckerar, *Electronic Materials Science and Technology*, Addison-Wesley, San Diego, CA, 1989.

Pozar, D. M., *Microwave Engineering*, Addison-Wesley, Reading, MA, 1970.

Rikoski, R. A., *Hybrid Microelectronic Circuits: Thick Film*, Wiley, New York, 1980.

Sergent, J. E., and A. Krum, *Thermal Management Handbook*, IMAPS, Washington, DC, 1998.

Stallings, W., *Data and Computer Communications*, 3rd ed., Macmillan, New York, 1988.

Taraseiskey, H., *Power Hybrid Circuit Design and Manufacture*, Marcel Dekker, New York, 1996.

Tummala, R. R., and E. J. Rymaszewski, *Microelectronic Packaging Handbook*, Van Nostrand Reinhold, New York, 1989.

Webster, J. G., ed., *Medical Instrumentation: Application and Design*, Wiley, New York, 1977.

Webster, J. G., *Encyclopedia of Medical Devices and Instrumentation*, Wiley, New York, 1988.

2

MATHEMATICAL FOUNDATIONS, CIRCUIT DESIGN, AND LAYOUT RULES FOR HYBRID MICROCIRCUITS

2.1 MATHEMATICAL FOUNDATIONS

Many different mathematical methods and concepts are used in the design and fabrication of hybrid microcircuits [1]. Many modern circuit elements are based on mathematical models [2], and without basic knowledge of the theory, it is difficult for circuit design engineers, material scientists, and technical personnel involved in the fabrication process to accomplish their jobs. In the first part of this chapter we present a combination of device physics background and technology, a review of existing device models, and more important, new and improved models compatible with the most advanced hybrid technology [3].

In thick-film technology, layers of resistive, dielectric, and conductive pastes are typically deposited on alumina substrate. The deposition of thick film is done by screen printing, which employs a fine mesh of either silk or stainless steel wire mesh to define the pattern of various films [4]. Thin-film circuit elements, by contrast, are formed by sequentially evaporating or sputtering resistive, dielectric, and conductive materials from a source or target onto an alumina substrate inside a vacuum chamber [5]. The fine lines of desired structures are created on the alumina substrate by either an additive or a subtractive method. In the additive method, films are deposited through a mask placed on or near the substrate. The subtractive method is performed using a photolithographic technique [6].

In a typical thick-film process the conductive paste is first screened to make interconnections, and consecutive layers for resistor terminations, bonding

Handbook of Thick- and Thin-Film Hybrid Microelectronics, By Tapan K. Gupta
ISBN 0-471-27229-9

pads, die attachment areas, and capacitor bottom electrodes are printed one after another after successive firing of each layer. After drying, the conductor film is fired at temperatures between 800 and 850°C to burn off the binder and sinter the conductor material [7]. Subsequent screenings and firings provide resistors, capacitors, and crossovers. Layering of conductor and dielectric pastes forms multilayer interconnections. Conductors, resistors, and capacitors are formed with this technology; discrete components are joined to this interconnect system electrically by solder or adhesive attachment.

The basic hybrid microcircuit fabrication is accomplished by adopting either thick- or thin-film technology or a combination of both. Thick- and thin-film technologies are best suited for low-volume custom circuits, complex arrays, and applications requiring medium to high power, high voltage, or stable high-tolerance passive components. Thick film has the added advantage of low product cost, low design cost, and low capital investment; thin films offer more circuit elements per unit area, more stability, and greater precision [8]. However, most passive circuit elements, in the form of chips, are the product of thin-film technology [9,10]. In both thick- and thin-film fabrication, process variation can result in resistor and capacitor values that exceed desired tolerance limits. To adjust the desired values precisely, elements are trimmed to a tolerance up to $\pm 0.2\%$ or better [11].

2.1.1 Factors Affecting the Value of a Resistor

Resistivity The choice of materials for thick-film resistor composition is a vital issue for hybrid circuit fabrication. Good resistor material should be such that it can go to a wide range of resistor values [12]. In fact, overall resistor properties are determined by parameters such as the bulk material resistivity [13], the size and shape of the resistor, and the interaction between resistive and end termination materials. The resistor value of a film (thick or thin) can be written as

$$R = \rho \frac{L}{A} \tag{2.1}$$

where ρ is the resistivity, L the length, and A the resistor area, which is equal to $d \times W$, where d is thickness and W is the width of the resistor. Substituting the values of A, equation (2.1) becomes

$$R = \frac{\rho}{d} \frac{L}{W} \tag{2.2}$$

where ρ/d is the sheet resistance (Ω/sq) and L/W is the aspect ratio or number of squares [14]. Figure 2.1 shows the typical rectangular geometry of a deposited resistor film with dimensions and two end terminations.

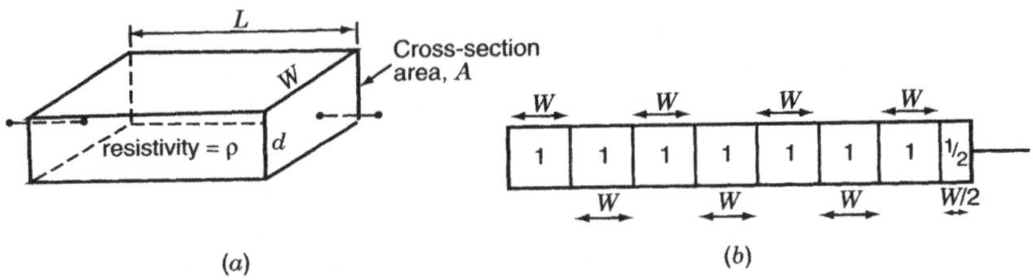

Figure 2.1 Deposited film resistor, showing dimensions: L = length; W, width; d, thickness. (a) Bulk resistivity; (b) total number of squares.

The total film resistivity ρ_{total} can be measured as the sum of three distinct resistivities:

$$\rho_{\text{total}} = \rho_{\text{ideal}} + \rho_{\text{residual}} + \rho_{\text{thickness}} \tag{2.3}$$

where ρ_{ideal} is due to the scattering of conduction electrons by thermal oscillations of the atoms in the lattice, ρ_{residual} is due to scattering of conduction electrons on impurities and other lattice defects [15], and $\rho_{\text{thickness}}$ is due to scattering of conduction electrons at the boundary of the film. Of these, ρ_{ideal} is temperature dependent (approximately proportional to the absolute temperature) and is zero at 0 K. The other two parameters, ρ_{residual} [16] and $\rho_{\text{thickness}}$, are independent of temperature; however, ρ_{residual} depends on temperature if the impurities or defects move out or are annihilated (as happens in annealing of deposited film). On the other hand, $\rho_{\text{thickness}}$ is dependent on film thickness when the thickness of the film approaches the mean free path length of the conduction electrons [17]. At room temperature the mean free path for copper is 400 Å, so that the electron passes about 150 atoms before colliding with one.

According to Matheson's rule, the residual resistivity $\rho_i(c)$, for a dilute solid solution is related to the concentration of the composition. $\rho_i(c)$ is temperature insensitive but very dependent on concentration, due to impurity atoms. For dilute alloys, where impurity atoms can be placed apart from each other, the $\rho_i(c)$ term can be approximated by [18,19]

$$\rho_i(c) = n_i(\Delta Z\, q)^2 \tag{2.4}$$

where n_i is the electron density per unit volume and $\Delta Z\, q$ is the difference between the ionic charges of the solvent and solute atoms.

Temperature Coefficient of Resistance The resistance of a film is dependent on temperature and arises from the interaction between the charge carriers and the lattice vibration mode [20]. A typical change in resistance of a film with temperature is shown in Figure 2.2. For a film resistor, the resistivity ρ as a function of temperature follows approximately as

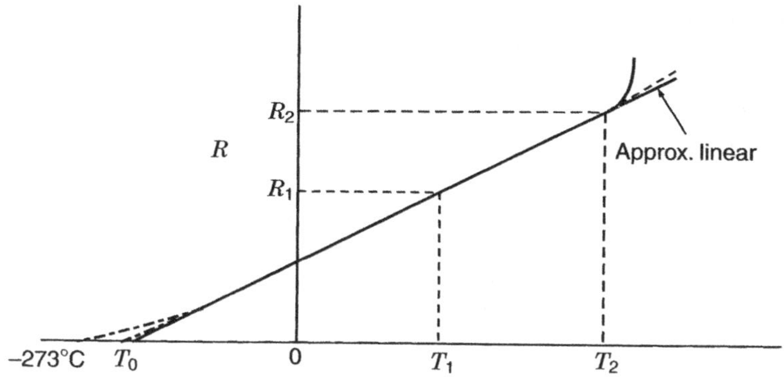

Figure 2.2 Change in a deposited film with temperature.

$$\rho_T = TA \exp\left(\frac{-E_a}{kT}\right) \tag{2.5}$$

where T is the temperature in kelvin (0 K = −273°C), A the material constant, E_a the activation energy, and R the Boltzmann constant.

In the linear range of Figure 2.2, the *temperature coefficient of resistance* (TCR) is defined as the amount of resistance change in a resistor (or resistor material) with temperature. The TCR of a resistor can be expressed mathematically as

$$\text{TCR} = \alpha = \frac{(R_2 - R_1)/R_1}{T_2 - T_0} = \frac{1}{R_1}\frac{\Delta R}{\Delta T} \tag{2.6}$$

TCR is commonly expressed in parts per million per degree Celsius (ppm/°C) and equation (2.6) can be written as

$$\frac{(R_{T1} - R_{T2}) \times 10^6}{R_{T2}(T_1 - T_2)} \qquad \text{ppm/°C} \tag{2.7}$$

where R_{T1} is the resistor value at temperature T_1 and R_{T2} is the resistor value at temperature T_2. In the usual case, test cycles for hybrid circuits are performed at +125°C and −55°C and the TCR at 125°C and −55°C can be expressed mathematically as [following equation (2.7)]

$$\text{TCR}_{125} = \frac{R_{125} - R_{25}}{R_{25}(125 - 25)} \times 10^6 \qquad \text{ppm/°C} \tag{2.8}$$

$$\text{TCR}_{-55} = \frac{R_{-55} - R_{25}}{R_{25}(-55 - 25)} \times 10^6 \qquad \text{ppm/°C} \tag{2.9}$$

where 25°C is assumed to be room temperature.

A typical recommended TCR value for a resistor in a hybrid circuit is approximately 100 ppm/°C. It has been observed that thick-film resistor pastes with sheet resistivity values below 3.5 kΩ/sq, usually show metallike positive TCRs, and above 3.5 kΩ/sq, TCRs are negative. The TCR is one of the important electrical properties of resistors. Since there are increasing applications for thick-film resistors with tight TCR tolerances and/or TCR tracking, there have been continuous efforts to reduce the TCR values of various types of thick-film resistors. On the other hand, the compositions of thin-film resistors generally include at least one transition metal, which dictates the properties of thin films, including TCRs [21,22].

TCR Tracking TCR tracking [23] is the numerical amount of mismatch between two resistors [24] at a given temperature [25,26]. The tracking factor T_f is the vertical spacing between any pair of TCR curves at a given temperature and has the dimensions of TCR. Figure 2.3 shows TCR curves of four resistor geometries of 100-Ω/sq sheet resistivity on the same substrate as a function of temperature and the vertical spacing between pairs of resistors T_f. The area bounded by a pair of curves and at a given interval of temperature is a direct measure of the total departure from ideal resistance tracking in ppm. If T_f is reasonably constant with temperature, the departure is simply the product $T_f \times \Delta T$.

Two resistors may be chosen such that they do not change much with temperature or both change such that $\Delta R_a = k\,\Delta R_b$. The latter case is closer to reality and implies that R_a tracks R_b in resistance value. It can be shown that

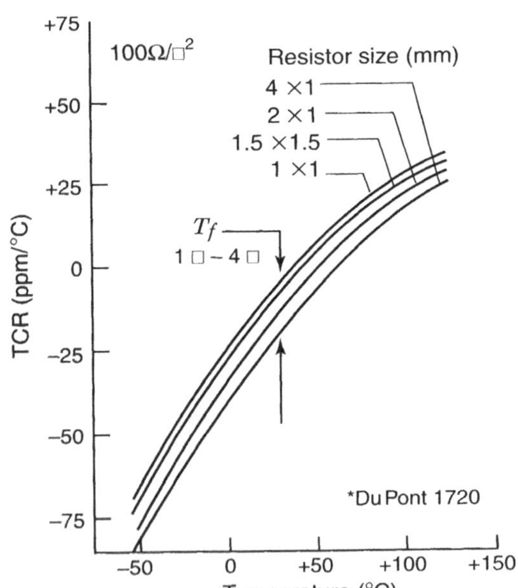

Figure 2.3 Relation between TCR and temperature.

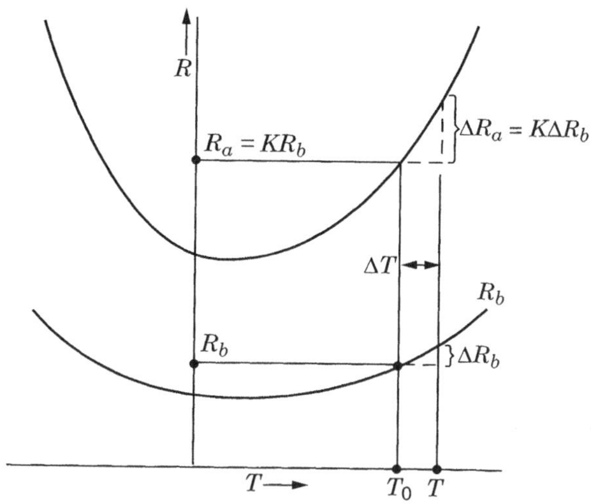

Figure 2.4 TCR-matched resistors R_a and R_b.

where R_a and R_b track resistance at equal temperatures, their TCR values are equal or matched. This is shown in Figure 2.4.

$$\frac{\text{TCR}}{T_0} = \lim_{T \to T_0} \frac{1}{RT_0} \frac{\Delta R}{\Delta T} = \frac{1}{R_b} \frac{dR_b}{dT} = \frac{k}{kR_b} \frac{dR_b}{dT} \tag{2.10}$$
$$V_o = iR_b$$

From the simple divider network of Figure 2.5, the tracking factor T_f can readily be applied to determine the stability of the output voltage $V_o = iR_b$ for a given range of temperatures (provided that the change in the ratio k is small over a temperature interval) and the value of $\Delta V_o / V_o$ is expressed as

$$\frac{\Delta V_o}{V_o} = \frac{k}{(k+1)} T_f \, \Delta T = \frac{R_a}{R_b + R_a} (T_f \, \Delta T) \tag{2.11}$$

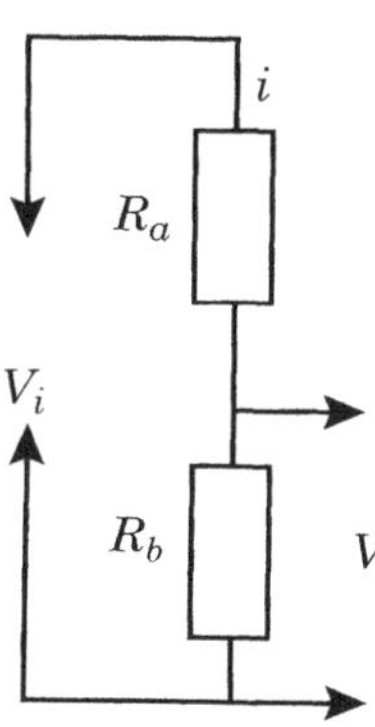

Figure 2.5 Voltage-divider network. i, current flowing through; V_o, output voltage; V_i, input voltage; R_a, R_b, resistors.

where $T_f = (\Delta V_o/\Delta T)(1/V_o)[1 + (R_b/R_a)]$ and $k = R_a/R_b$. Let $R_a = 1\ \text{k}\Omega$ and $R_b = 100\ \Omega$, $T_f = 50$ ppm/°C, and $\Delta T = 100$. Then

$$\begin{aligned} \frac{\Delta V}{V_o} &= \frac{50 \times 100 \times 10}{11} \\ &= 4545.45 \end{aligned} \tag{2.12}$$

Voltage Coefficient of Resistance The voltage coefficient of resistance (VCR) is defined as the fractional change in resistance per volt, with measurements usually being made at full-rated voltage and at one-tenth of that voltage [21,27]. So the VCR of a resistance can be expressed mathematically as

$$\text{VCR} = \frac{R_2 V_2 - R_1 V_1}{R_1 V_1 (V_1 - V_2)} \times 10^6 \qquad \text{ppm/V} \tag{2.13}$$

where V_1 and V_2 are the applied voltages and R_1 and R_2 are the resistances at the respective voltages. According to the definition, if the full-rated continuous working voltage is V_R, the equation for VCR will follow the relation as

$$\text{VCR} = \frac{R V_R - (0.1R) V_R}{(0.1R) V_R (V_R - 0.1 V_R)} \times 10^6 \qquad \text{ppm/V} \tag{2.14}$$

Again, VCR (%) can be expressed as

$$\text{VCR (\%)} = \frac{(R_2 - R_1)200}{(R_2 + R_1)(V_1 - V_2)} \tag{2.15}$$

R_1 and R_2 are the corresponding values of resistors when the voltages are V_1 and V_2. VCRs for resistors are believed to be due to the combined effects of contact resistance and the resistor structure [28].

Termination Effect The resistivity of a resistor is influenced by the contact resistance at the conductor/resistor interface [29,30]. The contact resistance is defined as

$$R_c = \frac{1}{(dJ/dV)_{V=0}} \tag{2.16}$$

where J is the current density, V the applied voltage, and R_c is in units of Ω·cm. For the Schottky barrier diode, the value of R_c changes to

$$R_c = \frac{kT}{eJs} = \frac{kT}{eAT} \exp\left(\frac{e\varphi}{kT}\right) \tag{2.17}$$

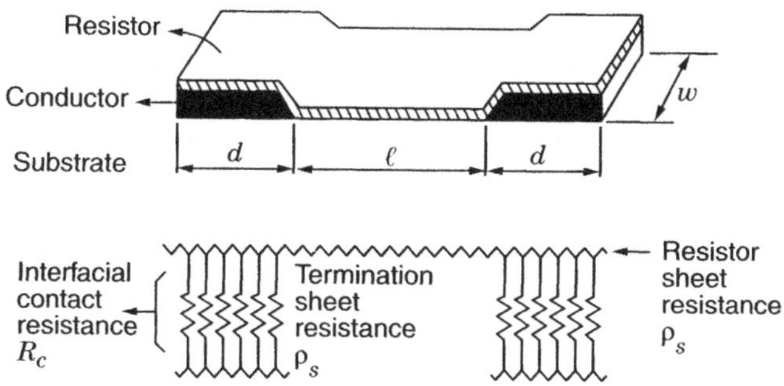

Figure 2.6 Schematic diagram of deposited resistor film and contact resistance. $l =$ length of the resistor; d, thickness; w, width; ρ_s, sheet resistance of the resistor.

where k is the Boltzmann constant, T the temperature in absolute degrees kelvin, $e = -q =$ charge on electrons, A is the Richardson constant, and φ the potential barrier.

To understand the essence of the science and to correlate the mechanism of the termination effect [31], a theoretical model is presented here. Figure 2.6 shows the schematic diagram and equivalent-circuit parameters of a deposited resistor film. Let us assume that the interfacial material has a uniform layer of resistive material characterized by its specific conductance G (S/sq). Let the sheet resistance of the resistor film be ρ_s and that of the termination be ρ_s'. The total resistance of the film resistor is made up of three parts: (1) the main body of the resistor, $R_f = \rho_s(1/w)$; (2) the parallel combination of the termination material and resistor material (R_p), equal to $[\rho_s'(2d/w)]/[(1 + (\rho_s'/\rho_s)]$, where d is the length of the termination material; and (3) the interfacial contact resistance at the conductor/resistor interface. This term can be calculated following the analysis given by Naguib [32] or Galla et al. [33]. In either case the result is

$$R_{\text{total}} = R_f + R_p + 2R_c \tag{2.18}$$

where R_p is the parallel combination of the resistors and R_c and R_f are the interfacial contact resistance at the conductor/film resistor interface and film resistance, respectively.

$$R_{\text{total}} = R_f + R_p + 2R_c = \rho_s \frac{1}{w} + \rho_s' \frac{2d}{w}\left(1 + \frac{\rho_s'}{\rho_s}\right)^{-1} + \frac{2}{w}\left(\frac{\rho_s}{G}\right)^{1/2} \tag{2.19}$$

where $2R_c = (2/w)(\rho_s/G)^{1/2}$. Again, $\rho_{s\,\text{total}}$ can be written as

$$\rho_{s\,\text{total}} = \rho_s + \rho_s' \frac{2d}{l}\left(1 + \frac{\rho_s'}{\rho_s}\right)^{-1} + \frac{2}{l}\left(\frac{\rho_s}{G}\right)^{1/2} \tag{2.20}$$

where G is the conductance of the material. So $R_{\text{total}} = \rho_{s\,\text{total}}(l_{\text{total}}/w)$. All three terms in equation (2.20) contribute to the aging effect of the resistor.

Structural Dependence of Resistivity The behavior of a resistor, which is a composite of two-phase alloys, behaves differently [34,35]. In this case the material is essentially a mixture of two different resistivities. From an electrical viewpoint it is a complicated network [36] of resistances in series–parallel combinations with one another. It can be shown that the resistance of such a material varies linearly with the volume fraction of either constituent. Thus, if the resistivities of two phases, α and β, are ρ_α and ρ_β, we can write

$$\rho_{\text{total}} = \rho_\alpha V_\alpha + \rho_\beta V_\beta \tag{2.21}$$

where V_α and V_β are the volume fractions of phases α and β.

Noise Noise in an electrical system is defined as random fluctuations of power [37], and accordingly, the noise within a film resistor will depend on the constituents of the resistor and the process parameters. In most cases the types of noises in film resistors are (1) thermal noise, (2) $1/f$ noise, (3) burst noise, (4) contact noise, and (5) current noise.

Thermal Noise The thermal noise power of a resistance R at a temperature T in the small frequency interval Δf can be written as

$$P_{\text{av}} = \frac{S_v(f)\,\Delta f}{4R} = kT\,\Delta f \tag{2.22}$$

where S_v is the spectral intensity of the thermal noise voltage of a resistance R at temperature T and is equal to $4kTR$, and k is the Boltzmann constant. The equation holds well in the lower frequency range [38].

$1/f$ Noise Flicker or $1/f$ noise has a $1/f^\alpha$ spectrum with α close to unity [39]. If a constant current I_0 is fed through the resistor, the open-circuit voltage spectrum will vary as I_0^2. The noise generated in the resistor is due to the fluctuations in the resistor value in time t and is given by [38]

$$R = R_0 + \delta R(t) \tag{2.23}$$

where R_0 is the average resistance, $\delta V(t) = I_0\,\delta R(t)$, $\delta R(t) = \delta V(t)/I_0$, and

$$S_v(f) = I_0^2 S_R(f) = \frac{V_0^2 S_R(f)}{R_0^2} \tag{2.24}$$

Then the $1/f$ noise expression in the resistor will be

$$\frac{1}{f} = \frac{S_v(f)}{K V_0^2 R_0} \tag{2.25}$$

where K is a constant for the material in question.

Burst or Popcorn Noise In the audio-frequency range, the magnitude of the burst or popcorn noise is larger than the conventional $1/f$ noise in the resistor [40] and is given by

$$S_I = A_1 I^2 + A_2 \exp(\beta I) \tag{2.26}$$

The first term represents the $1/f$ noise and the second term represents the burst noise. Burst noise has frequently been observed in highly resistive thick film [41]. The presence of burst noise in a trimmed thick-film resistor is often larger than $1/f$ noise fluctuations [42,43]. When burst noise exists, the noise waveform of the resistor consists of current noise superimposed on nonstationary random pulses of different amplitudes. The frequency spectrum lies in the range 30 Hz to 10 kHz. The total noise, including burst, shot, current, and thermal noise, is found to have a k/f^α relationship. The average value of α is found to have a value of 1.0 regardless of the integration time. The frequency f of the burst noise spectra is found to be 30 Hz $< f <$ 10 kHz. Below 30 Hz, burst noise seems to level off and is dominated by conventional current $(1/f)$ noise. Burst noise is of special interest to designers of telecommunications and data processing systems. Experimental data show that the burst noise frequency spectrum is dependent on (1) bias voltage, (2) annealing temperature, and (3) environmental temperature. Excess burst noise can be reduced through the use of thick-film resistor ink with a lower sheet resistance [40]. Resistors having blisters and/or craters show more burst noise, due to electric field spots in the resistors.

Contact Noise Contact noise arises from the interface of the resistor/contact material. The magnitude of contact noise depends on the interface material and the geometry of the film resistor. Noise voltage can be expressed mathematically as [44]

$$V^2 = i^2 R^2 = \frac{V_s^2 (R_1 + R_f)}{R_s + R_1 + R_f} + V_f^2 \left(\frac{R_s}{R_s + R_1 + R_f} \right)^2 \tag{2.27}$$

where

$$i^2 = \frac{V_s^2}{R_s^2} + \frac{V_f^2}{(R_1 + R_f)^2} \quad \text{and} \quad R = \frac{R_s (R_1 + R_f)}{R_s + R_1 + R_f} \tag{2.28}$$

V_s is the thermal noise voltage of R_s, V_f the excess noise voltage generated, R_s the resistance connected in series, and R_f the lower portion of the resistor. Then the square of V_f can be written as

$$V_f^2 = \left(1 + \frac{R_1 + R_f}{R_s} \right)^2 V^2 - \left(\frac{R_1 + R_f}{R_s} \right)^2 V_s^2 \tag{2.29}$$

The second term has a negligible effect on the total noise value.

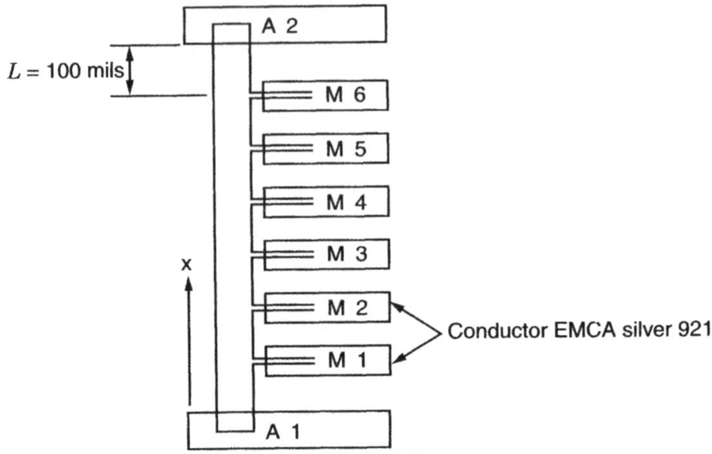

Figure 2.7 Resistor configuration [44]. (Reprinted from the September 1978 edition of *Solid State Technology*. Copyright 1978 by PennWell Publishing.)

Figure 2.7 is a resistor configuration in which contacts A1 and A2 carry dc biasing. A1 and A2, including M1 and M6, are used for noise measurement. Figure 2.8*a* shows the circuit used for measuring the excess noise generated between contact A1 and leg M1. In this circuit, the capacitance C is used to short circuit the resistor noise generated in the upper portion of the resistor. Since the resistance of each leg, R1, is much smaller than that of the upper portion of the resistor, R_f^2, this circuit can be drawn as in Figure 2.8*b* and then simplified as in Figure 2.8*c*.

Current Noise Current noise [45], which is observed in all resistances except those manufactured from bulk metallic alloy [46], is a form of excess noise observed above the thermal or Johnson noise level which occurs when a current is passed through a resistor. The root-mean-square (rms) value of the noise voltage V_n is expressed mathematically as

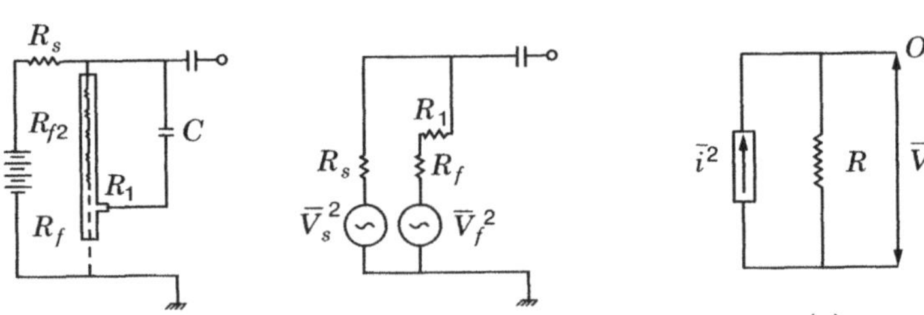

Figure 2.8 Noise equivalent circuit [44]. (Reprinted from the September 1978 edition of *Solid State Technology*. Copyright 1978 by PennWell Publishing.)

$$V_n^2 \propto \frac{V^\alpha df}{f^\beta} \tag{2.30}$$

where f is the frequency and the values of α and β are approximately 2 and 1.

The *noise index*, which is a measure of current noise, is given by [47]

$$\text{NI} = 20 \log K \tag{2.31}$$

where K is equal to V_n/V and V is the applied voltage. For an individual generator generating V_n rms volts (microvolts) of noise, the dc voltage occurring across the generator will be V/gl and equation (2.31) will become

$$\text{NI} = 20 \log \frac{K}{g(lw)^{1/2}} \tag{2.32}$$

where l and w are the length and width of the resistor, gl the number of generators along length l, and gw the number of generators across width w.

2.1.2 Mathematical Model for Thick-Film Deposition

In thick-film deposition the electrical circuit pattern is printed on a ceramic substrate [48]. The slurry (ink) is forced through the screen fabric opening with a squeegee. The thickness, uniformity, and flow of the ink depend on several parameters of the printing system [49]. To explain the mechanism of screen printing technology, a mathematical model is presented here [50].

The forces acting on the mesh during pull down are expressed as

$$f_m = 4\tau D(2d) + d\left(\frac{1}{M} + D\right)P_a \tag{2.33}$$

where $\tau = \eta(dw/dr)$ = shear stress at the edge of mesh opening, η is the viscosity of the slurry (ink), D the diameter of the mesh, d the wire diameter, P_a the atmospheric pressure, M the mesh per unit length, and dw/dr the paste (ink) velocity slope at the edge of the mesh opening where $w = -[(32/\pi)(Q/D^4)][(D^2/M) - r^2]$, and the value of τ and f_m turn out to be

$$\tau = 16\left(\eta \frac{V}{D}\right)\frac{a}{S+B} \tag{2.34}$$

and

$$f_m = 128\eta V d \frac{a}{S+B} + d\left(\frac{1}{M} + D\right)P_a \tag{2.35}$$

The pull-down force

$$F_m = Md^2\left[64\eta V + \frac{S+B}{2a}\left(\frac{1}{M} + D\right)\right]P_a \tag{2.36}$$

Now the fabric separation immediately behind the squeegee $= \tan\delta = 2a/(S+B)$, where δ is the snap-off angle [51]. The viscous flow of the ink through the screen mesh can be written as [52,53]

$$Q = \text{fluid volume flow} = \frac{D^2\pi v}{4} = \frac{\pi D^4 P_a}{128\eta 2d} \tag{2.37}$$

$$P_a \geq 64\left(\frac{d}{D^2}\right)\eta v \quad \text{and} \quad \eta V \leq \frac{1}{128}\left(\frac{D^2}{d}\right)\frac{S-B}{a}P_a$$

where S is the screen size, V the squeezee speed, B the length of the printed pattern, a the snap-off, and v the snap-off velocity of the screen.

2.1.3 Theoretical Model for Screen-Printed Film Thickness

Thick film is deposited on a substrate with the help of a screen [54]. The relative thickness of the wet film can be written as [55,56]

$$T_w = D\left(2 - \frac{D\pi}{2ma}\right) \tag{2.38}$$

where T_w is the relative thickness in mils (1 mil = 25.4 μm), D the fabric wire diameter in mils, m is $1000/M$ in mils (M is the mesh count per inch), and a is equal to cos(arctan $DM/1000$). The experimental data are found to fit well with [57]

$$T_w = HA \tag{2.39}$$

where H is the mesh thickness and A is the decimal equivalent of the percentage of open area. Rheological studies on thick-film paste offer a potentially useful means of formulating rheological criteria useful for screen printing. Experimental observation has demonstrated that the viscosity of the paste should be on the order of 100 P at a shear rate of about 600 s^{-1} [58,59].

2.1.4 Thick-Film Resistor Design

The empirical equation for the resistance of a thick-film resistor can be written as [60,61]

$$R = 10^{B_0} L^{B_1} W^{B_2} \tag{2.40}$$

The coefficients have the values $B_0 = 3.3$, 4.3, and 5.34; $B_1 = 1.1$; $B_2 = -0.98$; and L is the length and W the width of the resistor. The power

$$P = dLW \tag{2.41}$$

where d is the power density in $\mathrm{W/mm^2} = 0.01r + 0.03$ and the values of r can be as follows:

r (%)	d (W/mm^2)	d (W/in^2)
1	0.04	25.81
5	0.08	51.61

The length

$$L = \frac{K_L R}{10^{B_0}}\left[\frac{1}{(P/d)^{B_2}}\right]^{1/(B_1 - B_2)}$$

and the width

$$W = \left[\frac{K_W R}{10^{B_0}}\left(\frac{1}{L}\right)^{B_1}\right]^{1/B_2}$$

The values $K_L \approx 1.3$ and $K_W \approx 0.7$ are good approximations.

2.1.5 Theoretical Model for Thin-Film Thickness

Thin-film deposition is generally done by vacuum evaporation. Details of the deposition techniques are given in Chapter 7. The distribution of the evaporated material is given in terms of θ, φ, and r (Figure 2.9) [62]:

$$\text{mass deposited per unit area} = \frac{dm_r(\varphi\theta)}{dA_r} = \frac{m_e \cos\varphi \cos\theta}{\pi r^2} \tag{2.42}$$

where $dA_r = r^2 d\omega/\cos\theta$ and m_e is the total mass of the source material.

$$\frac{dm_r}{dA_r} = \frac{(m_e \cos\theta)}{4\pi r^2} \tag{2.43}$$

For point-source emission, equation (2.43) reduces to

$$\text{film thickness } d = \frac{1}{\rho}\frac{dm_r}{dA_r} \tag{2.44}$$

where ρ is the density of the film.

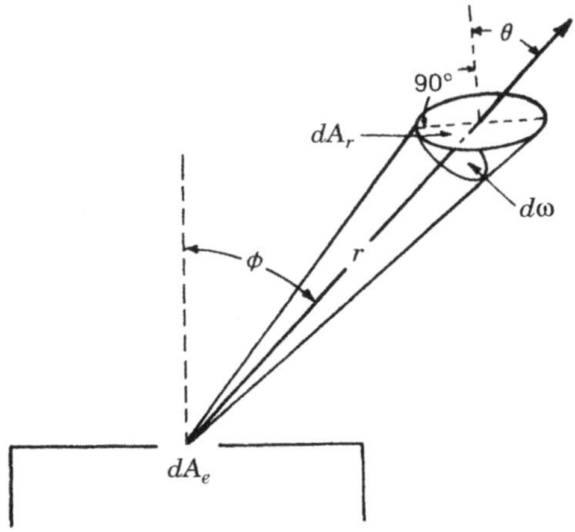

Figure 2.9 Surface element dA_r receiving a deposit from a small-area source dA_e.

Factors influencing the desired shape of the source and their amenability to different modes affect the complexity of the source and the external power supply. Attempts to meet these demands have led to numerous source designs [63]:

For a small source: $$d = \frac{m_e}{\pi \rho h^2 [1 + (l/h^2]^2} \tag{2.45}$$

For a point source: $$d = \frac{m_e}{4\pi \rho h^2 [1 + (l/h^2]^{3/2}} \tag{2.46}$$

where l is the distance from the center of the source material to the substrate and h is $r \cos\theta = r \cos\varphi = h$.

2.1.6 Dissipation Factor or Dielectric Loss within a Dielectric Material

When an electric field acts on any matter, it dissipates a certain quantity of its energy in the form of heat. This dissipation of electric power in a specimen is a measure of electric loss [64]. If the frequency of the applied voltage is higher, the loss of energy will be greater. The dissipation factor is a function of the frequency of the applied voltage, the temperature, and the dielectric material [65]. The dielectric losses of a ceramic substrate having capacitance C can be written as

$$P = V^2 \omega C \tan\delta \tag{2.47}$$

where ω is the angular frequency $= 2\pi f$, V the applied voltage, δ the dielectric loss angle, and $\tan\delta$ the loss tangent or dissipation factor. Figure 2.10 shows the current voltage relation through a dielectric material. When I_a and I_r are

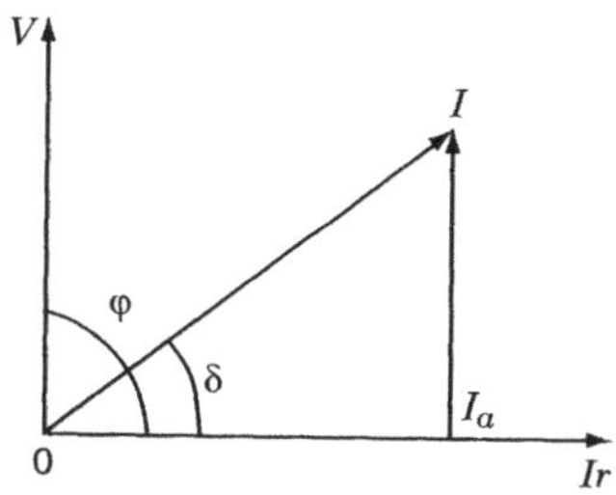

Figure 2.10 Current–voltage relation through a dielectric material.

the active and reactive components of the current I passing through the capacitor, w the width, L the length, and t the thickness of the dielectric, we have

$$P = V^2(2\pi f)\varepsilon_0\varepsilon_r \frac{wL}{t} \tan\delta \tag{2.48}$$

where $\delta = 90 - \varphi$ and $\tan\delta = I_a/I_r$. The quality factor

$$Q = \frac{1}{\tan\delta} = \cot\delta = \tan\varphi \tag{2.49}$$

$\varepsilon_0 = 8.854 \times 10^{-12}$ F/m, and ε_r is the relative permittivity.

The total loss in the dielectric [66] may be represented as the loss in an equivalent resistor connected across the capacitor. For a low-frequency circuit, a parallel equivalent circuit is used, whereas for high frequency, the equivalent circuit is regarded as a series circuit (Figure 2.11).

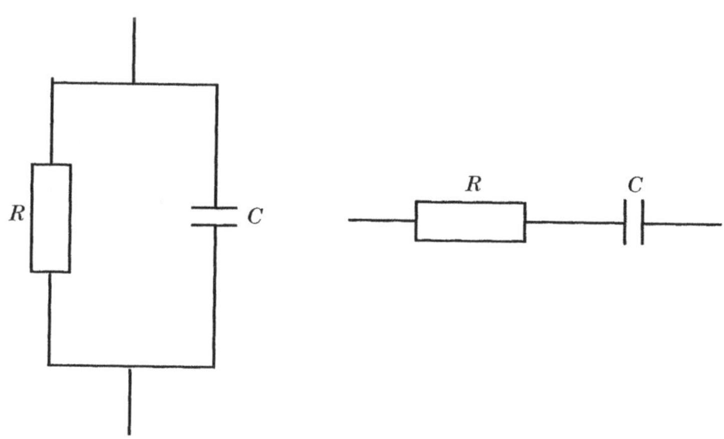

Figure 2.11 (*a*) Parallel equivalent circuit; (*b*) series equivalent circuit. *R*, resistor; *C*, capacitance [67].

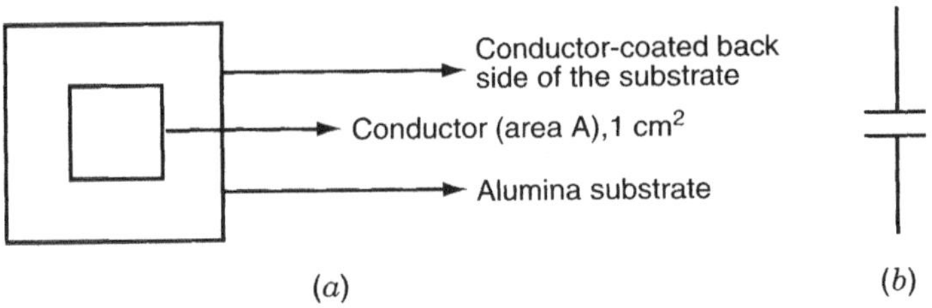

Figure 2.12 Calculation of stray capacitance (alumina substrate): (*a*) alumina coated with a conducting material; (*b*) equivalent-circuit capacitance.

For a parallel equivalent circuit,

$$\tan\delta = \frac{I_R}{I_C} = \frac{1}{R_p\omega C_p} = \frac{\text{capacitive reactance}}{\text{parallel resistance}} \tag{2.50}$$

For a series equivalent circuit,

$$\tan\delta = R_s\omega C_s = \frac{\text{series resistance}}{\text{capacitive reactance}} \tag{2.51}$$

Figure 2.12*a* shows alumina substrate coated with a conductor (area *A*), and the equivalent stray capacitance that will appear during electrical measurements is shown in Figure 2.12*b*:

$$\text{stray capacitance} = \varepsilon_0\varepsilon_r\frac{A}{t} \tag{2.52}$$

Let us assume that area A of the conductor $= 0.155\ \text{in}^2$ ($1\ \text{cm}^2$), t is the thickness of the alumina substrate $= 25$ mils $\simeq 625\ \mu\text{m}$, $\varepsilon_0 = 8.854 \times 10^{-12}$ F/m $=$ permittivity of the free space, and $\varepsilon_r =$ relative permittivity of alumina $=$ 9.6. Then

$$\text{stray capacitance} = 8.854 \times 10^{-12} \times 9.6 \times \frac{(0.01)^2}{625 \times 10^{-6}}\ \text{F} \simeq 13.6\ \text{pF} \tag{2.53}$$

Two conductors are running parallel on an alumina substrate (Figure 2.13) 10 mils (254 μm) wide and 200 mils (0.5 cm) long with an area 2000 mils^2 ($0.0127\ \text{cm}^2$). The thickness of the substrate is 25 mils (635 μm). Then the stray capacitance will be 0.18 pF. For alumina substrate, the rule of thumb is that

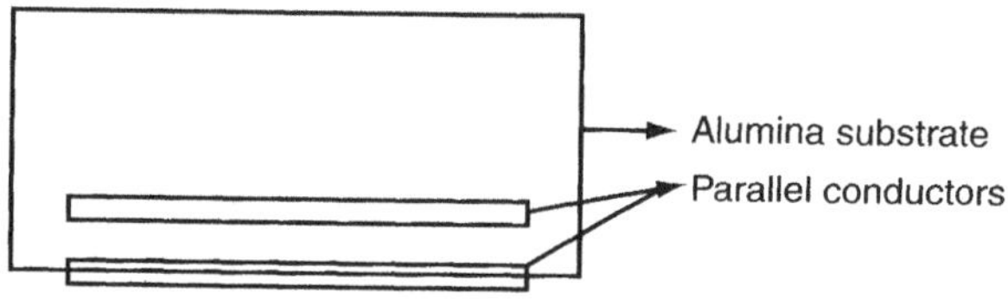

Figure 2.13 Capacitance coupling between parallel conductors.

capacitance = 1.0 pF/in. (0.4 pF/cm) with equal runs and spaces [67]. For a 10-mil (254-μm) conductor crossing a 10-mil (254-μm) conductor, the coupling capacitance will be approximately 0.15 pF. A typical dielectric constant of 9.4 corresponds to a capacitance of 0.0015 pF/mils^2 (2.3×10^{-6} pF/μm^2) for a 1.4-mil (35.56-μm) fired dielectric.

2.1.7 Inductors

When a conductor is looped into one or more turns in a cylindrical manner, the current in adjacent turns flows in the same direction and the magnetic fields reinforce each other [68]. The total flux of the turn therefore increases, due to the additional current induced by intercepting flux lines from any other turn. Such a coil forms an inductor [69]. In a multiple-layer solenoid (e.g., a bobbin winding), the flux from each turn is reinforced by interaction with flux from each turn and is reinforced additionally by interaction with flux from turns of the adjacent layers. The total lines of force are many times greater than the arithmetic sum [70]. The current causes an IR drop (following Ohm's law) in each turn, producing a difference of potential between turns. This potential difference creates an electric field, and thus a distributed capacitance C_D exists between turns. The C_D value for a single-layer solenoid is given by

$$C_D = \frac{\text{capacitance between a pair of adjacent turns}}{\text{total number of turns}}$$

and

$$C_D \text{ for a bobbin winding} = \frac{C_D \text{ of one layer}}{(N-1) \text{ layers}}$$

Inductors cannot be avoided in RF circuits, although their use can be avoided in low-frequency circuits. The designs are generally spiral, circular, or square. Current inductor designs involve the addition of one or more layers of ferromagnetic paste. Each ferrite layer is printed, dried, and fired, and the spiral coil is then printed on top of the ferrite layers [71]. Another type of inductor is a meander-type coil, and the inductance L can be calculated as [72]

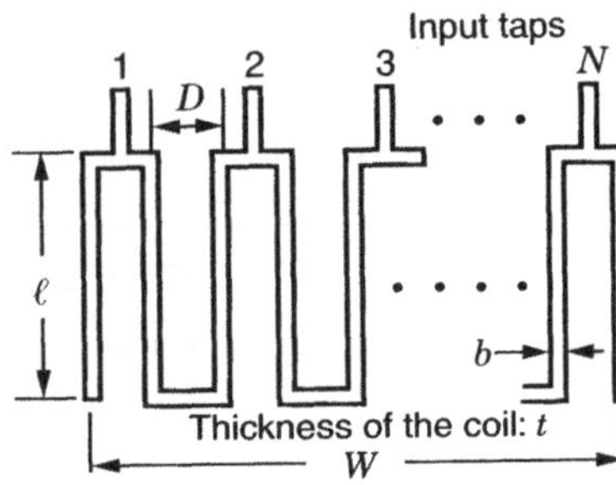

Figure 2.14 Schematic of a meander-type inductor [73].

$$L = \frac{\mu_c l N}{\pi} \log \frac{D - b}{a} \tag{2.54}$$

where $N = \left(\frac{W + D}{2D}\right)$ = number of windings, and μ_c is the permeability of the core material, b the width of the current flowing path, l the length of two parallel wires, N the number of windings, D the distance between two parallel wires, a the effective radius of the thick-film meander path, and W the total width of the meandering coil $= D(2N - 1)$. Figure 2.14 shows the pattern of a meandering inductance coil where a is the effective radius of the thick-film meandering path.

For flat spiral [74] inductors, the following equation can be used:

$$L = cn^2 aF \tag{2.55}$$

where c is a constant, n the number of turns, a the linear dimension of the coil, and F a factor depending on the ratio of outer to inner winding size.

For circular spiral inductors we can use the following equation derived by Burkett [75]:

$$L = \frac{0.8a^2n^2}{6a + 10c} \tag{2.56}$$

where $a = (d_o + d_i)/4$ mils, $c = d_o - d_i$, and $n = [0.5(d_o - d_i) - w]/(s + w)$ [w is the conductor width (mils), s refers to the surface area of the coil, and d_o and d_i refer to the outer and inner diameters of the coil]. The quality factor Q can be represented mathematically as

$$Q = 0.2\sqrt{\frac{f}{\pi \rho \mu_c}} d_o \frac{x^2 - 1}{7x - 13} \times 10^{-9} \tag{2.57}$$

where μ_c is the permittivity of the core material.

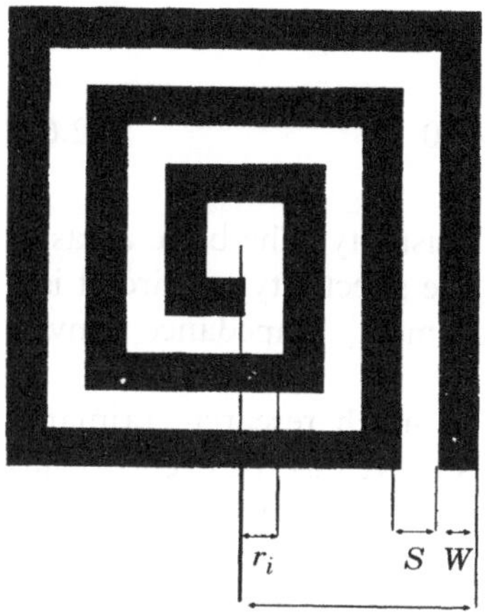

Figure 2.15 Square spiral inductor.

It has been observed that a square geometry results in maximum inductance for a given size, while a circular configuration yields the highest Q for a given inductance. The inductance L for a square spiral (Figure 2.15) is given by $L = 0.0216S^{1/2}N^{5/3}$ (nH), where N is the number of turns and S is the surface area of the coils in mils^2. A toroidal structure offers significant advantages if maximum inductance is wanted in a given volume. The value of the toroidal (Figure 2.16) inductance is calculated as [76–78]

$$L_{\max} = \frac{4\pi\mu_c/D}{V/d^2} \tag{2.58}$$

where d is the wire diameter, D the spacing of the printed coil, μ_c the permeability of the core material, and V the volume of the coils.

$$L = \frac{d_o(1+x)^2 n^2}{10(13-7x)} \qquad \text{nH} \tag{2.59}$$

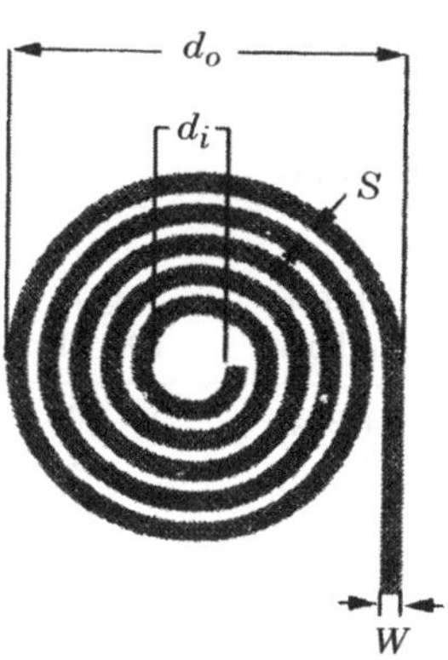

Figure 2.16 Circular inductor designed for hybrid circuits.

and

$$Q = 0.2\sqrt{\frac{f}{\pi\mu\rho}\frac{d_i^2 - d_o^2}{7d_i - 13d_o}} \times 10^{-9} \tag{2.60}$$

where $x = d_o/d_i$, f is the frequency, and ρ is the resistivity. The basic areas of applications for inductors in hybrid microcircuits are selectivity, ac circuit isolation, phase stripping (for correction and displacement), impedance conversion, pulse shaping, and phase distortion filtering.

Thick-film inductors are becoming the focus of much research, primarily because of developments in thick-film ferrite pastes as well as the ever-growing applications of thick-film technology to RF frequency [72]. Typically, the term *thick-film inductor* is used to refer to inductors fabricated mainly by a printed conductor coil made out of gold with or without some form of deposited magnetic material (ferrite). Gold is preferred over silver or copper for fine-line definition. The problem with silver is the metal electromigration, and special firing conditions are needed for copper conductor. Typical values of deposited thick-film inductors together with the number of turns and respective Q values are given in Table 2.1. Thick-film inductors have already been used successfully at high frequencies in RF tuned amplifiers and matching networks [79], TV filters [80], power electronic applications [81], and power amplifiers [70].

Thin-film technology generally uses aluminum as the metal conductor and silicon monoxide as the dielectric for the fabrication of microminiature inductors. The inductors are fabricated using vacuum deposition techniques, most often in a rectangular configuration, and the turns are of the same dimensions. The inductor coils are laid out in a single-layer pattern whose axis is normal to the substrate plane. A high inductance occupies a large surface area as a result of the increased number of turns, and is undesirable, especially when the device is to be mounted on a ground plane and operated at ultrahigh frequency and microwave ranges. The empirical equation relating to the inductance of the coil is given by [82]

$$L = \frac{0.009N^2 a a_1 F}{(b - 0.002)(a_1 + a)N(G + X)} \quad \mu\text{H} \tag{2.61}$$

TABLE 2.1 Characteristics of Various Thick-Film Conductor Materials

Material	Dc Resistivity, ρ (μΩ·cm)[a]	Skin Depth at 100 MHz (10^{-6} m)[b]	L (nH)	Number of Turns	Q at 100 MHz
Silver	1.59	6.40	20	1(3/4)	60
Copper	1.67	6.60	28	2(3/4)	45
Gold	2.35	7.86	44	3(3/4)	42
Aluminum	2.65	8.14	64	4(3/4)	36

[a] From *Handbook of Chemistry and Physics*, 55th ed., CRC Press, Cleveland, OH, 1974–1975.
[b] From D. M. Pozar, *Microwave Engineering*, Addison-Wesley, Reading, MA, 1990, p. 27.

where $X = 0.3(N-1)^2/N^2$, G and F are correction factors, a and a_1 are the dimensions of the core, b is the length of the inductor, and N is the number of turns. The self-resonant frequency f_0 is related to

$$f_0 = \frac{1}{2\pi(LC_d)^{1/2}} \quad \text{and} \quad Q = Q_0(1 - X^2) \tag{2.62}$$

where L is the inductance, C_d the distributed capacitance, and $Q_0 = 2\pi fL/R_s$ (R_s is the series resistance of the metal). For aluminum 0.0016 to 0.2 mil (400 to 50,000 Å) thick, skin effect is negligible in the frequency range 10 to 250 MHz [83].

Solid surfaces can be strongly reflecting or absorbing to the electromagnetic waves. These effects depend on the frequency of the incident radiation. At very high frequencies (above 100 kHz), the absorption coefficient is very high and the electromagnetic wave is very rapidly attenuated as it enters the metal surface. The damping distance, which we call the classical skin depth δ, is represented as

$$\delta = \frac{C}{k\omega} = \frac{c}{(2\pi\sigma\omega)^{1/2}} \tag{2.63}$$

where δ is the classical skin depth, $\omega = 2\pi f$ (where f is the frequency in hertz), σ is the conductivity, c the velocity of light, and k the extinction coefficient governing the damping of the wave medium. This phenomenon, at which current falls to about 37% of its value, is well known classically as the skin effect [84]. For high-frequency (above 100 kHz) currents, Joule heating occurs predominantly near the surface of the material, and the current induced decreases rapidly toward the interior.

2.1.8 Theoretical Model for Transport Properties during Hermetic Sealing

Hermetic sealing is a routine operation in the microelectronic industry. Some problems regarding the hermetic sealing test can be understood by considering theoretical models (e.g., transport properties of liquids [85] and gases). Let us assume free molecular flow and set up gas transport into and out of the package as [86,87]

$$Q = \frac{d}{dt}PV = F(P_u - P) \tag{2.64}$$

where P is the interior pressure, P_u the pressure at the upper stream, F the conductance, and V the volume of the transport material. The leak size L is defined as

$$L = FP_0 \tag{2.65}$$

where $P_0 = 1$ standard atmosphere. If $P_u = P_E$ for a period T and is then removed for a period t so that the tracer gas effuses back through the leak, the interior partial pressure can be represented by [88]

$$P = P_E\left[1 - \exp\left(\frac{-LT}{P_0 V}\right)\right] \exp\left(\frac{-LT}{P_0 V}\right) \tag{2.66}$$

For a laminar flow [89] approximation, the transport equation will be

$$Q = \frac{d}{dt} PVK(P_u^2 - P^2) \tag{2.67}$$

where K is equal to $\pi r^4/(16\eta l)$ (η is the viscosity of fluid in question, r is the radius, and l is the channel length).

In the helium test method, after pressurization the package is placed in a test chamber with liquid helium and the test chamber is evacuated. The leak rate R is given by

$$R = L\frac{P}{P_0} \tag{2.68}$$

On the other hand, in a laminar flow model, the radioisotope test can be described by the equation

$$\frac{P_2}{P_0} = \frac{1 - (1 - \alpha\gamma)/[(1 + \alpha\gamma)\exp(-t/\tau)]}{1 + (1 - \alpha\gamma)/(1 + \alpha\gamma)\exp(-t/\tau)} \tag{2.69}$$

where

$$\gamma = \frac{P_1}{P_t} = \frac{1 - [(\alpha - 1)/(\alpha + 1)\exp(-\alpha t/\tau)]}{1 + [(\alpha - 1\gamma)/(\alpha + 1\gamma)\exp(-\alpha t/\tau)]} \tag{2.70}$$

where $\alpha = P_t/P_0$, t is the dwell time, P_t the pressurization value, P_1 the pressure following pressurization, P_2 the final value, and $\tau = P_0 V/2L = \frac{1}{2}$ for molecular flow.

2.2 CIRCUIT DESIGN AND LAYOUT RULES

A hybrid microcircuit is a high-density electronic package [90]. Depending on the size of the substrate and the components, more than 45 to 50 components per square inch can quite easily be accommodated on a single ceramic substrate. To achieve such a high-density electronic package over a single ceramic substrate, resistor and conductor line widths must be tightly controlled and the spacing between the components minimized. So in designing a hybrid microcircuit the designer should seek to achieve a cost-effective way of accommodating smaller circuits [91]. The most important part of designing a hybrid microcircuit is a trade-off between circuit performance and cost. In all instances it is imperative that the set of specifications to be met be formalized at the inception of hybrid microcircuit design. In designing a hybrid microcircuit, several calculations are carried out. The calculations are simple, but the optimization process is tedious, particularly if several ink combinations are to be used for a particular hybrid microcircuit. These calculations include size and thickness of substrate, number of circuit layers, and shape, size, and thickness of resistors, capacitors, and sometimes, inductors [92].

The first step in hybrid circuit design is *partitioning* [93], the act of dividing an electronic subsystem into clearly defined subfunctions with a minimal interconnection interface. Successful development and application of a hybrid circuit start at the early stages of system design and require interactions of representatives of several technical disciplines on various management levels. In most applications requiring hybrid electronics implementation, the application engineer has to fabricate a prototype circuit such as a *breadboard or brassboard.* This is done to perform electrical evaluation of the circuit before its final design. Once the electrical functionality of the hybrid microcircuit has been evaluated successfully, the circuitry is transferred to the design engineer for miniaturization. To design a completely functional miniature version of a circuit, it is also important that the circuit designer understand the basic functional performance of the circuit elements.

Our purpose in this chapter is to define constraints and rules for hybrid microcircuit design in relative terms. The results of a good hybrid circuit design rest on three distinct factors: (1) the hybrid's ability to perform all the functions for which it is designed, (2) the ability of the circuit to perform all the functions without interruption for the operational life of the system, and (3) economical producibility. Unfortunately, most often, circuit designers give low priority to the producibility of the microcircuitry, resulting in a noncompetitive circuit. Figure 2.17 shows the flow diagram for the hybrid circuit design. In reality, hybrid microcircuits are not the least expensive circuit technology. However, for many applications they represent the most cost-effective approach and are an extremely flexible tool for realizing a wide variety of electronic functions built in on a single ceramic substrate. With miniaturization and ever-increasing density of devices, circuit designing plays an important role in the hybrid industry [94].

2.2.1 Hybrid Circuit Design Elements [95]

Development of a successful hybrid circuit and minimization of development cost depend largely on the design part of a circuit. The design engineer must have prior knowledge of the circuit under certain operating conditions and must follow certain rules and design sequences. This allows for maximum ordering of the compromises to be made and the adoption of available options

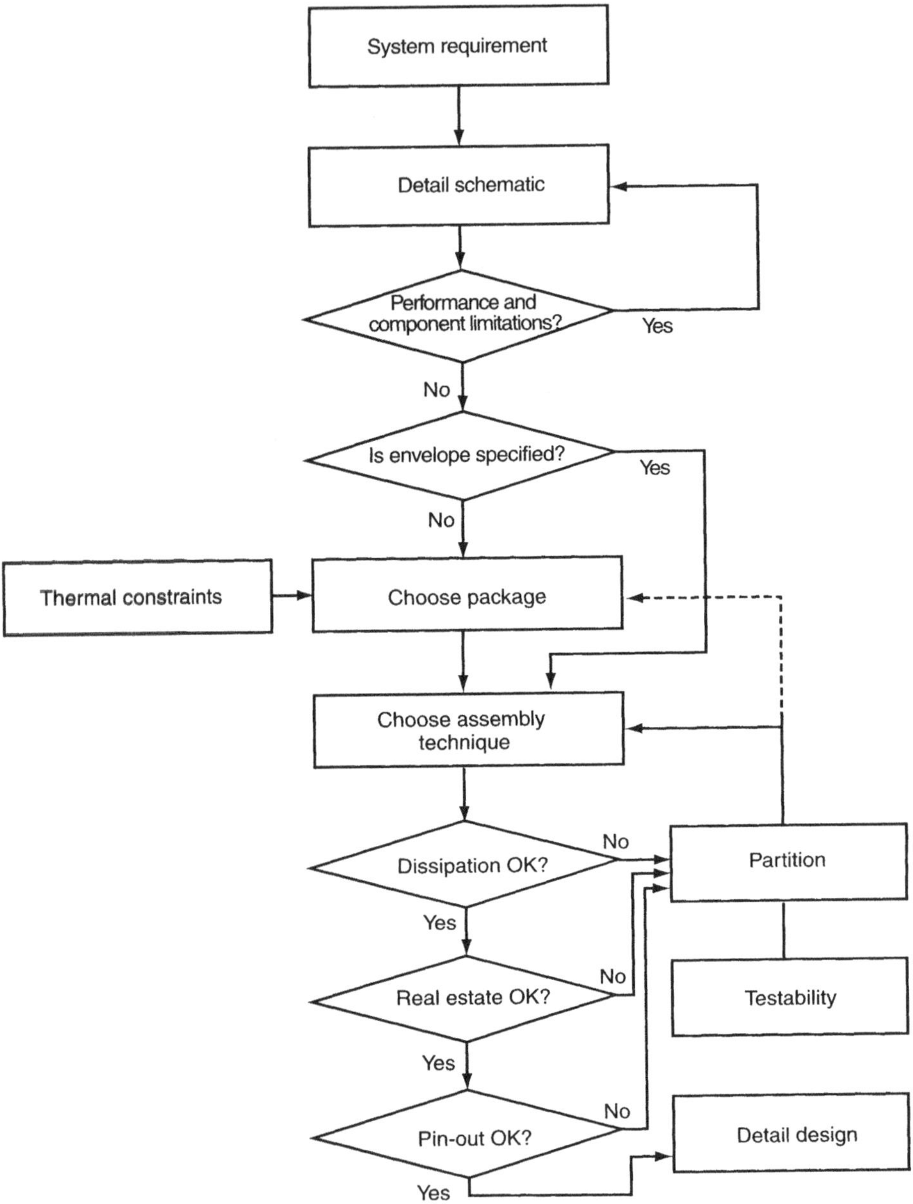

Figure 2.17 Flow diagram of a hybrid circuit design.

for design concept. Some design sequences are presented in Table 2.2, which indicates the performance or design parameters to be considered at each stage.

System Requirements The system specifications will identify several factors influencing the selection or determination of hybrid design criteria. The envi-

TABLE 2.2 Hybrid Circuit Design Sequence

State of Design	Parameters Considered
System requirements	Environment, size, weight, cost, reliability, power supplies, and signal
Circuit requirements	Complexity, I/O, count, form, fit, and function
Breadboard	Testability, functional verification, performance, and restrictions
Final circuit schematic	Component definition, pin out, layout, and restrictions
Component count and availability	Layout restrictions, module size, and component compatibility/availability
Process definition	Technology choice, temperature hierarchy, assembly technique, and packaging
Design review	Producibility and cost
Substrate layout	Size, pin out, and testability
Prototype fabrication	Process verification, performance, and producibility
Critical design review	Manufacturing resources, productivity, cost performance, and system integration
Manufacture	Capability/design rule update

Source: From [93].

ronmental conditions (e.g., temperature, humidity), system size, weight, power consumption by individual components, reliability, and signal source should be included in the specifications. In addition, the total cost of the materials, including labor, should be available as a result of the systems design activity.

Circuit Requirements Form, fit, and function are primary considerations at this stage of design. Partitioning into repetitive functions that can be configured in multiples represents an opportunity for cost reduction; conversely, increasing complexity within a packaged function may adversely affect producibility. The choice of thick or thin film or both may also affect the cost of the circuit.

Breadboard Demonstration of circuit functions in breadboard form serves to identify performance restrictions that affect eventual substrate layout. Testability and verification of functional performance are determined at this stage.

Final Circuit Schematic Final definition of the circuit schematic diagram will develop from a successful breadboard evaluation. At this stage of development, component definition must be considered, which includes performance specification and package compatibility with respect to alternative assembly processes. The availability of semiconductor circuit chips may well be the determining factor in the selection of process methods and/or the assembly sequence to be used.

Process Definitions Decisions concerning the use of thick- or thin-film technology (or both) for passive circuit elements can be made at this stage. The assembly techniques to be used (e.g., bonding, soldering) and the assembly temperature hierarchy (e.g., solder reflow, epoxy cure, package sealing) are influenced by component selection, package selection, and system requirements.

Design Review Component selection and thick- or thin-film decisions should be taken at this stage. The producibility of the processes selected should also be evaluated. Compromises made at previous stages should be reconsidered if they affect the ability to manufacture the product.

Substrate Layouts When component selection, processes (thick/thin), and producibility issues have been resolved, circuit layout on the substrate can begin. This involves coupling module size, pinout, and performance and testability requirements with established design rules.

Prototype Fabrications The inherent electronic performance improvement resulting from reducing the magnitude and complexity of interconnections must be balanced against possible interference due to the increased package density. Design rules will evolve with experience to increase the predictability of trade-off, but verification is best accomplished by prototype evaluation.

Manufacturing Technology Overview [95] Hybrid microcircuitry (HMC) represents a packaging approach for electronic components which is intermediate between the conventional printed board assembly of discrete components and monolithic circuit chips (devices). To achieve high package density on a PCB is very difficult, and accommodating discrete components is not possible in ICs. Hybrid circuits offer higher-density packaging because semiconductor dies are bonded directly to the substrate, and resistors are inherent in the films themselves. The substrate serves the same purpose in the hybrid circuit as the PCB does in the conventional circuit (i.e., it provides a rigid interconnecting structure). Due to the size of components used in PCBs [96], the lengths of component interconnects can also be a problem at high frequencies. On the other hand, the development cost and manufacturing time of IC technology are very high compared to those of HMCs, and the IC manufacturing yield drops off significantly as IC size decreases and the number of ICs increases. It is also true that discrete components such as resistors, large capacitors, and inductors are difficult to incorporate on ICs, whereas these are easily fabricated as films or added as discrete components in hybrid microcircuitry.

As the device size and number of discrete components increase in HMC, thermal dissipation problems frequently occur [97,98]. To meet this challenge, multilayer ceramic or low-temperature cofired ceramic (LTCC) substrates [99,100] with layers of screen-printable conductors and coefficients of thermal expansion matching that of silicon have been introduced successfully in HMC fabrication [101–103]. The basic hybrid electronic technologies are based primarily on thick and or thin film and assembled discrete circuit elements. Although thick films are indeed usually "thicker" than thin films, the terms no

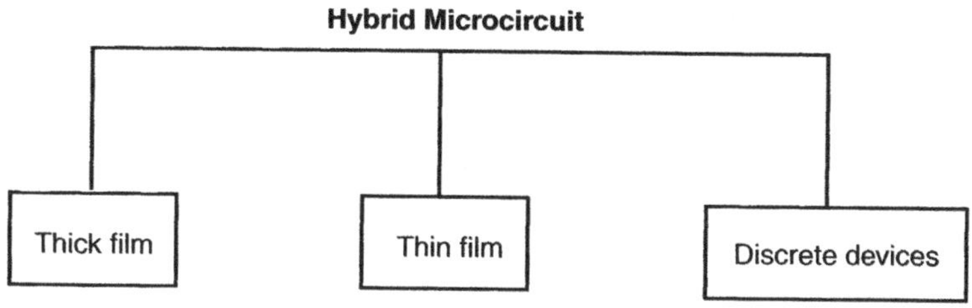

Figure 2.18 Hybrid circuit family members.

longer relate primarily to thickness but also to the processes used to deposit them. The term *thick film* is derived from the fact that the fired films are fairly thick, being typically 3 to 5 mils (76 to 127 μm). Thin films, on the other hand, are on the order of a few angstroms to 1 or 2 μm thick. Recently, however, these generic terms have come to be associated with their respective methods of film deposition. Thick film uses screen-printing technology for laying out circuits, whereas thin film generally uses vacuum deposition or sputtering technology. During vacuum deposition or sputtering, only 8 to 12% of the source material is deposited on the substrate [104]. So thin-film deposition is not a cost-effective way to fabricate a hybrid circuit. But sometimes it is difficult not to have thin-film circuitry within a hybrid microcircuit. Figure 2.18 shows various members of the hybrid microcircuit family.

2.2.2 Thick-Film Hybrid Circuit Design

Thick-film technology has undergone a remarkable transformation in the last 25 years, spurred on in part by the flexibility and adaptability it offers and in part by the speed with which design changes can be achieved. These inherent qualities of thick film have led circuit manufacturers to expect faster throughput and greater yields. Indeed, every aspect of thick-film technology has been geared toward the manufacture of circuits with throughput at lower cost and with greater reliability. Uses of improved screen printers with printing speeds of 118 to 157 mils/s (3.0 to 4.0 mm/s) are not uncommon today. High-resolution screens or masks and excellent inks have enabled some practitioners to develop thick-film fabrication to a high art. Automated parts handling, infrared drying, fast firing profiles, and high-speed soldering operations further developed thick-film technology. For high yields, design guidelines specify a minimum of 10 mils (254 μm) for line widths and spacing. However, with the developments of special pastes, an automated screen printing system, and use of finer mesh size, line widths and spacing as low as 2 mils (51 μm) have been achieved [105,106].

Thick-film hybrid microcircuitry is gaining wide acceptance because of its design flexibility and production economy [107]. A practical and useful design guide, however, must reflect not the best that can or will be achieved in the future, but rather, what a designer today can reasonably expect of machines and materials. The design guide is therefore based on the current production standards of many circuit manufacturers. However, there is no hard and fast

TABLE 2.3 Typical Guidelines for Discrete Component Size

	Maximum Area	
Component	mils^2	mm^2
Typical thick-film resistor	50 × 50	1.27 × 1.27
Typical thick-film capacitor	70 × 40	1.78 × 1.06
Diode die	20 × 20	0.5 × 0.5
Transistor die	25 × 25	0.63 × 0.63
Linear ICs	50 × 50	1.27 × 1.27
Resistor transistor logic	40 × 40	1.06 × 1.06
Diode transistor logic	40 × 40	1.06 × 1.06
Transistor–transistor logic	50 × 50	1.27 × 1.27
Medium-scale integrated circuit	75 × 75	1.9 × 1.9
Large-scale integrated circuit	120 × 120	3.0 × 3.0

Source: From [93].

rule for design engineers when the benefits of shrinking the minimum feature size far outweigh the limitations [108]. The number of devices that could potentially be accommodated on hybrid microcircuit substrate is strongly dependent on die size, pitch, and the dimensions of the passive circuit element [109]. The minimum feature size of the device has been reduced from 0.27 mil (7 μm) to less than 0.15 mil (0.4 μm).

Production Standards To set up a manufacturing facility and produce a wide range of working circuits, many factors are considered, particularly the limitations in the size of discrete components. As a first approximation, a designer should assume the discrete component sizes and proceed to design the circuit. A typical guideline for discrete component size is given in Table 2.3. As a rule of thumb, the total area of the discrete components should not consume more than 20% that of the substrate. The conductors, pads, and screened resistors should occupy 60 to 70% of the substrate area. The remaining 10 to 20% of the area should be left for maintaining gaps and additional spaces for power transistors and silicon-controlled rectifiers [110]. Given the approximate substrate area needed for the entire circuit, the number of packages required to implement the circuit design can be estimated once a definite package size has been selected.

Design Engineering* For convenience and to save area, use of different inks is recommended rather than long, thin or short, wide rectangular resistors. When the aspect ratio (L/W) is more than 4, a tophat design is preferred. Zigzag resistors with high aspect ratios can be designed, but with zigzag design, trimming is difficult and hot spots may develop. Although the power density recommended for thick film is 50 W/in^2 (7.75 W/cm^2), for all practical purposes, 25 W/in^2 (3.87 W/cm^2) is preferred [93,111,112]. Table 2.4 shows typical

*The material in this section, including the figures and tables, is from [93], with permission from ISHM, Reston, Virginia, now IMAPS, Washington, DC.

TABLE 2.4 Thick-Film Screened Resistor Sizes and Tolerance

	Dimension	Minimum mils	Minimum μm	Maximum mils	Maximum μm	Comments
Number of squares	L/W	1/3		1/10		Rectangular thick-film resistor should not exceed 1:3 to 1:4.
Length	L	30	762	60	1.5 mm	
Width	W 1% tolerance	20	508	40	1.0 mm	When the aspect ratio (L/W) is more than 5, a tophat design is preferable.
Conductor terminal (CT) distance beyond length of resistor	D_1	5	127	10	254	
CT distance beyond length of resistor	D_2	10	254	15	381	Ink of higher resistivity can be used.
Resistor overlap onto conductor terminal	D_3	10	254	15	381	Zigzag resistors can be made out, but trimming is difficult.
Conductor distance from resistor	D_4	15	381	20	508	Maximum power density about 50 W/in^2 (7.75 W/cm^2).
Edge of substrate distance from resistor	D_5	15	381	20	508	
Voltage stress		0.5 V/mil	0.02 V/μm	1 V/mil	0.04 V/μm	

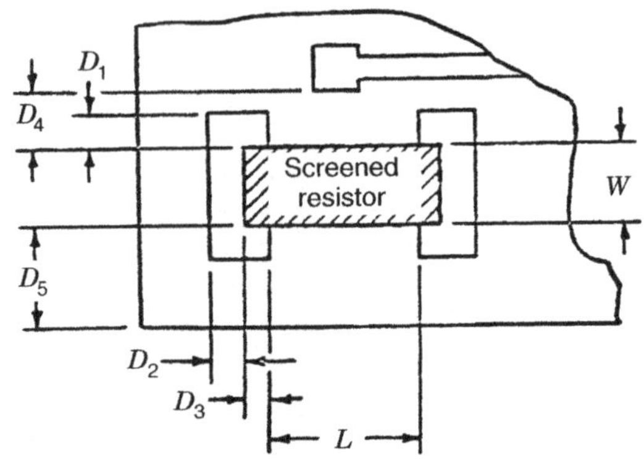

Figure 2.19 Thick-film screened resistor.

screened resistor size and tolerance, and Figure 2.19 shows a thick-film screened resistor.

Vacuum-deposited thin-film resistors, on the other hand, are designed to have a rated maximum power dissipation of 15 W/in^2 (2.2 W/cm^2). The corner square of a meander resister is taken to be half that of a regular square [113,114].

$$\text{minimum resistor film area} = A_{\min} = \frac{P_a}{P_{dm}} = \frac{L}{W} W^2$$

$$\text{width} = W = \left(\frac{P_a}{P_{dm}} \frac{R_s}{R}\right)^{1/2}$$

where P_a is the actual power dissipation in milliwatts, P_{dm} the maximum power dissipation density allowed = 15 W/in^2 (2.2 W/cm^2), R_s the sheet resistance, and R the resistance of the resistor [115]. The shape and actual size of a thin-film resistor are made by either an additive or a subtractive method. In a subtractive method, selective etching is done through photolithography. Table 2.5 shows thin-film etched resistor size and location. Figure 2.20 shows the shape of an etched thin-film resistor.

Figure 2.21 shows the minimum dielectric distance from lines and crossovers. The actual maximum and minimum values of dielectric line and crossover distances are tabulated in Table 2.6.

Table 2.7 corresponds to Figure 2.22 for the minimum dielectric distances that should be left over in laying out a thick-film hybrid circuit.

Table 2.8 is the tabulated data for the minimum lands for component mounting. Actual data are shown in Figure 2.23.

Table 2.9 shows the actual numbers for obstacle height and the minimum spacing that should be left between the wire bond and the obstacle. Figure 2.24 shows the locations of the wire bonds.

TABLE 2.5 Thin-Film Etched Resistor Sizes and Locations

		Minimum		Maximum	
	Dimension	mils	μm	mils	μm
Number of squares	L/W	0.5		25	
Length	L	300	7.62 mm	500	12.7 mm
Width	W	10	254	150	3.8 mm
Distance beyond resistor width	D_1	10	254	5	127
Conductor terminal width	D_2	2	50.8	5	127
Conductor distance from the resistor	D_3	10	254	15	381
Edge of substrate distance from the resistor	D_4	10	254	15	381
Voltage stress	—	0.5 V/mil	0.02 V/μm	1 V/mil	0.04 V/μm
Power dissipation	—	10 W/in^2	1.47 W/cm^2	15 W/in^2	2.2 W/cm^2

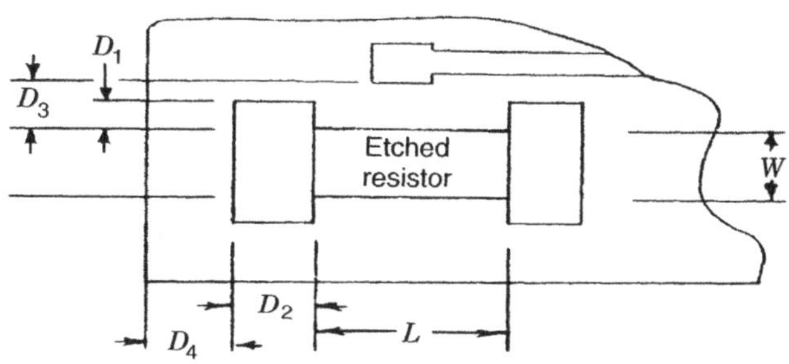

Figure 2.20 Thin-film etched resistor.

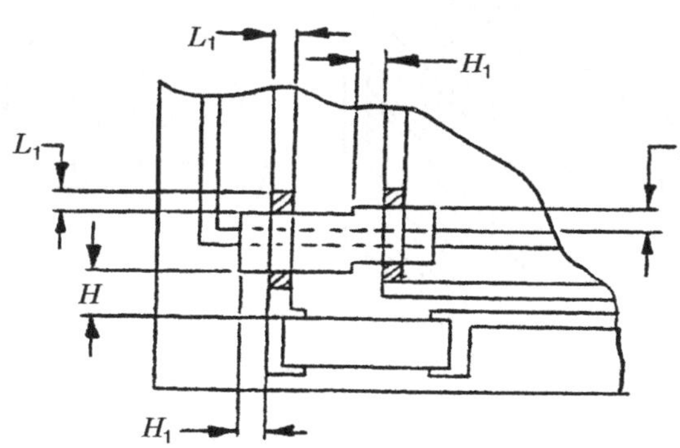

Figure 2.21 Minimum dielectric distance from other lines and crossovers.

TABLE 2.6 Dielectric Distance from Lines and Crossovers

		Maximum		Minimum	
	Dimension	mils	μm	mils	μm
Resistor distance to dielectric	H	50	1.2 mm	30	762
Conductor-line staggered distance	H_1	20	508	10	254
Step (typical on two-layer circuit, conductor lines overlap)	L_1	9 × 10	254 × 254	15 × 15	381 × 381
Dielectric crossover	H				
Distance extended beyond width (on either top or bottom conductor)	H_1	15	381	20	508

TABLE 2.7 Dielectric Spacing and Size

		Minimum		Maximum	
	Dimension	mils	(μm)	mils	(μm)
Dielectric distance to metallized land for solder or epoxy component attachment	W	15	381	20	508
Dielectric distance to edge of substrate, minimum size	W_1	10	254	15	381
Dielectric width when covering the run under bonded wire	W_2	10	254	15	381
For 10 by 10 mil (254 by 254 μm) conductor dielectric crossover area	A	15 × 15	381 × 381		

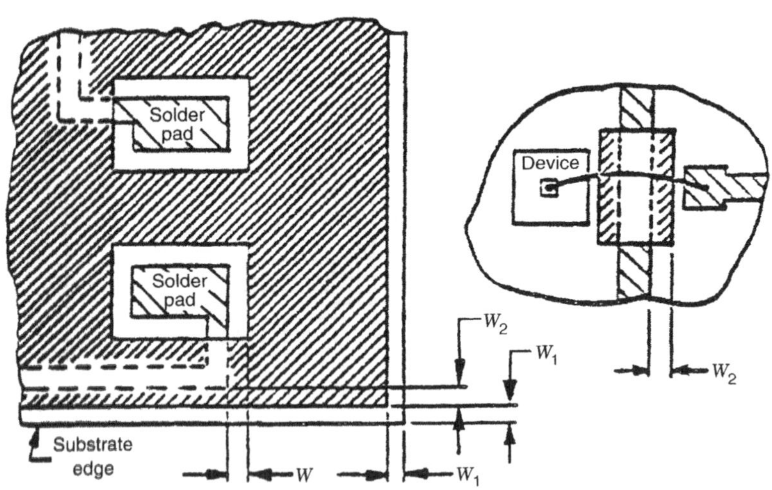

Figure 2.22 Minimum dielectric spacing and size.

TABLE 2.8 Lands for Component Mounting

		Minimum		Maximum	
	Dimension	mils	μm	mils	μm
Discrete leaded component	A	10	254	15	381
Metallized land overlap for epoxy or solder attachment, distance from leaded component	F	30	762	40	0.1 mm
Semiconductor chip	B	10	254	15	381
Die–bond land overlap for eutectic or epoxy attachment land distance from other metallization	E	10	254	15	381
Leadless component (chip capacitor or chip resistor) ceramic body overlaps onto land	C	10	254	15	381
Metallized end land overlaps onto land, for epoxy or solder attachment	D	15	381	20	508

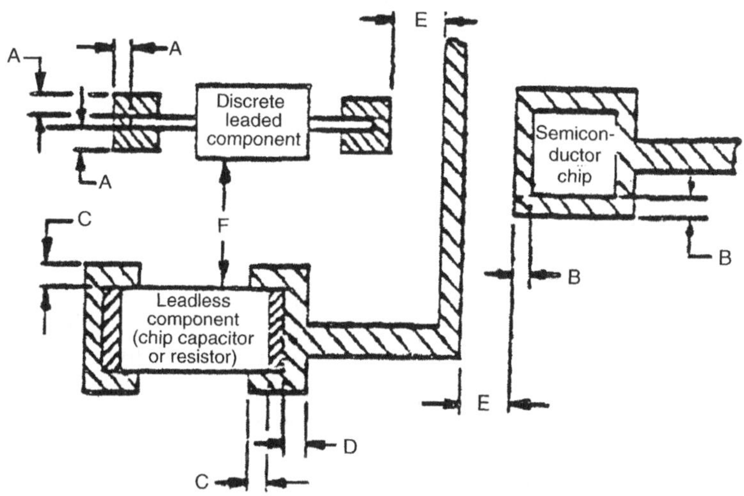

Figure 2.23 Minimum lands for component mounting.

TABLE 2.9 Locating Wire Bond Lands and Sizing

Obstacle Height		Wire Bond to Obstacle Minimum Spacing	
mils	μm	mils	μm
10	254	25	635
20	508	40	1 mm
40	1 mm	45	1.14 mm
60	1.5 mm	50	1.27 mm
80	2.0 mm	55	1.4 mm
100	2.54 mm	65	1.65 mm
150 and over	3.8 mm and over	70	1.78 mm
Package height from the bond level		15	381

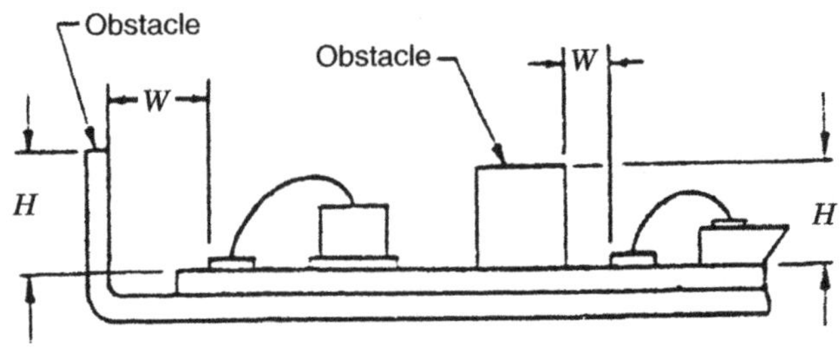

Figure 2.24 Locating wire bond and sizing components.

Table 2.10 and Figure 2.25 show layout features of the bonding wires and the cost and penalties that may result due to excessive space left between bonding wires.

Table 2.11 contains the data and Figure 2.26 illustrates those data for typical via size and location.

Figure 2.27 shows the minimum lead frame bonding, land size, and location. The actual dimensions are tabulated in Table 2.12.

Figure 2.28 indicates the suggested dimensions of a workable layout. Actual sizes and dimensions are tabulated in Table 2.13.

2.2.3 Basic Rules for Laying Out Hybrid Microcircuits

Layout begins with careful and thoughtful positioning of components to accommodate the maximum number of components per square inch without any deterioration in the electrical performance of the microcircuit. Laying out of hybrid microcircuits is the second important task for a hybrid circuit engineer [116,117]. Both uninterrupted operational life and economic producibility of the microcircuit are very dependent on proper circuit layout. To retain the manufacturing cost low without any compromise in quality, an effort should be made to minimize the number of bonds and thereby help to minimize the bonding difficulty. A prime factor that should be kept in mind is exact *placement of the components.* A good layout for volume production should be relatively immune to variations in part placement [93,67]. The guidelines and layout sequences that follow apply to both thick- and thin-film designs, although the final physical configuration will be different for each type [118]. A typical sequence of events for hybrid microcircuitry is as follows: (1) initial circuit formation, (2) layout preparation, (3) prelayout sketch, (4) preliminary layout, (5) preliminary layout review, and (6) final layout.

Thus we can conclude that the circuit can be made in thick- and/or thin-film form, and the designer must lay out the circuit properly so that it can be fabricated. The layout scale is generally 20 times the final size of the circuit, which can be used for preparation of the rubylith form. Sometimes, engineers adopt a layout scale 10 times the final size. To minimize the error between rubylith artwork and the final size of a circuit, 20× magnification is more desirable [119].

TABLE 2.10 Layout Features

	A min.		B min.		C max.		D min.		E min.		D max.	
Bonding Process	mils	μm	mils	μm	mils	μm	mils	μm	mils	μm	mils	mm
Gold wire bonding												
Thermocompression	4	100	4	100	30	762	25	635	10	254	100	2.54
Thermosonic	4	100	4	100	15	381	37	939	10	254	100	2.54
Aluminum wire bonding	2	50	4	100	30	762	25	635	10	254	100	2.54

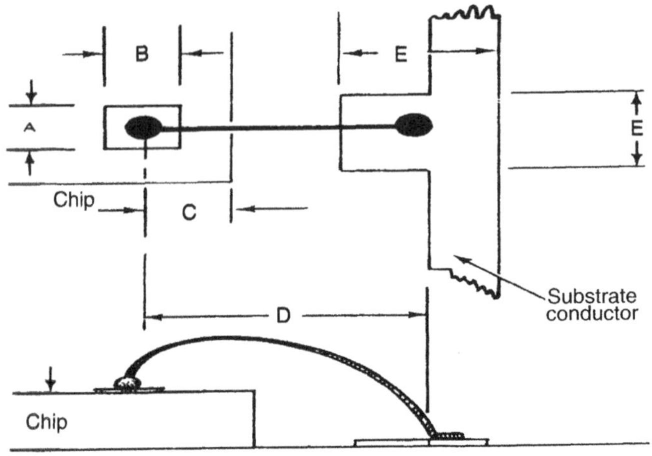

Figure 2.25 Layout feature.

TABLE 2.11 Minimum Via Hole Parameters

		A		B		C	
Parameters	Dimension	mils	μm	mils	μm	mils	μm
Length and width	$L \times W$	20 × 20	508 × 508	15 × 15	381 × 381	12 × 12	305 × 305
Spacing distance from mettallization	D	20	508	15	381	12	305
	D_1			15	381	12	305

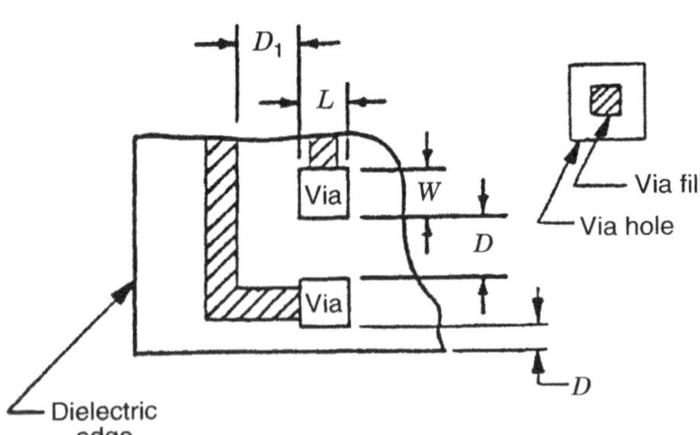

Figure 2.26 Via sizes and locations.

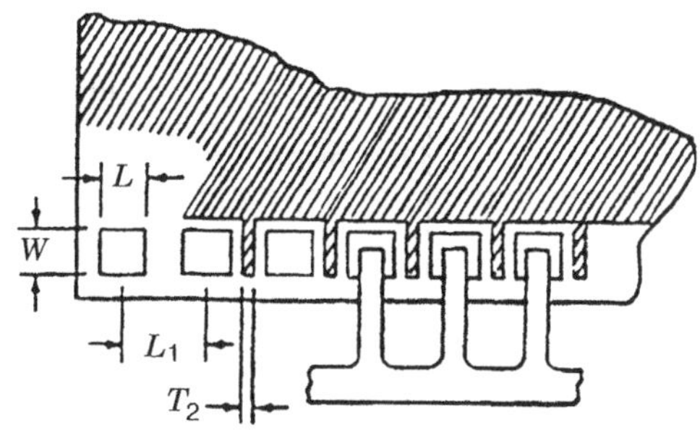

Figure 2.27 Minimum lead frame bonding lands, size, and location.

TABLE 2.12 Lead Frame Bonding

		Maximum		Minimum	
	Dimension	mils	mm	mils	mm
Land size for solder attachment	$L \times W$	60 × 60	1.5 × 1.5	50 × 50	1.27 × 1.27
Land spacing center to center	L_1	100	2.5	80	2.0
Screened dielectric between lands	T_2	15	0.38	10	0.254
Spacing between two solder attachments	—	40	1.0	30	0.762

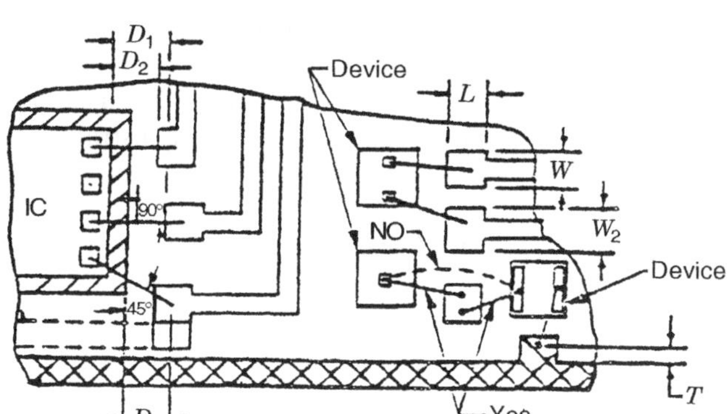

Figure 2.28 Wire bond land sizes and locations.

TABLE 2.13 Wire Bond Land Size and Locations

		Minimum		Maximum	
	Dimension	mils	μm	mils	μm
Wire bond land size					
Length	L	15	381	20	508
Width of single wire	W	10	254	15	381
Width of double wire	W_1	20	508	25	635
Locating angle die edge to center of wire bond site	<	45	1.1 mm	90	2.2 mm
Die edge distance to edge of land	D_1	50	1.3 mm	70	1.78 mm
	D_2	30	762	50	1.3 mm
Dielectric distance to center of wire bond site	D_3	15	381	20	508
Wraparound metallization distance to center of wire bond site	T	25	635	30	762

Conductor Layouts There should be no crossovers or a minimum number of crossovers between conductors, as every crossover requires an intermediate dielectric layer. The lengths of the connecting wires should not be more than 100 mils (2.54 mm) and should not have sharp bends or sags. For good electrical contacts, discrete components are placed on the conducting pads. The back side of a transistor, which is the collector, is laid down on a conductor pad, and a separate pad should be used for emitter connection. A minimum spacing of 5 mils (127 μm) [2 mils (50 μm) has been reported] should be maintained around the die on every side. This is necessary to allow the person involved in fabrication to see some metallization around the outside, which will make it possible to obtain a fillet to ensure a reliable die attachment. If a wire bond to the pad is required directly adjacent to a transistor die, some room will be needed to get the bonding tool down on the pad. A minimum of 15 mils (381 μm) between the edge of the chip and the edge of the pad should be allowed. No matter what kind of bond is being used—a wedge bond, ball bond, or ultrasonic bond—a minimum area of 10 by 10 mils (254 by 254 μm) on the bonding pads should be provided. These are the basic layout rules for design engineers. Table 2.14 shows the dimensional constraints for thick-film conductors and pads.

Resistor/Conductor Interfaces The conductors that interface with resistors also serve as contact pads for probing resistors during the trimming operation. The operations of probing and trimming, and limitations of equipment used, emphasize the need for standardized contact points for all resistors. Accurate readjustment of the probes after each resistor trim is a difficult process because of the nozzles, hoses, and fixtures. Furthermore, adjustments cannot be

TABLE 2.14 Dimensional Constraints for Thick-Film Conductors and Pads

Conductor Pad Sizes and Spacing	Minimum Dimension in.	Minimum Dimension cm	Nominal Dimension in.	Nominal Dimension cm	Maximum Dimension in.	Maximum Dimension cm
Conductor to edge of substrate for conductors ≤0.015 in. (0.38 cm) wide	0.010	0.025	0.020	0.05	None	
Conductor to edge of substrate for conductors ≥0.015 in. (0.38 cm) wide	0.010	0.025	0.0125	0.03	None	
Exit bonding pads (width and length)	—		0.0125	0.03	None	
Wire bonding pads [width and length—one or two wires on same pad (preferred)]	—		0.0125	0.03	None	
Wire bonding pads [width and length—one wire per pad (special)]	—		0.0125	0.03	None	
Conductor width (preferred)	—		0.020	0.05	None	
Conductor width (special)	0.005	0.012	0.010	0.025	None	
Conductor to conductor spacing (preferred)	0.010	0.025	0.020	0.05	None	
Conductor to conductor spacing (special)	0.005	0.012	0.010	0.025	None	
Conductor to resistor spacing (on untrimmed side of resistor)	0.010	0.025	0.015	0.038	None	
Conductor to resistor (on trimmed side of resistor)	0.015	0.038	0.020	0.05	None	
Upper and lower conductor widths at crossover junction	0.010	0.025	0.015	0.038	None	
Crossover conductor to connecting conductor overlap—length	0.010	0.025	0.0125	0.03	None	
Conductor crossover dielectric overlap	0.010	0.025	0.015	0.038	None	
Conductor to resistor spacing tophat configuration	0.030	0.076	0.035	0.089	None	

Source: From [93].

made under a microscope without removing nozzles and other attachments. The substrate is mounted on an XY table, which allows the substrate to be positioned with the substrate's long dimension in the trim direction or at 90° to it. Because of these impediments, conductor interfacing with resistors should be designed to provide 0.050 by 0.050 in. (1.27 by 1.27 mm) pads for probing.

Resistor Layouts The resistors are terminated with conductor pads and should overlap the conductor by a minimum of 15 mils (381 μm). The metal-

TABLE 2.15 Dimensional Constraints for Thick-Film Resistors

	Minimum Dimension		Nominal Dimension	
Resistor Size and Spacing	in.	cm	in.	cm
Resistor height	0.030	0.075	None	
Resistor width	0.060	0.15	0.080	0.2
Resistor-to-resistor spacing	0.015	0.038	0.020	0.05
Resistor-to-conductor overlap	0.010	0.025	0.0125	−0.03
Resistor-to-conductor overlap width	0.005	0.013	0.0075	0.2
Resistor length				
Ink >100 kΩ/sq	0.030	0.075	0.040	0.1
Ink <100 kΩ/sq	0.040	0.1	0.060	0.15
Resistor width				
Ink >100 kΩ/sq	0.030	0.075	0.040	0.1
Ink <100 kΩ/sq	0.040	0.1	0.060	0.15
Conductor pad length (beyond resistor)	0.005	0.013	0.010	0.025

Source: From [93].

lization should extend by at least 5 mils (127 μm) and a maximum of 10 mils (254 μm) beyond the resistor termination perpendicular to current flow. Table 2.15 illustrates the dimensional constraints for thick-film resistor layout.

Layout Guidelines for Resistors [67,93]

1. The side of the resistor that will not be trimmed should be located at least 20 mils (508 μm = 0.508 mm) from any screened area.
2. The trimmed side of the resistor should be located at least 40 mils (1.0 mm) from any other screened area.
3. Two resistor pastes with different values of sheet resistance should be separated by at least 10 mils (254 μm) when they are printed on a common pad.
4. Resistor pastes with the same sheet resistance are terminated on a common pad without any gap in between.
5. When a resistor termination pad also serves as a contact pad for a terminal or discrete component, the termination must be designed such that bleed-out of the resistor during firing does not affect the component add-on area on termination.
6. All resistors should be terminated in the X–Y direction.
7. High-power resistors should be evenly distributed on the substrate and should not be located near the substrate edge or close to active components.

8. Resistors screened from the same sheet resistance ink and similar geometry will track much better temperature than resistors of different sheet resistance ink and geometry.
9. All resistors should be located at least 30 mils (762 μm) from the edge of the substrate.
10. A closed loop of resistors should not be used.

One should remember that these guidelines are for optimum design of resistors and should not be considered as hard and fast rules.

Design Rules for Capacitors [67,93] The basic design rules for capacitors in both thick and thin films highlight the choice of dielectric materials. Dielectric pastes for thick films are available in the range 4 to 20 kΩ. Low-K dielectrics are very stable; high-K dielectrics are not as stable. Moreover, high-K materials have poor temperature coefficients [120,121]. On the other hand, thin-film dielectric materials are available in the range 4 to 25. Very recently, a dielectric

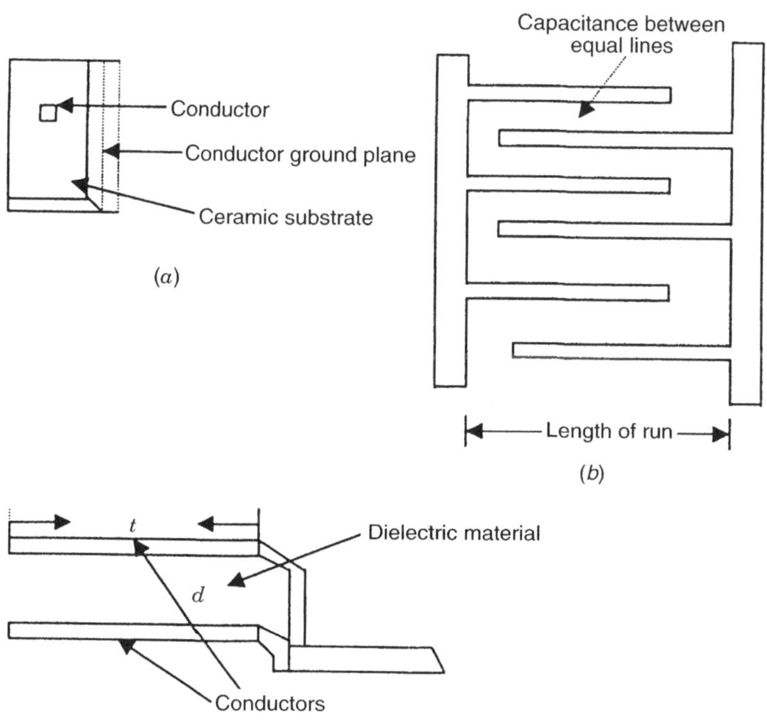

Figure 2.29 (a) Ceramic substrate carrying the conductors; (b) interdigital capacitor; (c) outline of a parallel-plate capacitor.

TABLE 2.16 Values of *d* and *t* for Capacitors [67]

	d	*t*
Thin film	>2 mils (50 μm)	5–7 mils (2000–3000 Å)
Thick film	0.5–2.5 mils (13–65 μm)	>10 mils (254 μm)

material higher than the dielectric value of silicon has been developed and has been deposited successfully by atomic layer chemical vapor deposition [122].

The capacitor value $C = \varepsilon_0 K(A/t) = 8.85 \times 10^{-12}$ F/m × K × meters (10^{-12} F = 1 pF), where A is the area in (meter)^2, t is the thickness of the dielectric in meters, ε_0 is about 8.85 F/m, and K is the dielectric constant of the dielectric material. For convenience, the substrate material in the hybrid circuit is used as a dielectric (the value of K for alumina is between 9 and 10). The lower part of the substrate, which is ceramic, is plated with metals and serves as a conductor ground plane (Figure 2.29). The dielectric material is placed in between the parallel-plate conductors, and the values of d and t are given in Table 2.16.

REFERENCES

1. J. J. Licari and L. R. Enlow, *Hybrid Microcircuit Technology Handbook: Materials, Processes, Design, Testing and Production*, Noyes Publications, Park Ridge, NJ, 1998, p. 363.
2. P. A. Sandborn and H. Moreno, *Computational Design of Multichip Modules and Systems*, Kluwer Academic, Norwell, MA, 1994, Chap. 2.
3. I. Brodie and J. J. Muray, *The Physics of Microfabrication*, Plenum Press, New York, 1982, Chaps. 1, 3, and 5.
4. A. Kosloff, *Screen Printing Electronic Circuits*, Signs of the Times Publishing, Cincinnati, OH, 1990, Chap. 11.
5. R. Glang and L. V. Gregor, Generation of patterns in thin film, in *Handbook of Thin Solid Films*, L. I. Maissel and R. Glang, eds., McGraw-Hill, New York, 1983.
6. M. J. Bowden, A perspective on resist materials for fine-line lithography, in *Materials for Microlithography*, L. F. Thomson, C. G. Wilson, and J. M. J. Frechet, eds., American Chemical Society Publishing, Washington, DC, 1994, p. 39.
7. P. J. Holmes and R. C. Loasby, eds., *Handbook of Thick-Film Technology*, Electrochemical Society, Ayr, Scotland, 1990.
8. J. E. Sergent, Design methods for hybrid circuits, in *Hybrid Microelectronics Handbook*, 2nd ed., J. E. Sergent and C. A. Harper, eds., McGraw-Hill, New York, 1995.
9. G. W. A. Dummer, *Fixed Resistors*, Pitman, New York, 1967.
10. G. Siddeal, Thin-film microelectronics, in *Thin-Film Microelectronics*, L. Holand, ed., Wiley, New York, 1965.
11. R. W. Berry, P. M. Hall, and M. T. Harris, *Thin-Film Technology*, Van Nostrand Reinhold, New York, 1968; R. C. Langford, *Proc. Electron. Comp. Conf.*, 1960, p. 66.

12. C. L. Wellard, Resistance and resistors, in *Handbook of Thin Solid Films*, L. I. Maissel and R. Glang, eds., McGraw-Hill, New York, 1983, Chap. 16.
13. R. A. Combs, ed., *The Electrical Properties and Applications of Thin Films*, Pitman, New York, 1967.
14. S. P. Murarka and M. C. Peckerar, eds., *Electronic Materials Science and Technology*, Academic Press, San Diego, CA, 1989, pp. 162 and 270.
15. W. Jones and N. H. March, *Theoretical Solid State Physics*, Vol. II, Wiley, New York, 1973, p. 710.
16. J. M. Ziman, Impurity scattering, in *Principles of the Theory of Solids*, Cambridge University Press, Cambridge, 1964, Chap. 7, p. 187.
17. D. I. Kennedy, R. E. Hoyes, and R. W. Alsford, *J. Appl. Phys.*, Vol. 38, 1967, p. 1986.
18. N. F. Mott, *Proc. Cambridge Philos. Soc.*, Vol. 32, 1938, p. 281.
19. S. P. Murarka and M. C. Peckerar, eds., Metallization, in *Electronic Materials Science and Technology*, Academic Press, San Diego, CA, 1989, p. 271.
20. S. A. Boctor, *Electric Circuit Analysis*, 2nd ed., Prentice Hall, Upper Saddle River, NJ, 1986.
21. J. F. Linder and M. A. Lee, Manufacturing technology for high ohms per square films, *Proc. Int. Symp. Microelectron.*, 1982, p. 192.
22. L. I. Maissel, Thin-film resistors, in *Handbook of Thin-Film Technology*, 2nd ed., L. I. Maissel and R. Glang, eds., McGraw-Hill, New York, 1983.
23. R. D. Jones, *Hybrid Circuit Design and Manufacture*, Marcel Dekker, New York, 1982, p. 46.
24. G. Rizzoni, Maximum power transfer, in *Principles and Applications of Electrical Engineering*, Irwin, Burr Ridge, IL, 1993, p. 102.
25. R. L. Cadenhead, TCR tracking: a definition, *J. Int. Soc. Hybrid Microelectron.*, Vol. 1, 1978, pp. 10–15.
26. R. C. Headley and R. E. Cote, Tracking of resistance and TCR in thick-film resistors: distinctions, definitions and derivations, *Int. Symp. Microelectron.*, 1979, p. 20.
27. G. W. A. Dummer, Resistive films, in *Materials for Conductive and Resistive Functions*, Hayden Book Co., New York, 1970, Chap. 18.
28. S. A. Halaby, Substrates/conductive films/resistive films, in *Materials of Thin-Film Devices*, Vol. 97, A. E. Javitz, ed., Electrotechnology Publishing, Menlo Park, CA, Sept. 1963.
29. J. Mayer and S. S. Lau, *Electronic Materials Science for ICs in Si and GaAs*, Macmillan, New York, 1990, pp. 32 and 365; J. Fisher and P. M. Hall, Termination materials for thin-film resistors, *Proc. IEEE*, Vol. 59, No. 10, Oct. 1971, pp. 1418–1424.
30. R. K. Willardson and A. C. Beer, *Semiconductors and Semimetals* Vol. 7, *Applications and Devices*, Part A, Academic Press, San Diego, CA, 1971, pp. 178–183.
31. L. Solymar and D. Walsh, *Lectures on the Electrical Properties of Materials*, 4th ed., Oxford University Press, New York, 1988.
32. H. M. Naguib, The characterization of thick-film resistors terminated with Pd/Ag conductors of various compositions, *Proc. Int. Symp. Microelectron.*, 1977, p. 53.

33. R. T. Galla, H. M. Greenhouse and W. C. Vergara, Evaluation of the interfacial resistance of thin film interconnections, *Microelectron. Reliab.*, Vol. 7, 1968, p. 185.
34. J. M. Ziman, Static properties of solids, in *Principles of the Theory of Solids*, Cambridge University Press, Cambridge, 1964, Chap. 4, p. 114.
35. M. H. Read and C. Altman, *Appl. Phys. Lett.*, Vol. 7, 1965, p. 51.
36. P. S. Castro, Microsystem circuit analysis, *Electron. Eng.*, Vol. 80, July 1961, p. 535.
37. D. A. Bell, *Noise and the Solid State*, Wiley, New York, 1985.
38. A. van der Ziel, *Noise in Solid-State Devices*, Wiley, New York, 1986.
39. H. Nyquist, *Phys. Rev.*, Vol. 32, 1928, p. 110.
40. A. Norrell, M. Grierson, and T. M. Chen, Characterization of burst noise in thick-film resistors, *Int. Symp. Hybrid Microelectron.*, 1978, p. 14.
41. W. H. Card and P. K. Chaudhuri, Characteristics of burst noise, *Proc. IEEE*, Vol. 53, 1965, pp. 652–653.
42. D. A. Bell, *Electrical Noise*, Van Nostrand Reinhold, New York, 1960, pp. 258–261.
43. T. M. Chen and J. G. Rhee, The effects of trimming on the current noise of thick-film resistors, *Solid State Technol.*, Vol. 20, 1977, pp. 49–53.
44. J. G. Rhee and T. M. Chen, Contact noise in thick film resistors, *Solid State Technol.*, Sept. 1978, p. 60.
45. W. C. Dunlap, Jr., Statistical mechanics for metals and semiconductors, in *Introduction to Semiconductors*, Wiley, New York, 1957, Chap. 5.
46. C. W. H. Bristow, W. L. Clough, and P. L. Kirby, The current noise and non-linearity of thick-film resistors, *Proc. Int. Symp. Microelectron.*, 1971, p. 7-6-1.
47. H. Rangchi, B. Huner, and P. K. Ajmera, A model for deposition of thick films by screen printing techniques, *Proc. Int. Symp. Microelectron.*, 1986.
48. D. C. Hughes, Jr., *Screen Printing Microcircuits*, Presco, Inc., Somerville, NJ, 1967.
49. T. C. Reissing, An overview of today's thick-film technology, *Proc. IEEE*, Vol. 59, No. 10, Oct. 1971; P. J. Holmes and R. G. Loasby, *Handbook of Thin-Film Technology*, Electrochemical Society, Pennington, NJ, 1976.
50. D. E. Riemer, The function and performance of the stainless steel screen printing ink transfer process, *Proc. Int. Symp. Micrelectron.*, 1986, pp. 826–831.
51. W. H. Banks and C. H. Mills, Tacky adhesion a preliminary study, *J. Colloid Interface Sci.*, 1953, pp. 137–147.
52. E. I. Radzimovski, *Lubrication of Bearings*, Ronald Press, New York, 1959, pp. 19–23.
53. I. H. Shames, *Mechanics of Fluids*, McGraw-Hill, New York, 1962, pp. 291–293.
54. N. R. Haskard, *Thick-Film Hybrids Manufacture and Design*, Prentice Hall, Upper Saddle River, NJ, 1988, Chap. 2.
55. C. A. Harper, ed., *Handbook of Thick-Film Microelectronics*, McGraw-Hill, New York, 1974.
56. S. G. Stahlnecker, Jr., and R. A. Wyant, New stencil screen printing development for thick-film printing, *Solid State Technol.*, Vol. 26, 1983, pp. 141–146.

57. R. M. Stanton, Rheology aspects of thick-film technology tack-and-paste transfer during screen printing, *Proc. Int. Symp. Microelectron.*, 1976, p. 112.
58. W. H. Banks and C. C. Mill, *J. Colloid Sci.*, Vol. 137, 1953.
59. M. Knudsen, *Ann. Phys.*, Vol. 25, No. 75, 1909, p. 999.
60. J. J. Licari and R. Enlow, Thick-film resistor design guideline, in *Hybrid Microcircuit Technology Handbook*, Noyes Publications, Park Ridge, NJ, 1998, Chap. 17.
61. D. F. Zarrow, A new approach to the thick-film resistor, *Proc. Int. Symp. Microelectron.*, 1979, p. 32.
62. R. Glang, Thickness distributed of evaporated film, in *Handbook of Thin Films Technology*, L. I. Maissel and R. Glang, eds., McGraw-Hill, New York, 1970, p. I-55.
63. L. Holland, *Vacuum Deposition of Thin Films*, Wiley, New York, 1961.
64. J. C. Anderson, *Dielectrics*, Chapman & Hall, London, 1964.
65. B. Tareev, Dielectric losses, in *Physics of Dielectric Materials*, Mir Publishing, Moscow, 1975, Chap. 3.
66. D. Gertenbery, Thin-film capacitors, in *Handbook of Thin-Film Technology*, 2nd ed., L. I. Maissel and R. Glang, eds., McGraw-Hill, New York, 1983, Chap. 19.
67. R. D. Jones, Substrate materials and properties, in *Hybrid Circuit Design and Manufacture*, Marcel Dekker, New York, 1982, Chap. 2.
68. C. W. T. McLyman, *Transformer and Inductor Design Handbook*, Marcel Dekker, New York, 1988, Chap. 8.
69. V. G. Welsby, *The Theory and Design of Inductor Coils*, Macdonald, London, 1964, p. 154.
70. F. W. Grover, *Inductance Calculations*, D. Van Nostrand, Princeton, NJ, 1962.
71. E. R. Hnatek, *A User's Handbook of Integrated Circuits*, Wiley, New York, 1988, p. 382.
72. A. E. Riad and J. C. Toscano, Wideband characterization and modeling of thick-film inductors, *Proc. Int. Symp. Micrelectron.*, 1987, p. 73.
73. M. Ukigai, T. Morzumi, and Y. Mida, A new digital variable delay device with hybrid microcircuit techniques, *Proc. ISHM*, 1987, p. 255.
74. J. R. Corkhill and D. R. Mullins, The properties of thick-film inductors, *Electron. Components*, May 1969.
75. F. S. Burkett, Improved design for thin-film inductors, *Proc. Electron. Comput.*, 1971, pp. 184–194.
76. E. Bellem, Thick-film chip inductors, *Proc. Int. Symp. Microelectron.*, 1971, p. 8-3-4.
77. F. E. Terman, *Radio Engineers Handbook*, McGraw-Hill, New York, 1950.
78. J. O. Renskers, Chip inductors—new breed, *Proc. Int. Symp. Microelectron.*, 1971, p. 8-2-1.
79. P. Barnwell, 100-MHz tuned amplifier with thick-film inductors, *Electron. Comput.* Vol. 13, No. 15, 1972, pp. 723–724.
80. J. Corkhill and M. Martin, Thick-film filter designed for television signals, *Electron. Eng. J.*, Vol. 43, No. 519, 1971, pp. 65–69.

81. W. Olshewski, The hybrid compatible transformer, *IEEE Trans. Comput. Hybrid Manuf. Technol.*, Vol. 2, No. 4, Dec. 1979, pp. 487–489.
82. R. E. Chaddock, The amplification of lumped element techniques to high-frequency hybrid integrated circuits, *Radio Electron. Eng.*, Vol. 44, No. 8, 1974, pp. 414–420.
83. B. D. Popovic, *Introductory Engineering Electromagnetics*, Addison-Wesley, Reading, MA, 1971; L. Solymar and D. Walsh, *Lectures on Electrical Properties of Materials*, 4th ed., Oxford University Press, Oxford, 1988, Chap. 1.
84. J. M. Ziman, *Principles of the Theory of Solids*, Cambridge University Press, Cambridge, 1964, p. 241; L. I. Maissel and R. Glang, eds., *Handbook of Thin-Film Technology*, 2nd ed., McGraw-Hill, New York, 1983, p. 2-70.
85. H. Lambs, *Hydrodynamics*, Dover Publications, Mineola, New York, 1945.
86. S. Ruthberg, Hermetic testing of large hybrid packages, *Proc. Int. Symp. Microelectron.*, 1982, p. 215.
87. D. A. Howland and C. A. Mann, The back pressurization technique of leak testing *Vacuum*, Vol. 15, 1952, pp. 347–352.
88. F. Gargione, System partioning for hybrid circuits, *Int. Symp. Microelectron.*, 1969, pp. 281–288.
89. R. J. Cornish, Flow in a pipe of rectangular cross section, *Proc. R. Soc. London*, Vol. A120, 1928, pp. 691–700.
90. M. Pecht, *Handbook of Electronic Package Design*, Marcel Dekker, New York, 1991.
91. Z. H. Meirksin, Design of thin and thick film, in *Thin- and Thick-Film Hybrid Microelectronics*, Lexington Books/D.C. Heath, Lexington, MA, 1976, Chap. 13.
92. R. W. Johnson, ed., *Hybrid Circuit Design*, in Modular Series in Hybrid Microelectronics, ISHM Publishing, Reston, VA, 1991.
93. D. D. Zimmerman and J. R. Dinitto, *Hybrid Microcircuit Design Guide*, ISHM/1402/IPC/855, Oct. 1982.
94. P. Moran, *Hybrid Microelectronic Technology*, Gordon & Breach, New York, 1984.
95. M. R. Hakard, *Thick-Film Hybrids Manufacture and Design*, Prentice Hall, Upper Saddle River, NJ, 1988, Chap. 3.
96. J. A. Scarlett, *PCB for Microelectronics*, 2nd ed., IMAPS, Washington, DC, 1990.
97. Y. Fukuoka and M. Ishizuka, Transient temperature rise for multichip packages, in *Thermal Management Concepts in Microelectronic Packaging*, R. T. Howard, S. S. Furkay, R. F. Kilburn, and G. Monti, Jr., eds., ISHM, Reston, VA, 1984.
98. A. J. Blodgett and D. R. Barbour, Thermal conduction module: a high-performance multilayer ceramic package, *IBM J. Res. Dev.*, Vol. 26, No. 1, 1982, pp. 30–36.
99. Horizons, DuPont, *Thick-Film News*, global ed., No. 14, Dec. 1999.
100. J. J. Licari and L. R. Enlow, *Hybrid Microcircuit Technology Handbook*, Noyes Publications, Park Ridge, NJ, 1999, Chap. 4.
101. A. L. Eustice, S. J. Horowitz, J. J. Stewart, A. R. Travis, and H. J. Sawhill, Low-temperature co-firable ceramics: a new approach to electronic packaging, *Proc. 36th Electron. Comput. Conf.*, Seattle, WA, 1986, pp. 37–47.
102. J. Matthey, Fine line gold conductor, *Thick-Film Products Bulletin*, JM1202.

103. Monograph Tektronics, Inc., *Design Methods for Hybrid Circuits*, Design Guides I and II, Hybrid Computer Operation, MTI, Beaverton, OR; J. E. Sergent and C. Harper, eds., *Hybrid Microelectronics Handbook*, 2nd ed., McGraw-Hill, New York, 1995.

104. R. F. Bunshah, ed., *Deposition Technologies for Film Coating*, Noyes Publications, Park Ridge, NJ, 1962.

105. J. J. Licari, Thin and thick films, in *Electronic Materials and Processes Handbook*, C. A. Harper and R. N. Sampson, eds., McGraw-Hill, New York, 1993, p. 8.66.

106. R. L. Geiger, P. A. Allen, and N. R. Strander, *VLSI Design Techniques for Analog and Digital Circuits*, McGraw-Hill, New York, 1990, p. 4.

107. G. V. Planer and L. S. Phillips, *Thick-Film Circuits*, Crane Russak, New York, 1972.

108. M. Pecht, Design for reliability concepts, in *Integrated Circuit Hybrid and Multichip Module Package Design Guidelines*, Wiley, New York, 1994.

109. H. W. Postlethwaite, Hybrid thick-film printed components, *Proc. IEEE Ind. Electron. Cont. Inst. Thick-Film Hybrid IC Tech. Symp.*, Mar. 22, 1968.

110. B. J. Baliga, *Modern Power Devices*, Wiley, New York, 1987; J. Millman and C. C. Halkias, Power circuits and systems, in *Integrated Electronics: Analog and Digital Circuit Systems*, McGraw-Hill, 1972, Sec. 18-14.

111. J. J. Staller, Introduction to thick-film hybrid circuits, *IEEE IC Tech. Symp.*, McGraw-Hill, 1972.

112. E. R. Hnatek, ed., *A User's Handbook of Integrated Circuits*, Wiley, New York, 1988, Chap. 13.

113. C. S. Yeh, Film resistor characteristics and design, *Proc. Int. Symp. Microelectron.*, 1976, pp. 26–33.

114. J. J. Licari and L. R. Enlow, Thin-film design limitations, in *Hybrid Microcircuit Technology Handbook*, McGraw-Hill, New York, 1998, Chap. 18.2.

115. J. J. Licari and L. R. Enlow, *Hybrid Microcircuit Technology Handbook*, McGraw-Hill, New York, 1998, p. 408.

116. F. W. Kear, *Hybrid Assemblies and MCMs*, Marcel Dekker, New York, 1993.

117. C. E. Jowett, *The Engineering of Microelectronics: Thin and Thick Film*, Macmillan, London, 1976.

118. M. Pecht, Layout compactness for high performance/large scale circuits, in *Advanced Routing of Electronic Modules*, CRC Press, Boca Raton, FL, 1995; J. J. Licari and L. R. Enlow, Hybrid design process, in *Hybrid Microcircuit Technology Handbook*, McGraw-Hill, New York, 1998, p. 363.

119. L. L. Hench and J. K. West, *Principles of Electronic Ceramics*, Wiley, New York, 1990, Chap. 6.

120. J. M. Herbert, Ceramic dielectric and capacitors, in *Properties of Dielectrics*, Gordon & Breach, New York, 1985.

121. A. E. Feuersanger, in *Preparation and Properties of High-Permittivity Thin-Film Dielectrics*, F. Vratny, ed., Electrochemical Society, Pennington, NJ, 1969, pp. 237–244; J. M. Herbert, ed., *Ceramic Dielectrics and Capacitors*, Gordon & Breach, New York, 1985.

122. B. Dance, Technology News, *Semicond. Int.*, Vol. 23, No. 14, Dec. 2000, p. 48.

RECOMMENDED READING

Anderson, J. C., *Dielectrics*, Chapman & Hall, London, 1964.

Brown, W. D., *Advanced Electronic Packaging*, IEEE Press, Piscataway, NJ, 1999.

Geiger, R. L., P. E. Allen, and N. R. Strader, *VLSI Design Techniques for Analog and Digital Circuits*, McGraw-Hill, New York, 1990.

Greenfield, W. E., *Introduction to Dielectric Theory and Measurement*, College of Engineering, Washington State University, Pullman, WA, 1972.

Grover, F. W., *Inductance Calculations, Working Formulas, and Tables*, Dover Publications, Mineola, NY, 1973.

Harnwell, G. P., *Principles of Electricity and Electromagnetism*, 2nd ed., McGraw-Hill, New York, 1949.

Kear, F. W., *Hybrid Assemblies and MCMs*, Marcel Dekker, New York, 1993.

Licari, J. J., and L. R. Enlow, *Hybrid Microcircuit Technology Handbook*, Noyes Publications, Park Ridge, NJ, 1998.

McLyman, C. W. T., *Transformer and Inductor Design Handbook*, Marcel Dekker, New York, 1988.

Pecht, M., ed., *Advanced Routing of Electronic Modules*, CRC Press, Boca Raton, FL, 1995.

Pecht, M., *Integrated Circuit, Hybrid, and Multichip Module Package Design Guidelines*, Wiley, New York, 1994.

Sandborn, P. A., and H. Moreno, *Conceptual Design of Multichip Modules and Systems*, Kluwer Academic, Boston, 1994.

Sergent, J. E., and C. A. Harper, eds., *Hybrid Microelecronics Handbook*, 2nd ed., McGraw-Hill, New York, 1995.

Taraseiskey, H., *Power Hybrid Circuit Design and Manufacture*, Marcel Dekker, New York, 1996.

Tareev, B., *Physics of Dielectric Materials*, Mir Publishing, Moscow, 1975.

Terman, F. E., *Radio Engineering Handbook*, McGraw-Hill, New York, 1960.

Von Hippel, A. R., *Dielectric Materials and Applications*, MIT Technology Press and Wiley, New York, 1954.

Walker, C. S., *Capacitance, Inductance, and Cross-Talk Analysis*, Artech House, Norwood, MA, 1990.

Zimmerman, D. D., and J. R. Dinitto, *Hybrid Microcircuit Design Guide*, ISHM and IPC Joint Publication, 1982.

3

COMPUTER-AIDED DESIGN AND PATTERN GENERATION TECHNIQUES

3.1 COMPUTER-AIDED DESIGN TECHNIQUES

The requirements to minimize development costs and optimize device circuit performance have led to a dramatic increase in interest in computer-aided design (CAD) techniques [1]. The development of modern computer software and hardware systems has helped to characterize and improve circuit design prior to circuit fabrication. Equivalent-circuit models have traditionally dominated design methods but require prior knowledge of circuits under certain operating conditions, making it very difficult to predict the characteristics of newly modified circuits and the device structures and configurations. Physical models have been shown to provide valuable insights into the operation of circuits and devices and in many cases have been used to improve circuit structures [2]. Two domain physical device models allow device and circuit operation to be investigated prior to fabrication and are intrinsically capable of modeling the construction of widely varying design through process parameters controlling steps from component generation to output for manufacturing. However, two-dimensional numerical simulations require extensive computer resources, making them unwise for use in most CAD applications [3].

The tools now available to hybrid microcircuit designers are very powerful and dynamic, and most require the use of large computers or, more recently, powerful graphically intense workstations. Investment in both hardware and software needed for current HMC design continues to be high, and it is crucial to remain competitive in the market [4]. As powerful and dynamic as these

Handbook of Thick- and Thin-Film Hybrid Microelectronics, By Tapan K. Gupta
ISBN 0-471-27229-9

tools may be, fierce competition in the marketplace has resulted in the evolution of user-friendly software with which the engineer can establish proficiency with a modest investment of time and effort [5].

Despite the sophistication and cost of both the hardware and software necessary to remain competitive in the field of HMC design, most major contributions to this field are based on relatively simple and basic innovations by engineers [6]. These innovations occur in circuit design, processing, and modeling as well as in the evolution of CAD tools themselves [7]. Because of the economic impact of innovations in the HMC design field, advancement potential and rewards for talented and ambitious engineers are essentially boundless. As in any high-technology area, a large amount of technical jargon has evolved which must be mastered by anyone wishing to be conversant with those working in the area [8].

3.1.1 Size and Complexity of Hybrid Microcircuits

The hybrid microcircuit, like its counterpart the IC, is not typically classified in terms of the number of devices or potential devices used in the design of the circuit and in terms of the feature size of the process. However, the complexity and number of potential devices accommodated in terms of area have gone beyond the limits [9]. Higher system performance, lower-cost material systems, and the cost of manufacturing systems are the driving forces behind the need for increased circuit density. To increase circuit density while keeping the size and weight of a module as small as possible is a challenge to HMC design engineers [10]. To address these problems, CAD engineers have used all the advantages of thick- and thin-film and IC technologies to make hybrid circuits with minimum feature size. The impact of shrinking the feature size warrants the design of a multilayer hybrid microcircuit. The pitch (the minimum of the sum of the minimum width of a feature and minimum spacing between similar features) is often nearly twice the minimum feature size. The feature size had shrunk to less than 0.4 mil (10 μm) in thick-film technology, under 0.08 mil (2 μm) in thin-film technology, and 0.04 mil (1 μm) or less than 0.04 mil (1 μm) in ICs [11].

Beyond the increase in device count, shrinking the feature size has other major benefits. As the size is decreased, the speed of the circuit increases linearly, so it is expected that smaller computers will work much faster [12]. However, there are some limitations associated with shrinking the feature size, including deterioration in matching characteristics, coupling noise, cost of aids, increased resistive path and power dissipation density, and overall processing complications.

HMC Design Process It goes without saying that an HMC design engineer should design a circuit with maximum device density while expanding minimum labor and physical resources in a short time frame. Furthermore, the production yield should be high, the process should be simple, and the substrate area should be small [13]. The conventional approach to circuit design often involves much iteration at the breadboard level. Preliminary designs are based on simple models of devices or subscripts. The preliminary design (e.g., 75% design rule) is used as an exercise for beginners. However, computer simulation using more accurate models is used to verify the performance of the preliminary design. Good design of the circuits and subscript models (placement of ICs, MOS, etc.) is critical. A performance verification after fabrication avoids the requirement of excessive computer time during simulation [14].

Once the preliminary design is found acceptable, actual layout takes place. A good overall floor plan reduces the cost and the design engineer's time. The floor plan contains all major busing and cell (subscript) placement information as well as I/O pad designations.

Following an acceptable computer simulation the entire circuit is approved for fabrication. In evolving processes or complicated designs, subcircuits and/or test structures are often fabricated. For large-scale HMC design, design automation and design verification are keys to success. As the circuit complexity and density increase, layout and checking of circuits are not possible manually and computer-based design automation tools are needed [15]. Design automation and design verification tools require computer-readable descriptions of the underlying circuit function as a structure to operate. A typical process flowchart for hybrid circuit design is presented in Figure 3.1. Before primary design and computer simulation, circuit element specifications are documented. After computer-simulated design of the circuit is accepted, circuit layout begins. The simulated circuit is passed for initial fabrication and then undergoes electrical testing and evaluation of circuit performance. Production starts once the circuit is accepted by the design and test engineers.

Layout Style Although the cost of handcrafted Mylar layouts of a circuit is higher, HMC engineers still use them because of the need to minimize the area required by high-volume HMC. Frequently, the basic form of the HMC is cut on paper prior to entry into a computer and to Mylar. The resulting geometrical layout is digitized, sometimes through use of a symbolic layout language but primarily with the help of an interactive cathode-ray tube (CRT) graphics editor. Most modern graphics workstations provide color displays. Handcrafted layouts can be entered directly into a computer in geometrical form through use of an interactive CRT graphics editor. A mouse is used in conjunction with a cursor to size and position geometric objects such as boxes on a high-resolution CRT display. A corresponding data file is kept in computer memory to describe the geometry displayed. A graphics package called *Magic* [16] developed at the University of California is currently very popular for CAD. It uses the painting idiom to create geometrical objectives on a color CRT. The user chooses a color (layout level) from a palette on the screen and paints areas on the screen by specifying two opposite corners of a rectangular field. The final result is the same as for other layout methods. A geometrical specification file is created in computer memory, saved, and ultimately trans-

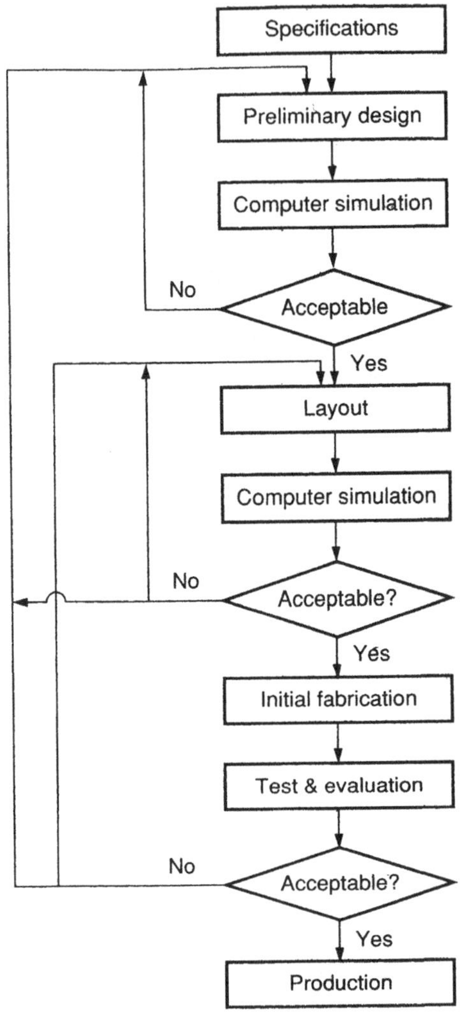

Figure 3.1 Block diagram of a conventional HMC design process [12].

ferred to the mask shop. For reasons of efficiency (fast editing response and minimum memory requirements), most industrial CAD tools provide conversions between their internal format and EDIF [17,18], which is defined as the interchange format. In addition, most CAD tools convert the internal format for submittal to the mask shop.

Graphical Symbolic Layout The parameterized layout representation described previously provides little insight into the geometrical relationship between circuit elements. This important insight can be provided by another symbolic form for hybrid microcircuit description called *graphical symbolic layout* [19]. *Sticks*, an earlier graphical symbolic layout method, combines circuit connectivity with layout topology information. The combination of connectivity and topological information is important in the generation of circuit layouts. It becomes a simple task for a designer to convert a circuit to full layout form once the graphical symbolic layout for a circuit is generated. The layout task has become simple for flattening connection lines and compacting the layout. If the graphical symbolic description has been entered into a computer, perhaps through an interactive graphics terminal, a symbolic compiler program can convert the symbolic layout to a full layout by expanding the line symbols according to a technology specification and then compacting the resulting layout [20].

CAD systems for hybrid design and layout have invariably been an afterthought. One might be attempted to think of CAD systems for hybrid design as the creative effort of marketing personnel who saw similarities between hybrid design and layout [21] and ICs and PCB design and layout. Hybrid circuits containing discrete components occupy considerably less area than that required in the conventional PCB/discrete approach and played a major role in the increasing demand for analog signal processing [12]. As in hybrid manufacturing, hybrid devices are typically built as an alternative process in which layers of materials are deposited on substrate. As in PCB layout, hybrids are typically built using discrete parts that can be placed and manipulated symbolically as a single entity [12,22]. Thick- and thin-film passive [23] components have reasonable tolerances, are easily trimmable, have acceptable temperature coefficients that can be tailored for tracking, and offer reasonable trade-offs between area required and component values. To provide a generalized solution to design and layout problems, the CAD system should focus on five areas: component generation, device placement, interconnection technology, layout analysis, and manufacturing output [24,25].

Parameterized Module Generation A parameterized module generator is a software procedure that can generate many different cell layouts, depending on the values that are specified when the generator program is executed. Many of the generators use input parameters to specify the width or number of bits. As an example, three designs might require an 8-bit adder for the first design, a 16-bit adder for the second design, and a 32-bit adder for the third design. Typical design style would use an interactive graphics editor to create each of these adder separately. Parameterized module generators use many of the powerful constructs of high-level programming languages to describe layout structure, to position subcells, and to fit the overall layout of a larger cell together. Parameterized variables are used with their values bound to a specific value when a module is generated [26].

The use of high-level programming language techniques also provides a disadvantage for many parameterized module generators: that layout cannot be visualized until the generation program has been coupled and linked to the layout. These potentially time-consuming steps may hinder the use of interactive layout in designing a module generator for a few cells. To circumvent this problem, there is ongoing research on ways to provide interactive graphical feedback as the geometrical structure of a cell is defined [27].

Component Generation To populate a substrate with components, copies are made from an online element library and placed in the component design file. Physical components to be used in a design are generated and stored in the library through an interactive process. Typical discrete devices such as chip capacitors and surface-mounted devices (SMDs) are constructed. Screened resistors, capacitors, and (sometimes) inductors are created as well. Construction of discrete devices consists of building a model to exact dimensions dictated by part satisfactions. Hybrid packs and resistors, however, are modifiable after creation and thus must be governed by a set of parameters defining design limitations and geometric construction.

Hybrid Pack Creation The unique feature of hybrid packs is the ability of their surface bond pads to "float" within delta parameters from the component die attachment point. Each bond pad with wire combination is constructed within constraints concerning wire length range and proximity to other wires. Once the peak is built, the die size and location of the connect pins within the die are permanent. In contrast, bond pads and wires may be modified during the interconnection phase to improve routability [28].

Resistor Creation The resistor geometry in a hybrid circuit, whether screen-printed thick-film design or preferentially etched vacuum-deposited thin-film design, depends on the operating characteristics of the circuit elements. For each design the variable parameters are resistance (ohms), resistivity (Ω·cm), power dissipation (watts), ink resistivity, resistor geometry (tophat or serpentine), and aspect ratio (under pretrim conditions). In 1977, LSI and VLSI CAD methodology was adopted to design complex thick-film multilayer hybrid microcircuits and modules [29]. The implementation of an efficient CAD standard cell library is enhanced by a powerful online design capability through a program called CALMA [29]. This has enabled programmers to create, edit, and manipulate components that are defined on many different levels. The physical geometry of the cells and I/O locations are specifically designed to take advantage of automatic IC placement and routing programs.

Device Placement Within CAD systems, parts can be placed on a hybrid circuit in different ways. No matter which placement routine is used, parts are always manipulated symbolically as single elements. Placement outlines may be built into parts and augmented by *side walks* that will ensure that sufficient space is left for manufacturing requirements (e.g., where a screen resistor is to be placed under a SMD part). If a device is to be placed on both sides of the substrate, two connected parts will be placed on opposite sides of the substrate if the via penalty is low. Specifying a high via penalty, while connections are made using a wraparound method, will result in parts being placed next to each other on the same side of the substrate. Once component placement is complete, prior to interconnection, the layout database can be converted to a three-dimensional representation.

An automatic routing tool uses a rip-up and retry strategy to complete all possible connections. Before actual routing is begun, device placement is evaluated. The designer is informed if there are any congested areas in the design, and the probability of success is estimated. To eliminate unnecessary reworking, the functional performance of a design must be tested before fabrication. CAD tools provided for functional analyses are layer trim simulation, actual/optimal track length analysis, and area calculations for material use. Other CAD tools developed are laser-trim simulation, track-length analysis, and area calculations and thermal analysis.

One of the challenging problems in hybrid circuitry is multilayer package design. As the speed and number of circuit elements are increasing, dense packaging structures evolved that shorten interconnect lengths and propagation delay [30,31]. To implement the requirements of high-speed multilayer package design, CAD tools should be developed to meet important electrical parameters (e.g., line impedance, propagation delay, signal rest time, attenuation, line ringing, and crosstalk [32,33]).

Modeling a package is the first step in CAD. After models are created, a network is constructed from the primitive package model. The success of CAD simulation is based primarily on how well the models have been constructed and the extent to which simulation has been conducted [34,35]. Figure 3.2 shows a CAD approach to the electrical analysis of multilayer packages. First, the physical parameters of the packaging systems are adjusted to provide the electrical requirements specified, and line width, line thickness, and dielectric thickness are selected for the characteristic impedance desired. Then electrical models of the package structure are constructed to represent the transmission lines, lead frames, vias, pins, and so on. The third step is to generate a computer database that represents the electrical equivalent of the package models and circuit I/O buffers. This is followed by a schematic entry which defines the chips, their approximate placement, chip-to-chip interconnects, and pin assignments. A place and route are then established, taking into consideration all critical line lengths. This is usually done in conjunction with the thermal requirements of the package. Once successful routing has been accomplished, CAD simulation is performed on selected critical transmission lines to verify performance. Parameters such as crosstalk, propagation delay, reflections, attenuation, and signal fidelity are observed in real time on a workstation graphics monitor. An analysis is made from these results, and if needed, iteration is performed by making necessary adjustments to the package structure or circuitry to obtain optimum performance.

To illustrate the CAD approach to solving problems in hybrid microcircuits, we have tried to be specific and have taken a simple case of packaging circuits [33]. A single strip line circuit cross section is shown in Figure 3.3, and Figure 3.4 is the equivalent-circuit diagram. The transmission line is 1 cm long with a

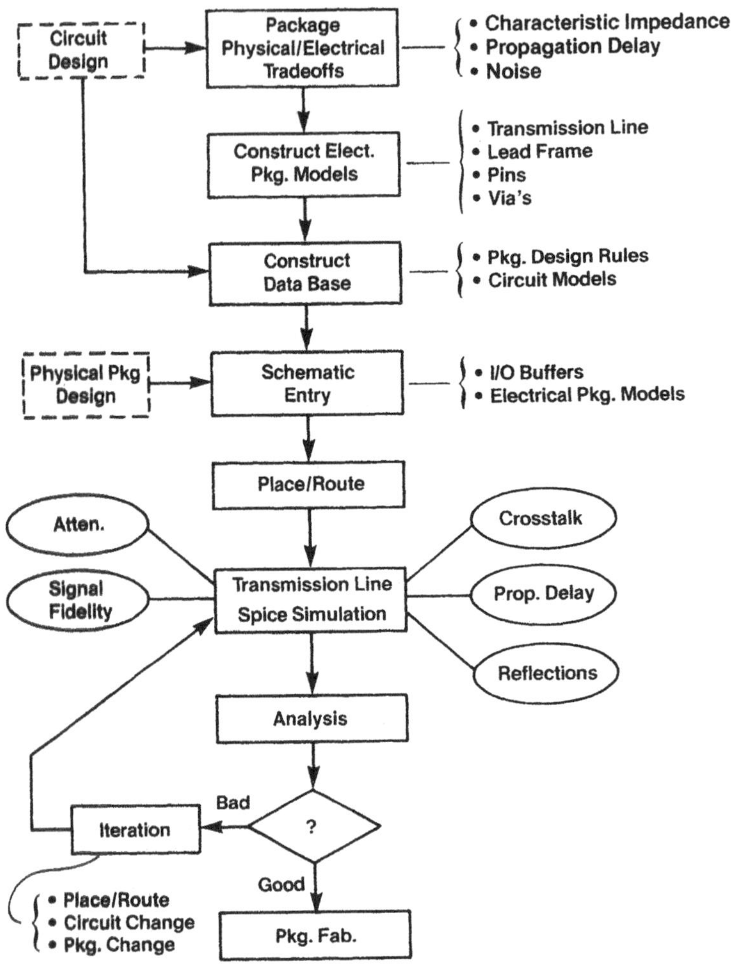

Figure 3.2 Electrical CAD approach.

600-Ω termination resistor. A circuit simulation approach is used here; therefore, the modeling problem [26] is to find an equivalent circuit to model the package elements. The transmission lines are modeled by an equivalent resistance (R), conductance (G), inductance (L), and capacitance C, per unit length [35]. The inputs to this model are: dielectric constant, 3.5; dielectric thickness, 3

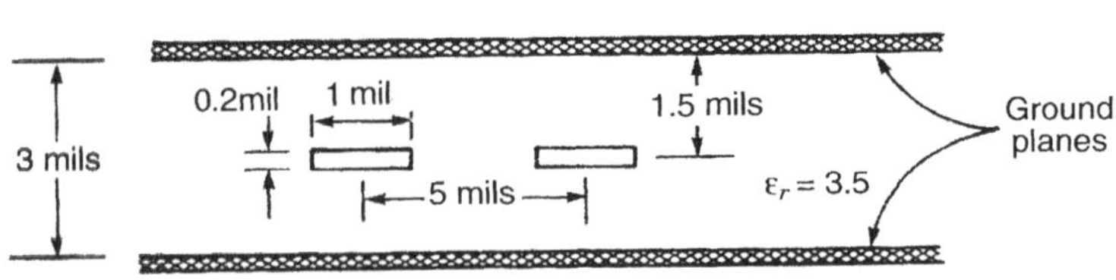

Figure 3.3 Physical structure (cross section) of the strip line (1 mil ≈ 25.4 μm).

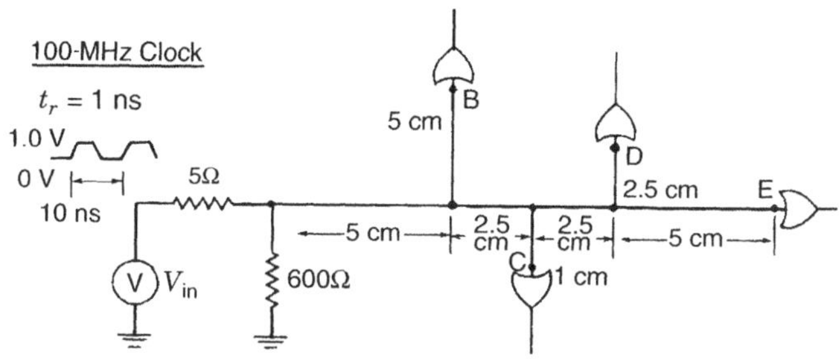

Figure 3.4 Single strip line circuit (1 in. ≈ 2.54 cm).

mils (76 μm); line width, 1 mil (25 μm); conductor thickness, 0.2 mil (5.1 μm); conductor offset, 1.5 mils (38 μm); and line resistivity, 1.6×10^{-6} Ω·cm. The results computed are $Z_0 = 52.5$ Ω, $C = \varepsilon_r^{1/2}/vZ_0 = 1.187$ pF/cm, and $L = CZ_0^2 = 3.00$ nh/cm. The R parameter can be obtained from the physical dimensions and the conductivity of the material as $R/l = \rho/tw = 1.26$ Ω/cm, where ρ is the resistivity, R the resistance, t the thickness; l and w are the length and width of the resistor deposited, assuming a no-loss dielectric and $G = 0$.

The next step is to calculate the physical length of the lump, or conversely, the number of lumps per centimeter. The following equation will give the minimum number of lumps per centimeter:

$$\text{lumps/cm} = \frac{1}{(1/N)[v/f_{\max}(\varepsilon_K)^{1/2}]} = 0.22$$

where N represents the lumps per wavelength, v the speed of light, $f_{\max}$ the highest frequency present in the propagating signal, and ε_K the dielectric constant of the package material. The result means that there should be at least 0.22 lump/cm. For these simulations, a conservative value of 4 lumps/cm is used.

The final step is to move onto the workstation and run the SPICE simulation [36–38]. The circuit is entered into the workstation graphically using standard SPICE elements, and Honeywell-designed transmission line symbols are implemented as subcircuits. The result is checked and expanded, after which the SPICE simulation is run. The run time is 4 minutes on an Apolo/Mentor DN420.

3.1.2 CALMA Online Design Process

In 1977, with an emphasis on low-cost fast-turnaround designs, GTE developed a comprehensive standard cell system for custom VLSI chips as well as a spin-off application for thick-film hybrid/module design [28]. CALMA's powerful online design capability has made this program successful [39]. It is a state-of-the-art second-generation interactive graphics system optimized for VLSI and hybrid circuit design. When the circuit layout is finalized, the

CALMA system generates a tape capable of direct operation of industry standard photo-tooling machinery. Thick- and thin-film layouts are designed online, and CALMA's pattern generation soft wire is used to write a magnetic tape capable of driving various commercial pattern generators. A CALMA online design process used for copper thick-film conductor design is shown in Figure 3.5.

The first step in the CALMA online design process is to design circuitry to minimize space. Once the circuit design is complete, the circuit components, including the substrate material, are chosen so that there will be no problem in the latter part of circuit testing (thermal, electrical, and quality). The next step is cell placement. The geometry of the circuit is also important from the standpoint of packaging. The next step is a review of the circuit, which is done primarily by the design engineer in consultation with a layout engineer. If they agree, the material is forwarded to the test engineer. Once the circuit passes all tests it is ready to be sent to the fabrication engineer. Next, the circuit is checked for violations of design rules. Finally, all engineering departments share their views before the circuit is passed for production.

3.1.3 Computer-Aided Engineering to Design Hybrid Microcircuits

At GTE, the capabilities of computer integration have been exploited by utilizing computer data from a graphic layout of an interconnect board. The layout of the board is designed on a graphic computer system, which can communicate with various miniframe computers, automated traffic exchange (ATE) systems, and component assembly equipment [41].

Efforts in the first area of hybrid design and layout are concentrated on the development of design rule checking (DRC) software. DRC software performs a series of geometric verifications of a graphic layout to ensure acceptable manufacturing yield. Computer DRC makes it possible to check the spacing between unconnected shapes of a mask and that between the shape of one mask versus that of another. Custom mask analysis files or DRC programs consist of Boolean operations and user-designated commands to be performed on preselected data. DRC will play an increasingly important role in the layout verification cycle of multilayer substrates as the complexity of thick-film multilayer substrates increases. The time required to complete the design of a circuit from a schematic to the finished product is costly, and DRC software is a valuable tool for implementing the job [42].

The second area of software integration is multilayer substrate material estimation. The software calculates paste quantities from a graphic layout of the interconnect board [43]. The flow diagram for the layout of such a board is shown in Figure 3.6. In step 1, the software algorithm performs a series of mask analyses. The second step of the software performs a fraction of the work of the data base graphic layout. In the third step, paste quantities (conductors, dielectric, passivation, etc.) are calculated. The result of this software application is reported to be highly accurate. It estimates the theoretical requirements of the amount of thick-film paste needed to print the substrate.

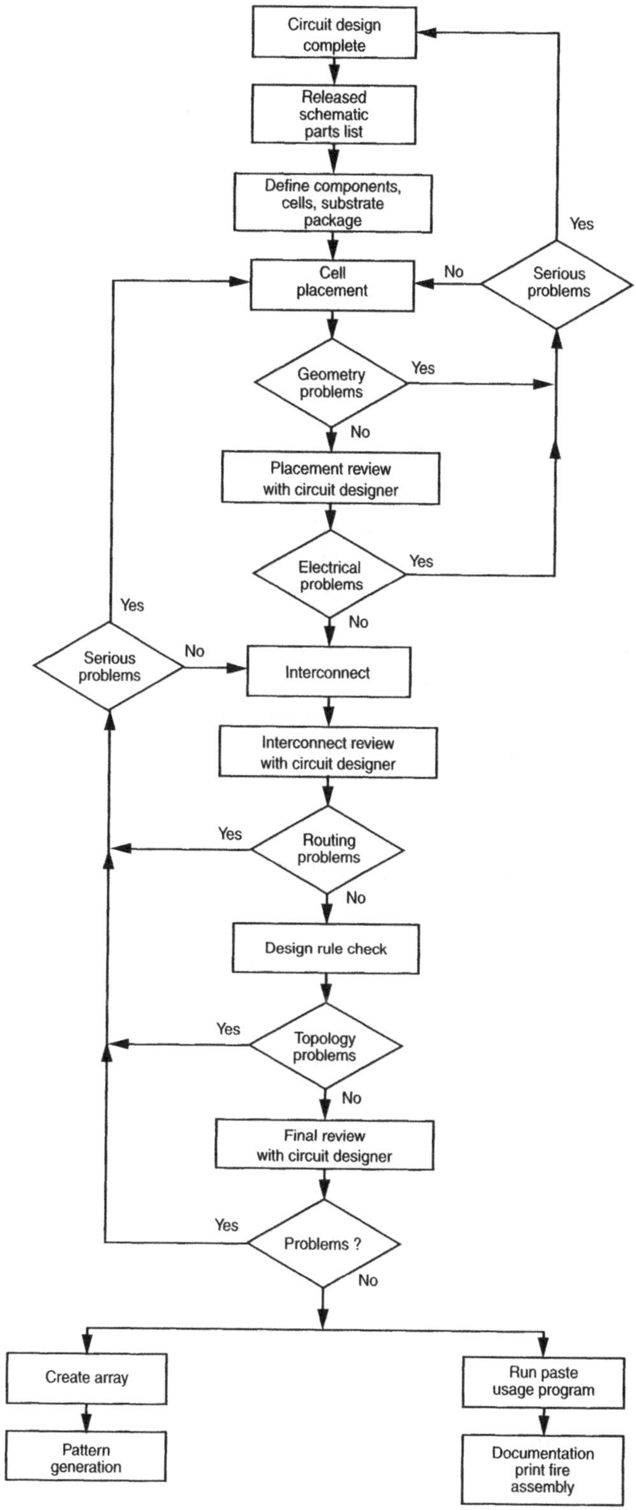

Figure 3.5 CALMA online design process [28,29].

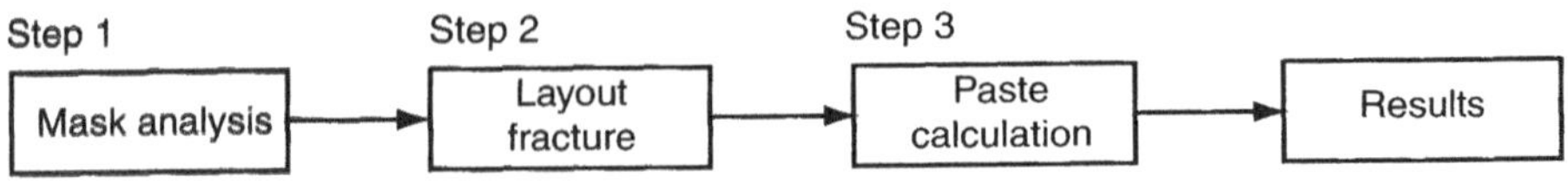

Figure 3.6 Flow diagram of the graphic layout of an interconnect board.

The third area of computer integration relates to microcircuit documentation. Custom menu-driven software uses operator-defined input to route connections between a chip's I/O pad and the corresponding bonding pad on the package to generate bonding diagrams. The software technique utilizes the design layout of the chip itself by first creating automatically a new structure consisting of the chip online, signal pads, and corresponding signal name. Using the *simplified* chip structure, the program will automatically prompt the user interactively for an IC package-type low-temperature cofired ceramic (LTCC) or DIP (leadless chip carrier or dual inline package), the number of package I/Os, and the size of the internal die attachment cavity of the package. Ceramic substrate layer drawings are another important aspect of microcircuit documentation as visual aids during manufacture. A complete documentation set of all ceramic board layers (conductors, via, and dielectric) can be generated automatically from the design layout.

The fourth area is high-speed bare-board testing of multilayer interconnect substrates. GTE has established a link between the CAD system and the tester through which the board net list can automatically be downloaded to the test system. Thick-film technology provides high-density chip and wire hybrid components and/or an interconnect substrate for components, resulting in a high-density module. Since the use of thick film is a custom process, various packaging options are possible, from plug-in modules to custom dual-inline packages to standardized packaging concepts. The common denominator in all these approaches is solderable thick-film substrate. Figure 3.7 illustrates the relationship of software to the networking of various computer systems in several processing operations.

So far we have been talking about different CAD systems and their usefulness in designing and testing a particular hybrid microcircuit. As production cost and competition increase, the demand for automation is also rising. As a result, almost all manufacturers have developed their own CAD system.

3.1.4 Circuit Layout Design

The miniaturization of a hybrid circuit generally rests on the design engineer. The engineer responsible for layout of the circuit sees to it that the circuit has not violated any circuit design rules. The fabrication engineer, on the other hand, is responsible for accommodating the circuit within a small area so that it is cost-effectve and the performance of the circuit does not affect the cost. There are 11 steps that are followed in designing a hybrid circuit [43]: (1) circuit partitioning, (2) package selection, (3) resistor design, (4) capacitor design, (5)

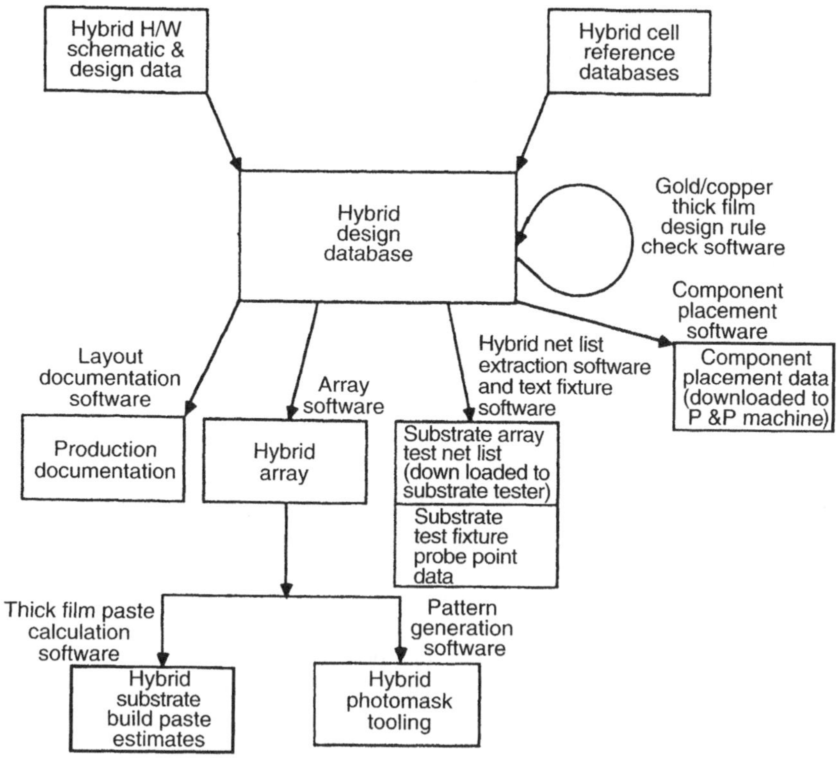

Figure 3.7 CAD system for a hybrid module [40].

substrate layout design, (6) artwork generation, (7) photomask generation, (8) die placement and bonding, (9) substrate fabrication, (10) package scaling, and (11) functional and environmental testing.

Circuit Partitioning [44,45] The most important step in hybrid microcircuit design is circuit partitioning: how much thick- and thin-film circuitry there will be in a particular hybrid microcircuit. The decision is based on the production cost, space required, and yield, and the design engineer should decide on the basis of the complexity and performance of the circuit which of the three parameters will be given preference.

Package Selection At this stage the power dissipation of the entire circuit [46] is calculated. For a better and safe design, the power dissipation of the entire circuit, including the surface-mounted discrete devices, should be 50 to 60% of the total dissipation capacity of the substrate [47].

Resistor Design and Layout [48] Once the total power dissipation calculations are complete, the size and shape of the resistors are calculated. Resistor design computer programs are now available for both thick- and thin-film circuits that will design resistors and generate output data consisting of minimum width and maximum length, or vice versa, depending on the shape of the resistors (simple rectangular, meander, or tophat type), trimmer length, and line drawing [49].

Capacitor Design [50] In hybrid circuit design the designer can choose two types of capacitors: chip capacitors or film-sandwich capacitors. Large capacitors are always deposited by depositing dielectric between conductors. Caution must be used in choosing different types of dielectric materials. The lower electrode is always made slightly larger than the bottom electrode as a dual precaution against voltage breakdown at sharp edges and loss of capacitance due to misregistration of pattern [43]. A computer-aided program is also available for capacitor design.

Inductor Design [51,52] Inductors cannot be avoided in hybrid circuits, especially in the high-frequency range. A computer-aided inductor design program is also available and can be used to design inductors.

Substrate Layout Design An interconnection pattern along with discrete components and a film resistor layout can be plotted by the computer. The layout obtained by the computer may not be optimal, but it will be less expensive and more efficient. After package selection, the designer chooses substrate size and then proceeds to the intermediate layout phase, which includes a final tailoring of circuit elements to fit the substrate area available. A typical flowchart for CAD hybrid microcircuit design is shown in Figure 3.8. The hybrid designer will proceed with partitioning the circuit, identifying the discrete components, conductor, resistors, capacitor printout, and layout design once the circuit schematic is received from the design engineer. Once this is accomplished successfully, the circuit is sent for artwork generation. Table 3.1 lists CAD systems that can be used to design hybrid microcircuit and multichip modules.

Design rules often depend on both process equipment and process design. They are needed when a new process is being developed. There is always a conflict between the circuit designer and the process engineer about the importance of the tighter and smaller design and the repeatability of the process. Despite their differences, both agree that design and layout rules are meant to prevent unworkable, unreliable, or difficult-to-implement constructs. As a matter of fact, the goal is to devise design and layout rules that are simple, constant in time, and applicable to many processes. The Mead Conway approach is to characterize the process with a single scalable process-dependent parameter called *lambda* that is defined as the maximum distance by which a geometrical feature on any layer can stray from another feature, due to overetching, misalignment, distortion, over- or underexposure, and so on, with a safety factor included. Lambda is thus equal to maximum misalignment of a feature from its intended position on the wafer, and two features on different

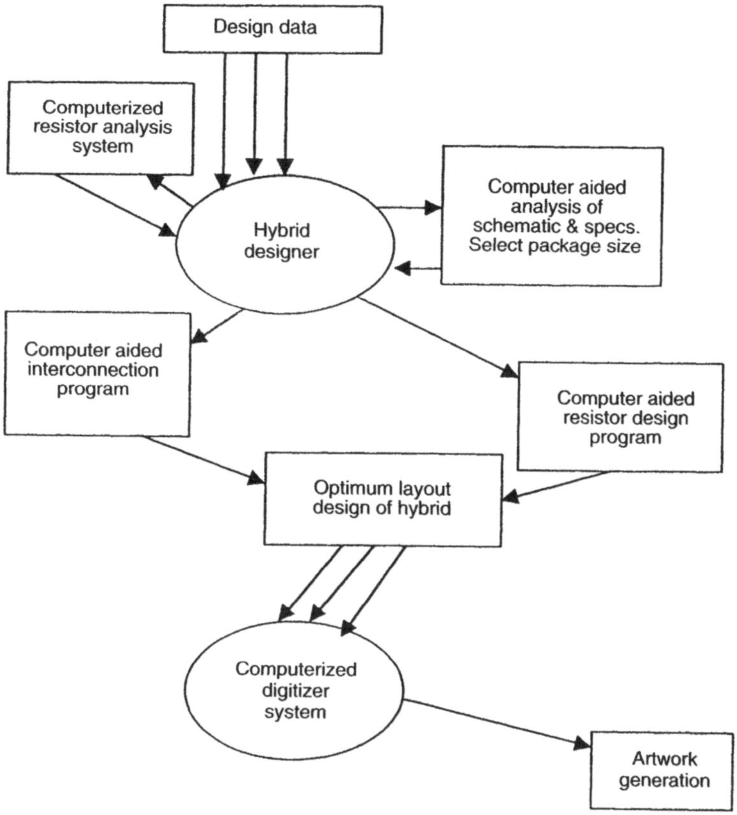

Figure 3.8 Flowchart for CAD hybrid microcircuit design [43].

mask layers can be misaligned on the wafer by a maximum of 2λ. The purpose of defining lambda is to make the design itself independent of both process and fabrication locations and to allow the design to be rescaled at a future date when the fabrication tolerances have shrunk [15].

3.2 PATTERN GENERATION TECHNIQUES

Utilization of the deposited films (thick/thin or both) in microelectronics is predicted on the ability to distribute the deposited material in a well-defined pattern. The trend toward more complex and higher-density hybrid microelectronic circuits has stimulated considerable development, with resulting improvements in techniques for generating film patterns. Most methods commonly used for generating film patterns fall into two categories: additive and subtractive [53]. Figure 3.9 is a flow diagram of several pattern generation processes. The pattern generation process is generally divided into two sub-

TABLE 3.1 CAD Systems Available for Hybrid Circuit Design

Components	Package	Platform	Operating System
Cadence design system	Allegro MCM	DEC, HP, IBM, SUN	Unix
Caduces	Auto hybrid	IBM PC	DOS
Cooper and Chyan Tech, Inc.	Spectra PS	HP, IBM, SUN	HP-U, IBM, SUN, OS
Daises (an Integraph company)	MCM engineering	Integraph SUN	Unix, Integraph, OS, SUN OS
Harris EDA	Funesse MCM	SUN	SUN OS
HEM Data Corp.	Snap master for Windows	IBM PC	Windows #. X
Mentor Graphics Corp.	MCM standard	HP, SUN	Unix
Recall-Redac Ltd.	MCM design suite	DEC, IBM, SUN	Unix
Zuken, Inc.		CR 3000	HP 300, 400, 700; SUN

Source: From [12]; reproduced with permission of The McGraw-Hill Companies.

groups: additive and subtractive processes. The first is performed by applying a physical barrier (mask) of appropriate shape and size in close proximity to the substrate during film deposition. The second encompasses techniques of forming the desired film pattern after deposition by selective removal of portions of the film. This can be accomplished by either a wet or a dry process. Perhaps the simplest pattern generation process is wet chemical etching. But wet

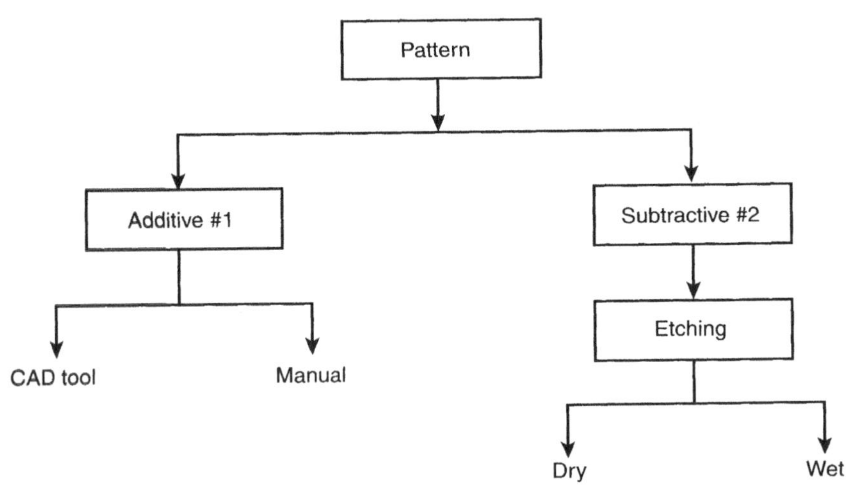

Figure 3.9 Flow diagram of various pattern generation processes.

etching usually suffers from one major drawback, *isotropy*, which means that during etching the etchant etches both horizontally and vertically. This leads to undercutting. On the other hand, dry etching systems (e.g., plasma-etching process) generally achieve *anisotropy* etching and usually minimize undercutting [54].

3.2.1 Additive Processes

In addition to deposition of thin films of several materials and their oxides, the utilization of films in hybrid circuits is predicted on the ability to distribute the material in well-defined pattern for resistors, conductors, and capacitors. As the dimensions of microcircuits are shrinking and the number of components per square inch is increasing, the trend toward more complex and improved techniques for generating a fine pattern is increasing. In an additive process a physical barrier of appropriate shape and size commonly referred to as a *mask* is interposed between a particle stream and a surface on which a particle can condense. Portions of the evaporate beam are thus intercepted and prevented from condensation on corresponding areas of the substrate. The circuit is formed on the substrate either with the help of a CAD/CAM direct writing system or manually by placing an appropriate metal mask in contact with the substrate inside a vacuum chamber.

Most commonly used masks are made of metal. Metal masks can be fabricated by two methods: machining techniques and etching techniques. In the first method, milling machines with suitable cutting tools can be used to make a desired pattern on the metal. This method is quite economical and can yield line-width tolerance ± 0.0002 in. (5 μm). Electron-beam machining has also been used for mask making but is not used much because of the irregular edge of the mask [55]. However, the second method, which uses photochemistry, can obtain higher resolution [56]. The resolution in this method depends primarily on the type of photochemical solution (more generally, called *photoresists*), optical systems, and the light source used to bring chemical reactions.

CAD/CAM System The micropen writing system [57] has been used successfully to produce thick-film conductors, resistors, and capacitors, which are comparable to conventional screen-printing processes. Fine line widths [about 0.5 mil (12.7 μm)] and spacing [about 0.5 mil (12.7 μm)] for a complex circuit pattern have been achieved on a large substrate area [6 by 6 in. (15 by 15 cm)]. Satisfactory results were obtained for lower pen writing speeds, up to 2 in. (5.1 cm) per second [58]. This CAD/CAM writing process has several advantages: (1) elimination of most of the normal prototype design to test turnaround time required in a screen-printing process; (2) the potential to write finer lines; (3) the ability to make thicker conductors, resistors, and capacitors (compared to thin-film circuits) and forming localized multilayer structures; (4) the potential to repair defective lines; and (5) the ability to fabricate microstrips, coaxial cables for microwave, and digital signal line impedance applications.

Manual System The manual system goes with the attachment of an appropriate metal mask onto a substrate inside a vacuum chamber. Metal, metal oxides, and metal alloy are evaporated either from a resistive heater or sputtered onto the substrate through the metal mask. Both the substrate and the mask should be optically flat and should not have an airgap in between. More separation between the substrate and the mask might result in a deformed or blurred replica of the mask over the substrate. Accurate replication of the mask pattern by the deposited film is contingent on two conditions. First, the mean free path of the evaporant particles must be compared to mask-to-substrate spacing to avoid random condensation caused by intermolecular collisions. Second, the sticking coefficient of the impinging vapor must be close to unity to prevent reevaporation and lateral spreading under the mask. When the mask is in intimate contact with the substrate, these prerequisites are usually satisfied [59]. Since the area of the metal is a measure of the conductor, resistor, or capacitor value of the device formed on the substrate through the metal mask, care must be taken to produce an exact replica of the mask. This is achievable only when the mask and substrate are in close proximity.

The intricacy and precision of the film pattern made by evaporation in the additive process also depend on the mechanical properties of the mask material and the accuracy of mask fabrication. The choice of mask material is governed by the desire to produce a mask that will maintain planarity and whose dimension will not change during deposition. Metals and metal alloy are suitable in the sense that they have a low thermal expansion coefficient and mechanical strength and amenability to a mechanical forming process. Patterns are made on a metal mask by several machining processes, including an automated milling machine [60] and arc erosion [61,62]. Masks of higher resolution are fabricated by selective etching of metals or metal sheets. Metals such as copper [63], stainless steel [64], molybdenum, and nickel are in very common use in metal masks.

3.2.2 Subtractive Processes

The subtractive process is commonly used to generate film patterns by selective removal of portions of the film. This is accomplished by the use of organic polymers called photoresist or lacquers that protect the film in the desired configuration while superfluous and unprotected regions are etched by either dry or wet chemical etching processes. The dry etching system is generally anisotropic and is used primarily for submicrometer pattern generation.

Perhaps the simplest pattern generation technique in routine use is wet chemical etching. The wet etching system is isotropic, which leads to undercutting of the resist and shrinkage of the etched lines. Undercutting is a particular problem in making near-micrometer lines. The substrate is covered completely with photoresist, and the processes involved can be categorized as follows: (1) spinning, spraying, or dipping of photoresist; (2) soft bake; (3) exposure under a photomask to a special light source; (4) developing; (5) hard baking; (6) etching in a chemical solution (wet etching) or by dry etching; (7) striping of the remaining hard-baked PR; and (8) cleaning.

Dry Etching System

Ion-Beam Etching Ion-beam anisotropic etching (IBAE) technology meets the challenges of the modern electronics industry, which demands increased product density, yields, and tighter tolerances [65]. However, it is not a cost-effective etching system and so does not dominate the electronic industry. The most cost-effective process, wet chemical etching, however, produces lifetime limiting defects due to contamination, undercutting of films, chemical reactions with other materials, and surface roughening and pitting [66].

Fiber optics will play an important role in the way data are transmitted to and from the office, and home across the world. The most important development in fiber-optic component development is the hybrid optical chip. Development of the IBAE system has helped to etch V-grooves in silicon, and control circuitry for diode lasers has provided a rapid method of alignment for single- and multimode fibers. IBAE technology has helped to develop micromachining technologies. A number of biomedical devices based on thin-film metals, polymers, glass, and silicon microstructures are manufactured by the IBAE method. These structures are common in pregnancy and drug detection kits, where specialized reagents are used to detect a specific drug in a blood or urine sample.

In the ion-beam etching system, an ion beam is used to sputter-etch material exposed by a mask (typically, a photoresist) to obtain a desired pattern. The process is performed in a vacuum chamber, where initially a vacuum of $>10^{-6}$ torr is attained, and a typical pressure of 10^{-4} torr is attained during the gas flow. Ions that strike the exposed material with sufficient energy will dislodge atoms or molecules. The number of atoms etched during the sputtering process is referred to as the *sputter yield*. The substrate temperature does not exceed 100°C during the entire etching process. Experimental observations show that a typical etch rate for gold is 0.005 mil (1200 Å) per minute and 0.0008 mil (200 Å) per minute for photoresist (at $V_{\text{beam}} = 500$ eV and current density = 1 mA/cm^2) [67]. In some applications, multiple iterations of applying the photoresist pattern and etching may be required to achieve the desired depth of etching. Ion-beam etching is used to fabricate highly sophisticated patterns [68] which require long operational lifetimes as well as precise performance specifications. These patterns are needed for commercial disk drive products, military and commercial communication components, microelectronic circuits, and sensor products for automotive, medical [69], and aerospace applications.

Plasma Etching Plasma is a fully or partially ionized gas composed of ions, electrons, and neutrons. A plasma is produced when an electric field of sufficient energy is applied to a gas, causing the gas to break down and become ionized. Free electrons that are released and gain kinetic energy from the applied electric field initiate the plasma. When the applied voltage is larger than the breakdown potential, sustained plasma is formed. In this discharge, free electrons gain energy, which they lose by collision with gas molecules. A very small amount of energy is transferred by elastic collisions because the inelastic collision rate is relatively high. This results in a wide range of energetic metastable species. Typical electron temperatures of 9000 to 10,000°C are achieved, although thermal temperatures are very low (50 to 100°C). The electron concentrations in the plasma for dry etching are relatively low, typically on the order of 10^9 to 10^{12} cm^{-3}. At a pressure of 1 torr, the concentrations of gas molecules are 10^4 to 10^7 times higher than the electron concentrations. Plasma-assisted dry etching is a low-temperature process (the average gas temperature is kept between 50 and 100°C) [70,71].

Typical etching rates are between 0.0012 and 0.004 mil (300 and 1000 Å), depending on the etchant gas, the materials to be etched, and the presence of the additive gases. The degree of anisotropy depends on the ratio of its etch rates [72]. Many types of plasma etching reactors are available. Nonmagnetically confined cold cathode dc (and low-frequency ac) glow discharge (also called gas discharge) is probably the simplest plasma generation technique used in plasma etching systems. The plasma thus generated is used as a source of high-energy ions for sputtering and ion plating (*atomic pinning*) [73]. The exciting frequency varies from 5 kHz to 56 Hz, and the operating pressure varies from 1 millitorr to 1 torr. Electrode arrangements can be internal, external, capacitive, or inductive, and total etch rates can vary from 0.0004 to 0.4 mil (10 to 10,0000 Å) per minute. Reactive ion etching, which is a reactive sputter etching system, is very much in use in the modern electronic industry. The system is different from plasma etching system in a few respects. It frequently applies an additional potential to the substrate electrode. The ground electrode is the chamber and usually has a larger area combined with lower operating pressures (5.0 to 100 millitorr). This leads to significantly higher plasma sheath potentials (20 to 5000 V) at the substrate surface [71,74]. The early reactive ion etcher, designed to hold the substrates on six vertical facets in the form of a hexagonal prism, is called a *hexagonal cathode etcher*. The instrument can etch many substrates simultaneously in a *batch process*. Later, for better line width control, higher selectivity, and better uniformity, a single-substrate etching process is introduced. Automated machines have been introduced to perform cassette-to-cassette operation. Etch rates for aluminum and photoresist etching are given in Table 3.2.

TABLE 3.2 Etch Rates

Material	Gas	Etch Rate [mils (Å)/min]
Aluminum	Chlorine content	0.002 (500)
Photoresist	Fluorine content	0.006 (1500)

Source: From [75].

3.2.3 Photolithography*

Llitho/lithos is a Greek word meaning *stone*. *Lithography*, on the other hand, is the art or process of printing from a stone or metal plate by a method based on the repulsion between grease and water. In a similar way, we can define *photolithography* (*photo* here representing photons or light quanta) as the art or process of printing from a mask made of glass or metal using a method based on the repulsion between the light quanta (photons) coming out of a special radiation source (especially in the ultraviolet region) and a light-sensitive polymer called *photoresist*. This process produces the fine lines and small structures needed for microminiaturization of hybrid microcircuit elements. Smaller microelectronic components are generally faster, lower in power, and cheaper. Photolithography, the most universal microcircuit fabrication step, plays a major role in gating progress toward further miniaturization. Optical lithography predominates in manufacturing because it is the lowest cost and simplest method [76]. The definition of the pattern in the photoresist (PR) layer is achieved by exposing the PR to a suitable form of patterned radiation, such as ultraviolet light, electrons, x-rays, and ions.

Photons generally interact with materials somewhat differently than electrons, which are again different from ions, as the momentum of incoming radiation increases from photon to electron to ion. The incoming radiation excites electrons in the PR material and these generally produce the chemical changes developed by a dry or wet chemical process in many forms of lithography [77]. Ions generally do not scatter transversely in nearly the same degree that lighter electrons do, and they lose energy over much shorter distances. Hence for a given incident energy, ions penetrate the material much less but deliver much more dose per incident particle. The exposure dose (energy) is an important parameter in photolithography. The incoming radiation energy should be capable of yielding an irradiation intensity of at least 64.5 mW/in^2 (10 mW/cm^2) at the substrate surface. The PR contains radiation-sensitive groups, which respond chemically to the incident radiation, forming a latent image of the circuit pattern, which can subsequently be developed by solvent or plasma treatment to produce a three-dimensional real image [78]. Figure 3.10 is a flow diagram of the formation of images with positive and negative photoresists.

Photoresist processing is a highly empirical art. Methods of resist application to substrate range from spinning to spraying, flowing, and rolling. However, whirler coating has become the standard technique in thin-film circuitry, as it yields the most uniform and reproducible thickness. A uniform surface thickness of PR is another important criterion for PR processing. PR thickness can be controlled by controlling the viscosity of the resist material. The PR-coated substrate is baked at low temperature (70 to 85°C). The purpose of this

*Some of the material in this section is excerpted with permission from L. F. Thomson, C. G. Wilson, and J. M. J. Frechet, *Materials for Microlithography*, ACS Symposium Series 266. Copyright 1984 American Chemical Society.

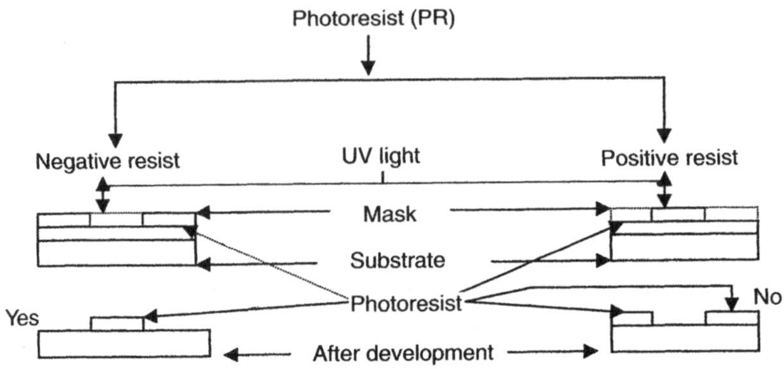

Figure 3.10 Flow diagram showing the formation of images with positive and negative photoresists.

operation is to drive out residual solvents, which if left in the film may weaken the adhesion of the PR to the substrate. Incompletely baked samples may also have variable exposure requirements, while excessive baking may cause fogging due to thermally activated cross-linking [79]. Soft-baked PR-coated substrate is exposed to appropriate radiation with a proper mask. Exposure will cause a change in the solubility of the PR. Following exposure, the PR-coated substrate is washed with a solvent called a *developer*, which preferentially removes the resist areas of higher solubility. This step is called *development*. Postbaking (100 to 120°C) of the developed PR is generally performed to evaporate residual solvents to enhance the chemical stability of the polymer coat and to promote its adhesion to underlying film. For pre- and postbaking and for process developments, manufacturers recommend certain cycles, but experience with a particular PR system under investigation is the best guide in this matter. At this stage the exposed substrate with open areas is ready to go for etching. The final step in the resist processes is called *stripping*. During this process the cross-linked polymer film is removed from the substrate using the solution recommended, followed by cleaning.

Two types of photoresist are available: positive and negative [80,81]. They differ in their responses to light and in their solubility behavior. Negative resist is less soluble in developer after being exposed to light and yields a negative mask pattern; the opposite behaviors are observed in positive resists. A mask for selective etching is designed accordingly. These light-sensitive materials change their chemical structure when exposed to light. The photosensitivity of a photoresist depends on the concentration of light-active centers present in the photoresist. When a light quantum is absorbed by such a center, it becomes inactive to light [82]. Negative resists adhere well on substrate but have lower resolution than that of positive resists. Lines 0.00001 in. (0.25 μm) wide separated by 0.00001 in. (0.25 μm) of space are possible with positive resist, whereas typical negative resist can produce lines about 2 μm wide with about 2 μm of spacing.

TABLE 3.3 Composition of Commercial Photoresists

Negative PR	Positive PR
(a) *Resin:* form of cyclized rubber (b) *Sensitizer:* bisaryl diazide (c) *Solvents:* toluene, xylene, acetone	(a) *Resin:* novolac (b) *Sensitizer:* diazonaphtha quinone (c) *Solvents:* ethylene glycol monoethyl ether, methyl cellulose
On exposure, the *aryl* group loses N_2 and forms free radicals. The result is *cross-linkage*, and *nitrene* is formed as an intermediate material.	On exposure, the N_2 group dissociates and forms *ketene*. The result is called *chain scission*.

Table 3.3 shows the composition of negative and positive photoresists. These are proprietary items and exact compositions are not known. The degree of exposure energy is adjusted by controlling the energy impinging on the resist (a product of the intensity of the source and the time of exposure). The exposure time of the photoresist is a critical step in the resist processing procedure. During processing, substrates are processed individually. It is important that the film be exposed quickly with the correct light source for the correct amount of time. Overexposure will result in undercutting; underexposure will result in overcutting and scalloped edges, which give poor definition. Figure 3.11 is a typical flow diagram for photoresist processing.

Photoresist Chemistry Photoresist chemistry [78,83] has advanced so much that what seemed to be impossible a year ago is now being accomplished by scientists and engineers. Advanced resists with high-NA (numerical aperture) scanners and resolution enhancement and 200-nm features using i-line (365-

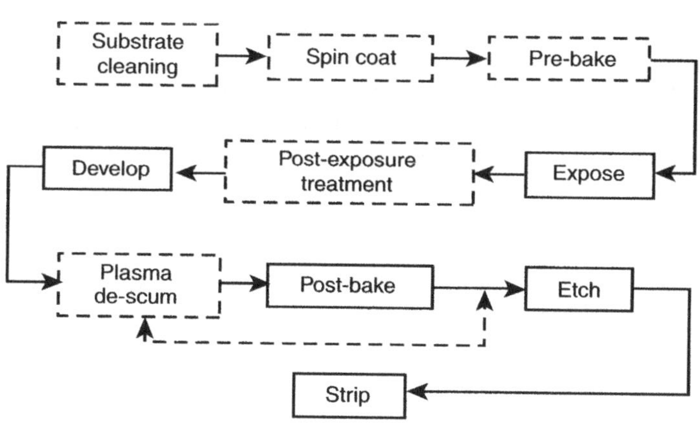

Figure 3.11 Photoresist processing steps. (Reprinted from *Materials for Microlithography*, ACS Symposium Series 266, with permission from the American Chemical Society.)

TABLE 3.4 Manufacturers of Resist Materials

Supplier	Positive Resist	Negative Resist
Eastman Kodak	Micropositive 800 series	Microresist 700 series
JT Baker	Baker 1-PR	
American Hoechst	AZ-1300, 4000 series	
Shipley	AZ-1400, 2400 micropositive	
Phillip A. Hunt	Waycoat HPR-2000 series	Waycoat HNR and HR series
Macdermind	Ultramac PR 73, 74	
Allied Chemical	Acculith	
E.M. Chemicals	Selectilux P-series	Selectilux N

nm) and sub-100-nm contact holes using 248-nm lithography are no longer a dream. At 193 nm, chemically amplified (CA) resist technology is poised to reveal 80-nm features, and attaining even finer dimensions at 157 nm are possible [80]. To achieve subwavelength lithography results, resist technologies such as phase-shifting masks, scanner optics, and inspection techniques are pushed to its absolute limit. Most important, the fundamental structure of resist must change from i-line's Novolac-based resists to CA hydroxystyrene-based resist at 248 nm, to cyclic olifin or acrylate-based resist at 193 nm, to an as-yet-undetermined polymer family at 157 nm. Although such steps are technically feasible, the challenge for many companies will be getting up to speed very rapidly in two or three distinct materials. Advanced resists require much engineering ingenuity and extensive optimization of polymer chemistry, photoacid generators, and various additives. Table 3.4 shows some suppliers and manufacturers of photoresist materials.

Imaging Methods There are three major imaging methods of transferring a pattern on a mask to a photoresist-coated substrate [84]: contact printing, proximity printing, and projection printing. The optical method should be capable of transforming with high accuracy an image of varying topography from a mask plate or reticle to a resist-coated substrate. The optics used for this process must not only be of high optical quality but must be able to fix the substrate accurately at the focal plane of the image. The optical system must be able to align the image properly. Figure 3.12 shows the schematic of three optical lithographic techniques: contact, proximity, and projection [84].

Contact Printing Contact printing is one of the oldest photographic methods. The mask containing the circuit pattern is first positioned (aligned) correctly and clamped to the resist-coated substrate. The drawback of this process is that the mask is in intimate contact with the resist surface during printing. As a result, the mask has to be abandoned after a few uses.

Proximity Printing To minimize the number of defects on a mask due to contact printing, a gap of about 4 to 1.0 mil (ca. 100 to 25 μm) is left

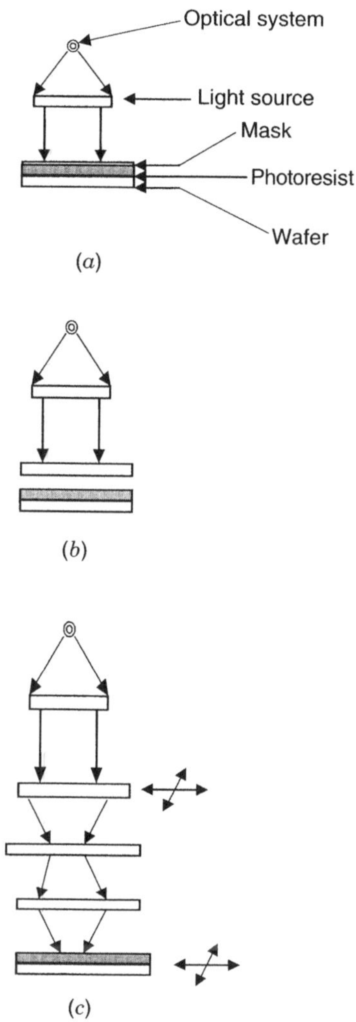

Figure 3.12 Schematic of three optical lithographic techniques: (*a*) contact printing; (*b*) proximity printing; (*c*) projection printing.

between the mask and the resist-covered substrate surface. Due to this gap, resolution is reduced. This method needs optically flat mask and substrate surfaces.

Projection Printing In projection printing, the mask and substrate are well separated and the image of the mask is projected on the PR-coated substrate with the help of lens and mirrors. Several types of projection printing techniques are available, and they can be classified into three major groups: 1 : 1 projection scanners, reduction step-and-repeat projection aligners, and non-reduction step-and-repeat projection aligners.

Figure 3.13 Production-oriented electron-beam lithographic system developed by IBM. (Courtesy of IBM, Yorktown Heights, New York.)

Advanced Lithography

Deep Ultraviolet Conventional photolithography employing wavelengths greater than 360 nm [deep ultraviolet (DUV)] is considered to be deep UV lithography. This lithography is quite mature and is being used in industry. Resists based on polymethyl methacrylate (PMMA) belong to this category [85–88].

Electron Beam Computer-controlled electron beams have been used for many years to generate high-resolution patterns in polymeric resist. Electron-beam lithography has been the advanced technology pursued most actively during the past two decades. It is capable of exhibiting submicrometer resolution, and its sensitivity to electron irradiation at 10 to 30 kV is on the order of 6.45×10^{-6} C/in^2 (10^{-6} C/cm^2) [89–91]. Figure 3.13 shows a production-oriented electron-beam lithographic system developed by IBM.

X-ray Resists X-ray lithography is an extension of near-contact photolithography for replication of 0.04 mil (1 μm) and small geometry. Typical wavelengths that have been used for resist exposure include 0.16 mil (4.2 Å) ($Pd_{L\alpha}$), 0.2 mil (5.4 Å) ($Mo_{L\alpha}$), and 0.32 mil (8.1 Å) ($Al_{k\alpha}$) [92,93].

Ion-Beam Lithography Ion-beam lithography is a technology that is still in its development stage. It has many potential advantages, notably the absence of proximity effects, which have spurred a great deal of interest [94,95].

Photoresist Material Parameters According to the requirements of a particular circuit, one has to select a particular type of resist over the others. The material parameters [96,97], which are the decisive factors for this, are described next.

Resolution The resolution of a photoresist is defined in terms of a modulation transfer function (MTF) which specifies the consistent ability to print minimum-size images under conditions of reasonable manufacturing variation. While evaluating the resolution of the resist, the feature size is the most difficult to resolve and has always been the ultimate benchmark for a particular resist material [98]. Resolution is also a factor for thick PR film. The positive resist shows higher resolution with the same thickness as that of a negative resist. Thick resist coating is better for step coverage, produces fewer defects, and is favorable for dry etching.

Contrast A measure of contrast is given by the value of the quantity γ. For negative resist, contrast is related to the rate of cross-linked network formation at a constant irradiation dose. For positive resist this is related to the rate of chain scission and change of solubility at a constant irradiation dose [99]:

$$\text{critical resist MTF} = \frac{I_{100} - I_0}{I_{100} + I_0} = \frac{10^{1/\gamma} - 1}{10^{1/\gamma} + 1}$$

where I_{100} is the minimum exposure energy for 100% film loss, I_0 the maximum energy for zero exposure, and γ the contrast (Figure 3.14). The MTF of an exposure system is known for various feature sizes, and knowledge of the γ value of the resist allows a prediction of the smallest features that are printable effectively with that system [99].

Swelling The feature size of the pattern created in the PR will alter if the resist volume increases during the development step (swelling) as a result of penetration of the resist material by developer solution. Such swelling of the PR is usually noted with negative resist. Positive resist is recommended when the feature size is less than 0.12 mil (3 μm).

Proximity Printing Resist resolution is seen to be lowered by proximity printing (when the substrate and mask do not touch each other). However, the

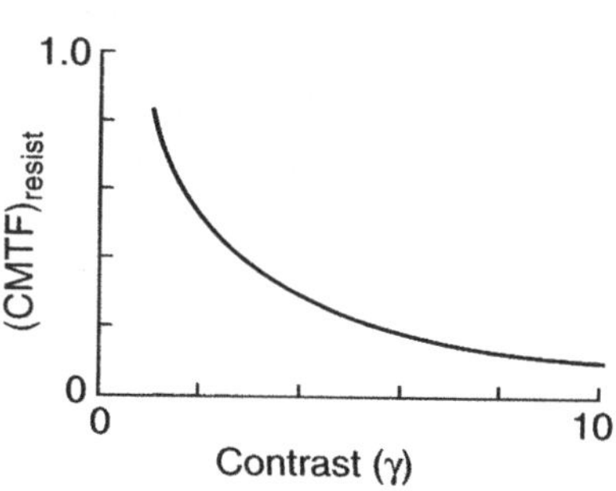

Figure 3.14 Relationship between critical resist MTF ($CMTF_{resist}$) and contrast (γ) [84].

defects are minimized in proximity printing. For proximity printing,

$$\text{resolution } L_{\min} = 1.5[(S + D)\lambda]^{0.5}$$

where S is the spacing between the mask and the resist, D the thickness of the resist, and λ the wavelength of light used for irradiation.

Sensitivity The sensitivity of a PR is defined as the input energy required to cause a specified design of chemical response, which results in the resist pattern desired:

$$\text{sensitivity} = \varphi = \frac{\text{number of photo-induced events}}{\text{number of photons absorbed}}$$

Recent Developments in Lithography The 1999 ITRS road map needs are accelerating, while optical lithography continues to extend capability. Further progress in accelerating the road map is limited by the industry's mask-making capability. Mask writing and inspection equipment must be developed and will be able to progress at the same speed as that for lithographic materials and equipment. In addition, both technical and business solutions must be found for repair technology. The mask industry needs major focus, paradigm shifts, and worldwide collaboration to overcome challenges together with rapid escalation of mask costs [100].

The first lithography tools employed masks held in contact with the workpiece and there was the possibility of mask feature damage. To remedy this, proximity printers were used. The mask was separated from the PR-coated substrate by about 1.0 mil (25 μm). The next major development in this field occurred in the mid-1970s when the Perkin-Elmer Corporation introduced the full substrate projection system. The high-quality image of the mask is projected like a slide photo onto PR-coated substrate. Resolution was better than that obtained with a proximity printer. For transmission lenses, chromatic aberration becomes a problem, as does astigmatism. To remedy these problems, the exposure field size was reduced so that it fills only a potion of the PR-coated substrate. The system using reduced field size is called a *wafer stepper*. Lithography steppers in the middle to late 1980s used a 426-nm g-line and achieved a feature size of 0.04 to 0.03 mil (1.0 to 0.7 μm). The move to the shorter 365-nm i-line wavelength in the early 1990s enabled feature sizes of 0.02 to 0.013 mil (0.5 to 0.35 μm). For 426- and 365-nm optical lithography, borosilica glasses are used with fused silica for photomasks. The borosilicate glasses are transmissive to the 350- to 450-nm bandwidth and have well-demonstrated properties for this application [101].

Technological advances required of even shorter wavelengths [248 nm (KrF laser) and 193 nm (ArF laser) for line widths below 0.25 μm] bring myriad challenges for engineers and material scientists. As a start, mixed oxide composition glasses such as borosilicates are too opaque for use in this wavelength range. Projection optics and photomask materials for 248 and 193 nm must be constructed from fused silica, which is highly transparent in this region. The transparency of optical materials for deep UV (DUV) is of obvious importance, but understanding interactions of materials with light is essential. Fused silica is a classic material that has been used extensively in high-energy gamma irradiation, x-ray window applications, and in DUV with excimer laser exposure [101]. The sidewall-chrome alternating aperture (SCAA) mask, a recent development, will be a challenge for the industrial lithographic process in the future [60]. In alternating aperture phase-shifting mask (alt-PSM) processes, phase trenches are etched through previously opened apertures in an opaque chrome layer. In the SCCA mask system, on the other hand, the phase level is etched first and then transparent apertures are formed in a deposited conformal chrome layer. SCCA masks are also more stable mechanically than alt-PSM structures, where quartz walls are recessed more than 90 nm underneath chrome edges to minimize trench wall effects. SCCA masks can now be fabricated with phase-shifting critical dimensions as small as 280 nm, which have projected with 4× lithography as 70-nm lines [102].

Critical levels for the first generation of 0.005-mil (0.13-μm) devices will use 248-nm lithography, but second-generation devices will probably require three to four critical levels that may only be patterned using 193-nm lithography. With the transition from 248-nm to 193-nm lithography comes a shrinking process window and increasing need to optimize the overall lithography cells to produce the best image, also known as *process image integration* [103]. According to an observer [104], several factors are still keeping the 193-nm system from being a lithography tool for the masses. Figure 3.15 is a scanning electron micrograph of a 130-nm transistor patterned with 193-nm lithography.

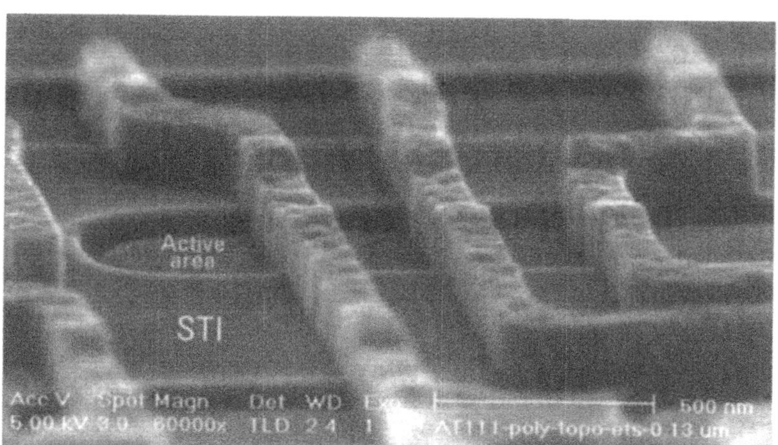

Figure 3.15 Scanning electron micrograph of 130-nm transistors patterned with 193-nm lithography [103]. [Copyright 2001. This photo is the property of IMEC, Leuven, Belgium (www.imec.be). It may not be used without express written permission by IMEC. All rights reserved.]

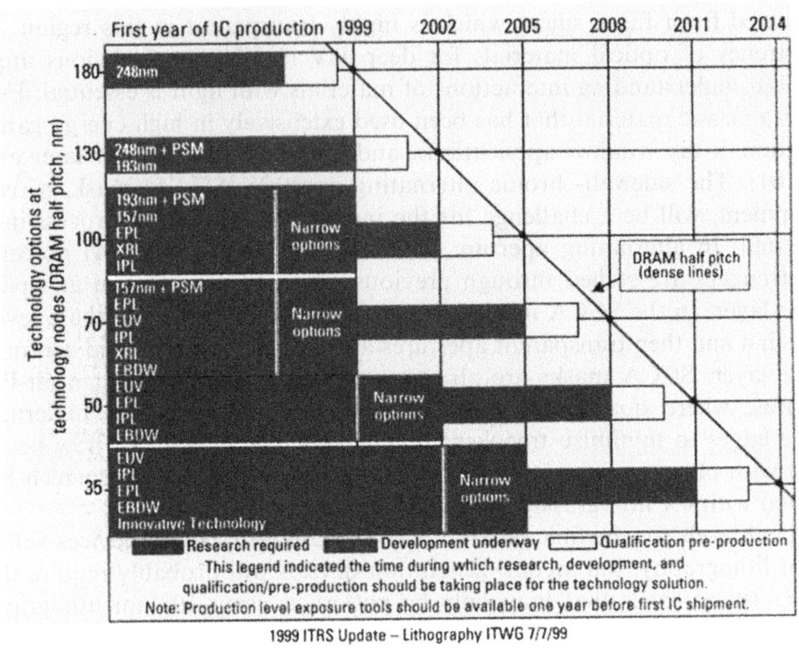

Figure 3.16 Many alternatives exist for nodes 100 nm and below, including 193 nm, 157 nm, electron projection, proximity x-ray, ion projection, extreme ultraviolet, and electron-beam direct write [100]. (© 2000 by International SEMATECH.)

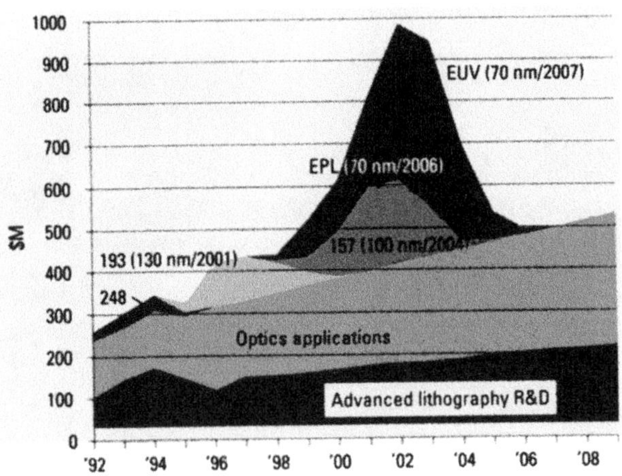

Figure 3.17 Total R&D spending for lithography may exceed $1 billion in 2002 [100]. (© 2000 by International SEMATECH.)

Lithography is the semiconductor industry's key enabler. It increases the speed and density of microchips directly while making possible about one-half of chip productivity improvements in each year by steady decreases in minimum feature size. The potential solutions road map (Figure 3.16) shows that the options at the 130-nm node have been narrowed to 248 nm + PSM and 193 nm. For 100 nm and below, there are many alternatives. They include 193 nm, 157 nm, electron projection, proximity x-ray, ion projection, extreme ultraviolet, and electron-beam direct write.

The lithography industry has a vested interest in extending optical lithography as far as it can go because its reliability has been outstanding [105]. The enormous challenge of device size [106] and the rapid growth and demand to push the lithography below 50 nm could cause total R&D spending to approach $1 billion in 2002 alone (Figure 3.17).

REFERENCES

1. D. L. Andrews, M. A. Thornton, M. Conrad, et al., Computer-aided engineering and design, in *Advanced Electronic Packaging with Emphasis on MCM*, W. D. Brown, ed., IEEE Press, Piscataway, NJ, 1999.
2. J. J. Licari and L. R. Enlow, Hybrid design processes, computer-aided design, in *Hybrid Microelectronics Handbook*, Noyes Publications, Park Ridge, NJ, 1999, Chap. 5.
3. I. Sutherland and C. Mead, Microelectronics and computer science, *Sci. Am.*, Sept. 1977.
4. R. Grabowski, *CAD Manager's Handbook*, Delmar, New York, 2001; R. Tummala and E. Rymaszewski, *Microelectronic Packaging Handbook*, Van Nostrand Reinhold, New York, 1989, p. 114.
5. C. E. Jowett, *The Engineering of Microelectronic Thin and Thick Film*, Macmillan, London, 1976.
6. M. S. Haskard and K. Pitt, Computer-aided hybrid layout (CAHL), in *Thick-Film Technology and Applications*, IMAPS, Washington, DC, 1997.
7. W. D. Brown, ed., Computer-aided engineering and design, in *Advanced Electronic Packaging*, IEEE Press, Piscataway, NJ, 1999, Chap. 8; K. Lee, M. Shur, T. A. Fjeldly, and T. Y. Herdal, *Semiconductor Device Modeling*, Prentice Hall, Upper Saddle River, NJ, 1993.
8. R. R. Johnson, The significance of wafer-scale integration in computer design, *Proc. IEEE Int. Conf. Comput. Des.*, Rochester, NY, Oct. 1984, pp. 101–105.
9. F. W. Kear, *Hybrid Assemblies and MCMs*, Marcel Dekker, New York, 1993.
10. B. A. Latin, VLSI design methodology and problems of the 80's for microprocessor design, *Proc. Caltech Conf. VLSI*, Jan. 1979, pp. 247–252.
11. D. L. Peltzer, Wafer-scale integration: the limits of VLSI? *VLSI Design*, Vol. IV, No. 5, Sept. 1987, pp. 43–47.
12. R. L. Geiger, P. E. Allen, and N. R. Strader, *VLSI Design Techniques for Analog and Digital Circuits*, McGraw-Hill, New York, 1990.

13. R. D. Jones, *Hybrid Circuit Design and Manufacture*, Marcel Dekker, New York, 1982, Chaps. 4 and 6.
14. D. Lee and O. A. Palusins, Modification of Spice for simulation of coupled packaging interconnections, *Trans. IEEE Comp. Hybrid Manuf. Technol.*, Vol. 15, No. 14, 1978, p. 491.
15. E. D. Fabrics, *Introduction to VLSI Design*, McGraw-Hill, New York, 1990, p. 2.
16. J. K. Ousterhout, G. T. Hamachi, R. N. Mayo, W. S. Scott, and G. S. T. Taylor, Magic: a VLSI layout system, *Proc. 21st Des. Automat. Conf.*, June 1984, pp. 152–159.
17. EDIF, *Electronic Design Interchange Format, Version 200*, ANSI/EIA-548-1988, Electronic Industry Association, Berkeley, CA, 1988.
18. D. S. Harrison, P. Moore, R. L. Spickelmier, and A. R. Newton, Data management and graphics: editing in the Berkley design environment, *Proc. Int. Conf. CAD*, 1986, pp. 24–27.
19. J. D. William, "Sticks": graphical compiler for high-level LSI design, *AFIPS Conf. Proc.*, Vol. 47, June 1978, pp. 289–295.
20. R. Zinszer, H. De Man, and K. Croes, Technology-independent symbolic layout tools, *Proc. Int. Conf. Comput.-Aid. Des.*, Sept. 1983, pp. 12–13.
21. R. L. Geiger, P. E. Allen, and N. R. Strader, Device modeling and circuit simulation, in *VLSI Design*, McGraw-Hill, New York, 1990, Chaps. 3 and 9.
22. A. Vladimirescu and S. Liu, *The Simulation of MOS ICs Using Spice 2*, UCB/ERL M80/7, University of California–Berkeley, Engineering Research Laboratory, Feb. 1980.
23. D. A. Weins and S. C. Tasker, A parameterized hybrid CAD system, *Proc. ISHM*, Minneapolis, MN, 1987, p. 491.
24. J. Sergent, H. Chiles, and P. Molkenthin, CAD systems pervade every area of hybrids production, *Electron. Packag. Prod.*, Mar. 1985.
25. S. Fu and C. Yang, A complete aided layout system for multilayer hybrid circuits, *Proc. Int. Symp. Microelectron.*, 1985; G. W. Pan, J. A. Prentice, S. K. Zahn, A. J. Staniszewski, W. L. Walter, and B. K. Gilber, The simulation of high-speed high-density digital interconnects in single-chip packages and MCM, *Trans. IEEE Comp. Hybrid Manuf. Technol.*, Vol. 15, No. 4, 1992, p. 465.
26. R. Amantea, An introduction to modeling and simulation, *RCA Rev.*, Vol. 46, Sept. 1985, pp. 281–288.
27. P. Six, L. Claesen, J. Rabasy, and H. De Man, An intelligent module generator environment, *Proc. 23rd Des. Automat. Conf.*, June 1986, pp. 730–735.
28. J. R. Sims, R. G. Creter, and A. H. Leville, An approach to CAD of multilayer hybrids, *Solid State Technol.*, Vol. 6, 1979, p. 65.
29. J. R. Sims and P. G. Creter, Application of custom microelectronics technology to government communication systems, *Proc. Int. Symp. Microelectron.*, Atlanta, GA, 1986, p. 809.
30. H. B. Bakoglu, Transmission lines, in *Circuits, Interconnections, and Packages for VLSI*, Addison-Wesley, Reading, MA, 1990, Chap. 6.
31. P. Penfield and J. Rubinstein, Signal delay in RC tree networks, *Proc. IEEE 18th Des. Automat. Conf.*, 1981, pp. 613–617.

32. R. Ansari and B. Liu, Multilayer signal processing, in *Handbook of Signal Processing*, S. K. Mitra and J. F. Kaiser, eds., Wiley, New York, 1993.

33. A. Feller, H. R. Kaupp, and J. J. Digiacamo, *Cross-Talk Reflections in High-Speed Digital System*, Spartan Books, Washington, DC, 1965, pp. 511–525.

34. F. J. Court and T. A. Lane, Electrical CAD analysis for multilayer package design, *Int. Symp. Hybrid. Microelectron.*, 1986.

35. R. Sainati, Characteristic impedance and coupling programs, memo, Internal Honeywell, Golden Valley, MN, Oct. 25, 1985.

36. A. Vladimirescu, K. Ahang, A. R. Newton, D. O. Pederson, and A. Sangiovanni, *Spice Version 2G User's Guide*, Department of Electrical Eng. & Computer Science, University of California–Berkeley, 1981.

37. J. Keown, *Pspice Circuit Analysis*, 2nd ed., Macmillan, New York, 1994.

38. M. C. Herniter, *Schematic with Pspice*, Macmillan, New York, 1994.

39. J. J. Licari and L. R. Enlow, *Hybrid Microcircuit Technology Handbook*, Noyes Publications, Park Ridge, NJ, 1999, p. 374.

40. D. J. Mannion, Integration of computer-aided engineering to the design, manufacturing, and testing of hybrid microelectronics, *Proc. Int. Symp. Microelectron.*, 1988, p. 426.

41. M. Yin, Layout verification of VLSI designs, *VLSI Des.*, July 1985, pp. 30–38.

42. J. R. Sims, P. G. Creter, and A. Leveille, An approach to computer-aided design of multilayer hybrids, *Solid State Technol.*, Oct. 1979, pp. 95–100.

43. V. M. Shukla, Computer applications to hybrid microelectronic circuit design, *Proc. Int. Symp. Microelectron.*, 1972, p. 2B-5-1.

44. P. W. Sandborn and H. Merreno, *Conceptual Design of Multichip Modules and Systems*, Kluwer Academic Publishers, NY, 1994. F. W. Kear, *Design and Development in Hybrid Assemblies and MCMs*, Marcel Dekker, New York, 1993.

45. D. D. Zimmerman and J. Dinto, Electronic system design considerations, in *Hybrid Microelectronics Design Guide*, ISHM/1402/IPC-H-855, Oct. 1982, Chap. 3.

46. B. S. Matisoff, *Handbook of Electronics Package Design and Engineering*, Van Nostrand Reinhold, New York, 1982.

47. G. D. Giacomo, *Reliability of Electronic Package and Semiconductor Devices*, McGraw-Hill, New York, 1997.

48. J. J. Licari and L. R. Enlow, Layout guidelines common in thick and thin film, in *Hybrid Microelectronics Circuit Technology Handbook*, Noyes Publications, Park Ridge, NJ, 1999, p. 408.

49. H. C. Frankel, *Thin-Film Hybrid Microcircuit Design and Layout*, Integrated Electronics Division, Electronic Component Laboratory, Technical Report ECOM-2890, U.S. army Electronics Command, Fort Monmouth, NJ, Sept. 1967, DA task 1, TO24401A-348-00-53.

50. H. J. MacLeod, The variation with frequency of the power losses in dielectrics, *Phys. Rev.*, Vol. 20, Jan. 1963, pp. 53–73; D. S. Cambell and J. A. Hayes, *Capacitive and Resistive Electronic Components*, Vol. 8, Gordon & Breach, Amsterdam, 1994.

51. J. E. Sergent, Discrete passive components for hybrid circuits, in *Hybrid Microelectronics Handbook*, 2nd ed., J. E. Sergent and C. A. Harper, eds., McGraw-Hill, New York, 1995, Chap. 8.
52. L. W. Wilson, Transformer and inductive devices, in *Passive Electronic Components Handbook*, C. A. Harper, ed., McGraw-Hill, New York, 1997, Chap. 3.
53. R. Glang and L. V. Gregor, Generation of patterns in thin films, in *Handbook of Thin-Film Technology*, 2nd ed., L. I. Maissel and R. Glang, eds., McGraw-Hill, New York, 1983, Chap. 7.
54. W. S. Ruska, Plasma etching, in *Microelectronic Processing*, McGraw-Hill, New York, 1987, p. 213.
55. M. Kaminsky, *Atomic and Ionic Impact Phenomena on Metal Surfaces*, Academic Press, San Diego, CA, 1965.
56. G. W. W. Stevens, *Microphotography*, Chapman & Hall, London, 1957.
57. C. E. Drumheller, Micropen, Inc., Pittsford NY, Dynamic pen control CAD/CAM thick-film writing system, *Proc. Int. Symp. Microelectron.*, 1982.
58. D. Nayak, L. Hwang, and A. Reisman, Evaluation of a CAD/CAM system for direct writing of thick-film conductors, resistors, and capacitors, *Proc. Int. Symp. Microelectron.*, 1986, p. 775.
59. L. V. Gregor, Thin-film processes for microelectronic application, *Proc. IEEE*, Oct. 1971, pp. 1390–1403; R. Glang and L. V. Gregor, Generation of pattern in thin films, in *Handbook of Thin Film Technology*, 2nd ed., L. I. Maissel and R. Glang, eds., McGraw-Hill, New York, 1983, Chap. 7, p. 4.
60. R. B. Delano, Jr., U.S. Govt. Contract, Nobstr 77508, Final Summary Report, 1961, p. 73.
61. F. Jaques and J. Schmidt, *Mod. Plast.*, Vol. 38, No. 4, 1960, p. 109.
62. A. E. Lessor, R. E. Thun, and L. I. Maissel, *IEEE Spectrum*, Vol. 1, No. 4, 1964, p. 72.
63. H. Fuchs and K. Heine, *Microminiaturization*, Vol. 5, 1966, p. 335.
64. F. A. Nun and D. S. Cambell, *J. Sci. Instrum.*, Vol. 40, 1963, p. 337.
65. R. Bakish, ed., *Electron and Ion Beam Science and Technology*, Electrochemical Society, Pennington, NJ, 1974.
66. W. M. Stauss and T. E. Lizotte, Ion beam etching solutions for microcircuitry and packaging, *Vac. Technol. Coat.*, Mar./Apr., 2000, p. 43; S. P. Murarka and M. C. Peckerar, Wet etching, in *Electronic Materials Science and Technology*, Academic Press, San Diego, CA, 1989, p. 500.
67. W. M. Stauss and T. E. Lizotte, Ion beam etching solutions for microcircuitry and packaging, *Vacuum Technol. & Coating*, p. 43, Mar./April 2000.
68. M. Madou, *Fundamentals of Microfabrication*, CRC Press, Boca Raton, FL, 1997.
69. T. Lizotte, *Micromolding for Medical Applications*, Prototyping Technology International, 1997, pp. 17–21; M. Caulton, Film technology in microwave integrated circuits, *Proc. IEEE*, Vol. 59, No. 10, Oct. 1971, p. 1481.
70. S. K. Ghandhi, *VLSI Fabrication Principles*, Wiley, New York, 1983, p. 499.
71. S. M. Sze, *Semiconductor Devices: Physics and Technology*, Wiley, New York, 1985.

72. R. H. Burton, R. A. Gottscho, and G. Smolinsky, Dry etching of Gr. III–IV compound semiconductor, in *Materials Processing: Theory and Practices*, F. F. Y. Wang, ed., North-Holland, Amsterdam, 1983.
73. D. M. Mattox, Plasma technology, *Vac. Technol. Coat.*, Vol. 1, No. 8, Oct. 2000, p. 14.
74. B. Chapman, *Glow Discharge Processes, Sputtering, and Plasma Etching*, Wiley, New York, 1980, p. 370.
75. D. Bollinger, S. Ida, and S. Matsumoto, *Solid State Technol.*, Vol. 27, 1984, p. 167.
76. A. N. Broers, Practical and fundamental aspects of lithography, in *Materials for Microlithography*, L. F. Thomson, C. G. Wilson, and J. M. Frechet, eds., ACS Symp. Ser. 266, American Chemical Society, Washington, DC, 1984, p. 11.
77. T. E. Everhart, Fundamental limits of lithography, in *Materials for Microlithography*, L. F. Thomson, C. G. Wilson, and J. M. Frechet, eds., ACS Symp. Ser. 266, American Chemical Society, Washington, DC, 1984, p. 5.
78. M. J. Bowden, A perspective on resist materials for fine lithography, in *Materials for Microlithography*, L. F. Thomson, C. G. Wilson, and J. M. J. Frechet, eds., ACS Symp. Ser. 266, American Chemical Society, Washington, DC, 1984, p. 49.
79. T. K. Gupta, Effect of temperature on the photosensitivity of a photoresist (AZ 1350J), *Eur. Polym. J.*, Vol. 17, 1981, pp. 1127–1131.
80. L. J. Peters, Resists join the sub-λ revolution, *Semicond. Int.*, Feb. 2000, p. 4.
81. J. Bowden, *CRC. Crit. Rev. Solid State Sci.*, Vol. 8, 1979, p. 223.
82. A. Charlesby, *Atomic Radiation Polymers*, Pergamon Press, Oxford, 1960; B. F. Griffin and P. R. West, Photopolymers Principles, Process and Materials, SPE Tech. Conf. Ellenvile, NY, Nov. 8–10, 1982, p. 185.
83. J. E. Guilet, S. K. Lili, and H. C. Ng., Photochemistry of ketone polymers in solid phase, in *Materials for Microlithography*, L. F. Thomson, C. G. Wilson, and J. M. J. Frechet, eds., ACS Symp. Ser. 266, American Chemical Society, Washington, DC, 1984, p. 165.
84. S. K. Wolf and R. N. Tauber, Lithography II, in *Silicon Processing for the VLSI Era*, Vol. I, Lattice Press, Sunset Beach, CA, 1986, p. 471.
85. B. J. Lin, *J. Vac. Sci. Technol.*, Vol. 12, No. 6, 1975, p. 1317.
86. Y. Mimura, T. Oncubo, T. Takeuchi, and K. Sekikawa, *Jpn. J. Appl. Phys.*, Vol. 17, No. 3, 1978, p. 541.
87. H. Hiraoka, Functionally substituted novolok resin, in *Materials for Microlithography*, L. F. Thomson, C. G. Wilson, and J. M. J. Frechet, eds., ACS Symp. Ser. 266, American Chemical Society, Washington, DC, 1984, p. 339.
88. T. Iwayanagi, T. Kohashi, and S. Nonogaki, *J. Electrochem. Soc.*, Vol. 127, 1980, p. 2759.
89. T. Tada, *J. Electrochem. Soc.*, Vol. 128, No. 8, 1981, p. 1791.
90. A. Novembre, L. Masakowski, J. Frackoviak, and M. Bowden, Photopolymers: principles, process, and materials, *Proc. Reg. Tech. Conf.*, Mid-Hudson section, SPIE, Ellenville, NY, 1982.
91. M. Dole, *The Radiation Chemistry of Macromolecules*, Vol. II, Academic Press, San Diego, CA, 1973.
92. G. N. Taylor, *Solid State Electron.*, Vol. 23, No. 5, 1980, p. 73.

93. E. Spiller and R. Feder, X-ray lithography, in *X-ray Optics*, J. H. Queizer, ed., Springer-Verlag, Berlin, 1977.
94. W. L. Brown, T. Venkatssan, and A. Wagner, *Solid State Technol.*, Vol. 24, No. 6, 1979, p. 60.
95. K. Gamo, in *Microcircuit Engineering '80*, R. P. Kramer, ed., Delft University Press, Delft, The Netherlands, 1980.
96. C. G. Wilson, Organic resist materials, theory and chemistry, in *Introduction to Microlithography*, L. F. Thomson, C. G. Wilson, and M. J. Bowden, eds., ACS Symp. Ser. 219, American Chemical Society, Washington, DC, 1983.
97. M. C. King, Principles of optical lithography, in *VLSI Electronics, Microstructure Science*, N. G. Einspruch, ed., Academic Press, San Diego, CA, 1981.
98. L. F. Thomson, C. G. Wilson, and M. J. Bowden, eds., *Introduction to Microlithography*, Advances in Chemistry Series, American Chemical Society, Washington, DC, 1983, p. 172.
99. S. Wolf and R. N. Tauber, *Silicon Processing for the VLSI Era*, Vol. II, Lattice Press, Sunset Beach, CA, 1986, p. 413.
100. J. Canning, Challenges of next-generation lithography, *Semicond. Int.*, Sept. 2000, p. 107.
101. C. Smith and B. G. Ackerman, Advanced lithography needs rigorously tested fused glass, *Solid State Technol.*, Jan. 2001, p. 111.
102. M. D. Levenson, Mask design with lower cost, *Solid State Technol.*, Jan. 2001, p. 104.
103. L. Peters, Ramping the 0.13-μm generation, *Semicond. Int.*, Jan. 2001, p. 58.
104. A. Hand, Higher repetition rate makes ArF laser more production-ready, *Semicond. Int.*, Jan. 2001, p. 40.
105. A. E. Braun, 100 nm: the undiscovered country, *Semicond. Int.*, Feb. 2000, p. 79.
106. F. J. Belcourt and T. A. Lane, Electrical analysis for multilayer package design, *Proc. Int. Symp. Microelectron.*, Atlanta, GA, 1986, p. 802.

RECOMMENDED READING

Brown, W. D., ed., Computer-aided engineering and design, in *Advanced Electronic Packaging*, IEEE Press, Piscataway, NJ, 1999, Chap. 8.

Chua, L. O., and P. M. Lin, *Computer-Aided Analysis of Electronic Circuits*, Prentice Hall, Upper Saddle River, NJ, 1975.

Fabrics, E. D., *Introduction to VLSI Design*, McGraw-Hill, New York, 1990, p. 2.

Grabowski, R., *CAD Manager's Guidebook*, Delmar, Albany, NY, 2001.

Haskard, M. R., and K. Pitt, *Thick-Film Technology and Applications*, IMAPS, Washington, DC, 1997.

Johnson, R. W., ed., *ISHM Modular Series in Hybrid Microelectronics*, ISHM, Reston, VA, 1991.

Jones, R. D., *Hybrid Circuit Design and Manufacture*, Marcel Dekker, New York, 1981.

Licari, J. J., and L. R. Enlow, *Hybrid Microcircuit Technology Handbook*, Noyes Publications, Park Ridge, NJ, 1998.

Maissel, M. I., and R. Glang, eds., *Handbook of Thin-Film Technology*, 3rd ed., McGraw-Hill, New York, 1983.

Pecht, M., ed., *Integrated Circuit, Hybrid, and Multichip Module Package Design Guidelines*, Wiley, New York, 1993.

Sandborn, P. A., and H. Moreno, *Circuit Design of Multichip Modules and Systems*, Kluwer Academic, Boston, 1994.

Sergent, J. E., and C. A. Harper, eds., *Hybrid Microelectronics Handbook*, 2nd ed., McGraw-Hill, New York, 1995.

Sutherland, I., and C. Mead, Microelectronics and computer science, *Sci. Am.*, Sept. 1977.

Taraeiskey, H., *Power Hybrid Circuit Design and Manufacture*, Marcel Dekker, New York, 1996.

Thomson, L. F., C. G. Wilson, and J. M. Frechet, eds., *Materials for Microlithography*, Advances in Chemistry Series, American Chemical Society, Washington, DC, 1984.

4

THICK-FILM FUNDAMENTALS

Engineers and scientists adopt materials and energy to society's needs. Since all disciplines in engineering and science involve the design of technical products, and all engineering disciplines encounter materials, it is essential to learn the properties and behavior of the types of materials used in a particular engineering product. Similarly, thick-film hybrid microcircuitry, which represents a packaging approach for electronic components intermediate between the conventional assembly of discrete components mounted on a printed board and monolithic integrated circuits is dependent on materials [1]. A thick-film hybrid microcircuit has components such as resistors, capacitors, and conductors, which are screen printed on alumina substrate [2,3]. The film resistor and the conductor used in a thick-film hybrid circuit are composed of mixtures of particulates of metals, metal oxides, and fine glass powders suspended in organic binders. The dielectric paste (ink), on the other hand, is composed of insulating materials such as oxides of metals and semiconductors and fine glass powders suspended in organic binders. Proper handling of these materials depends on knowledge of their physical, chemical, mechanical, and electrical properties. The effect that one material may have on another material within a hybrid circuit is of extreme concern [4]. Therefore, material substitutions that are being contemplated should be reviewed to determine whether a material selected is susceptible to malfunction under harsh conditions. Nonconforming materials can result in failure of hybrid circuits caused by one or more of the following: metal migration [5], corrosion or oxidation [6], ionic contamination [7,8], improper adhesion [9], intermetallic formation [6,10], and differences in thermal

Handbook of Thick- and Thin-Film Hybrid Microelectronics, By Tapan K. Gupta
ISBN 0-471-27229-9

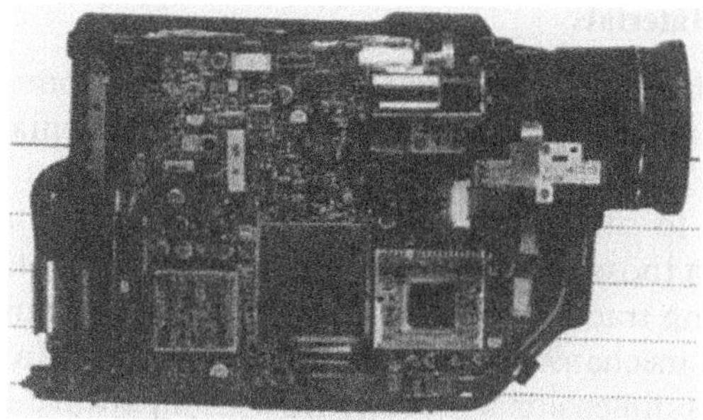

Figure 4.1 Thick-film material used in a Panasonic Palmcorder [13]. (Reproduced with permission from the May/June 1994 issue of *Advanced Packaging*™ magazine. Copyright 1994 PennWell Corporation. All rights reserved.)

coefficients of expansion [4,11]. Thick-film process cost is relatively low compared with those of its thin-film counterpart and monolithic circuits. Although screen-printed capacitors are included in thick-film circuits, the electrical characteristic of thick-film capacitors is not outstanding and the capacitance density is quite low compared to that of thin-film capacitors. Inclusion of thick-film inductors is avoided as far as practicable, although it is difficult in high-frequency circuits [12]. Figure 4.1 shows a home video product, the Panasonic Palmcorder. DuPont and Matsushita have tailored thick-film materials for mass-produced hybrids [13].

4.1 THICK-FILM SUBSTRATES

The passive substrate on which a thick-film hybrid circuit is built and fired serves as a mechanical base and as an insulating material [14,15]. Additionally, the thick-film substrate, as a passive material, insulates conductors screened on it. An ideal substrate, then, is physically rigid and should have high thermal conductivity, low electrical conductivity, and good thermal stability throughout both high (ca. 850°C)- and low-temperature (ca. 55°C) operation cycles. The substrate should be low in cost and readily available in quantity.

An important parameter in the selection of insulating substrate materials is dielectric loss [16]. Ion migration [17] and space-charge polarization [17] are the two important sources of dielectric loss [17] in ceramic materials. At frequencies above 1010 Hz, vibration and deformation losses at room temperature are small; thus ceramics are usually preferable for hybrid microcircuits. Also, interelectrode capacitance becomes significant as the distance between conductive paths decreases [18,19]. Thermal mismatch of the hybrid microcircuit substrate with conductors, resistors, dielectric pastes, and surface-mounted discrete devices should be taken into account [20]. Recently, thick-film substrate with a solderable thin-film back side has been used, which has shown many technical advantages [21].

4.1.1 Substrate Materials

Ceramic Thick-film substrates are usually produced from ceramic materials [22–24] such as alumina, beryllia, magnesia, thoria, zirconia, or combinations of these materials [25,26]. Ceramics have a closer coefficient of thermal expansion (CTE) match to silicon, higher thermal conductivity, and higher dimensional stability than those of most laminate materials. This stability allows finer patterning for wiring traces than on laminates. The importance of dimensional stability is not just mechanical. The electrical properties of a ceramic substrate are as important as dimensional stability [27]. Comparative data for the thick-film substrates used in hybrid circuits are given in Table 4.1. The screened and fired films depend heavily on the electrical properties (e.g., sheet resistance, dielectric constants, power density, and dielectric loss at high frequencies) of the substrate material. Fortunately, most ceramic materials show low losses at higher frequencies. In general, dielectric constants of most ceramics are less sensitive to frequencies, although there are exceptions [28]. Their dielectric constants may be higher than those of flex or laminate materials, leading to slower propagation speeds, but high-speed transmission has little value if the signal dies. For high-speed communication, the interconnect lines in the package are essentially transmission lines, and the impedance properties of those transmission lines can change significantly with anisotropic thermal expansion. Many laminate materials have an anisotropic CTE. Ceramics have been used for much higher performance ICs, modules, and special devices, such as surface acoustic wave filters [29].

Alumina The most commonly used ceramic substrate material for thick-film hybrid microcircuit is alumina (Al_2O_3) containing 4 to 6% glasses. Selection of this material is based on a compromise of thermal, electrical, and mechanical properties together with economic considerations [30]. Substrates are generally used in the *as-fired* state in a wide variety of sizes and shapes with standard 1% dimensional tolerance [31]. The use of alumina should be carefully engineered in consideration of substrate thickness and location in order to anticipate potential stress cracking problems in substrate processing and packaging [32]. Alumina offers excellent combinations of the electrical and mechanical properties required for hybrid microcircuits. The industrial standard material is 96% alumina mixed with 4% glass. Alumina substrates of 99.5% purity can be specified when low loss at microwave frequencies is important. The grain size is approximately 0.12 to 0.15 mil (3 to 5 μm), and the surrounding glassy grain boundary reacts with the thick-film binder glass to give a significantly high bond strength. Alumina substrates provide high physical strength, good thermal stability, and excellent dielectric properties and show excellent film adhesion for most thick-film pastes [33]. But alumina has a trigonal structure and a differential expansion coefficient both parallel and perpendicular to the C axis (anisotropy), which manifests in nonisotropic stress [34,35]. Alumina substrate is available from a variety of vendors with a tolerance of $\pm\frac{1}{2}\%$ in linear dimen-

TABLE 4.1 Comparative Data on Thick-Film Substrates in Hybrid Circuits

Substrate Material	Density $(g/cm)^3$	Flexural Strength $(dyn/cm^2 \times 10^7)$	Tensile Strength $(dyn/cm^2 \times 10^7)$	Young's Modulus $(dyn/cm^2 \times 10^{10})$	Thermal Expansion (ppm/°C)	Thermal Conductivity $[cal/(s \cdot cm \cdot °C)^{-1}]$
Al_2O_3, 90–96%	3.5–3.8	275–345	69–193	276–317	6.5–8.0	0.03–0.06
BeO						
98%	2.8	172–193	—	310	6.5	0.25
99.5%	2.85–2.88	207	138	351	8.0	0.35
Steatite	2.4–2.8	117–152	59–90	103–110	6.0–11.1	0.009–0.01

Substrate Material	Loss Factor at 25°C, 1 MHz	Dielectric Strength (V/μm)	Volume Resistivity (Ω·cm)	Dielectric Constant K at 25°C, 1 MHz	Power Factor Loss Tangent	Hardness (S)
Al_2O_3, 90–96%	0.00085–0.0048	8.6–13.4	10^{13}–10^{15}	9.0–10.0	0.0003–0.0015	>9
BeO						
98%	0.001	14.0	10^{14}	6.5	0.0001	9
99.5%	0.002	14.0	10^{14}	6.7	0.0003	9
Steatite	0.005–0.021	8.2–9.0	10^8–10^{11}	5.9–6.3	0.0008–0.0035	7.5

Source: From [48]; reproduced with permission of The McGraw-Hill Companies.

sion and 4 mils/in. (40 μm/cm) in camber. In volume these substrates cost less than 5 cents/in^2 (0.77 cent/cm^2). Hole patterns are also available at small extra cost.

Beryllia Beryllium oxide (BeO)–based ceramics are in many ways superior to alumina-based ceramics [36]. In the area power handling, beryllia, with its tenfold higher thermal conductivity, is a very good substitute for alumina substrate [37]. It is also attractive because of its unique combination of electrical, physical, and chemical properties [38]. Beryllia has a lower dielectric constant and thermal expansion than alumina. However, BeO is somewhat weaker than alumina, and the Occupational Safety and Health Administration (OSHA) has declared beryllia both a potential carcinogen and a toxic substance [39]. Current and proposed environmental regulations create concern about long-term use of BeO. Beryllia materials are fabricated in much the same way as alumina materials. However, the price of beryllia substrate is considerably higher than that of alumina substrate.

Aluminum Nitride A new ceramic material, aluminum nitride (AlN), has been developed for electronic applications [40]. AlN is relatively inexpensive, nontoxic, and has high thermal conductivity and a low CTE, matching that of silicon [41]. It has good electrical and mechanical characteristics as a substrate material. The demand for power modules is growing. Thus the need for an acceptable BeO substitute is getting more acute [42]. Fundamental studies and material characterization methods have indicated that both thick- and thin-film hybrid circuits can be fabricated on AlN substrates without difficulty. These hybrids are qualified to handle environmental conditions as specified in MIL-STD-883C [43]. To achieve higher clock speeds (gigahertz) and increased performance, a hermetic packaging system that provides mechanical support, matched interconnects, high I/O density, and effective thermal management is necessary. Research results show AlN to be promising as a thick-film substrate material [44]. MCM-C/D test vehicles have been fabricated successfully on the top of a cofired AlN pad/ball grid array substrate [45]. AlN offers thermal paths in all directions; and a designer has considerable flexibility in solving thermal problems. It also offers interconnect solutions in wirebond, flip-chip, and tape automated bonding [46]. AlN substrates have been used for many years to replace alumina using direct-bond copper metallization. The computer industry uses AlN as a heat sink in alumina copper–tungsten pin grid arrays. In the automotive industry, BeO substrates have been replaced in ignition modules by AlN. AlN offers additional flexibility in multilayer and power hybrid circuits [47,48].

Steatite Steatite (silicate of magnesium) and forsterite have sometimes been used for thick-film hybrid circuits [49,50]. They have a lower dielectric constant than that of alumina (6.1 and 6.4, respectively, compared with 9.3 for 96% alumina). They also have low tensile strength and low thermal conductivity.

Use of these materials is recommended where the lower dielectric constant is not a major factor in any particular circuit. Forsterite is one of a very few relatively high expansion ceramics. Because of an absence of alkali in the vitreous phase, the material remains a good electrical insulator at high temperatures [40].

Special Material Experimental studies on low-carbon-steel substrate coated with an electronic-grade glaze show good prospects for use of this material as a potential candidate for thick-film substrate. The difficulty with this substrate is that the glaze is a low-alkali-content glass frit, which creates problems during thick-film processing at moderately high temperatures (ca. 850°C). Special conductor, resistor, and dielectric pastes have been developed which are compatible with this porcelain steel material [51], whose advantages include high mechanical strength, easy merchantability, excellent thermal properties, and low cost. RCA has developed a porcelain [52] enamel steel system that can cope with leaded and surface-mounted components on substrates up to 18 by 24 in^2 (45.72 by 60.96 cm^2) in size [53]. The major disadvantage of this substrate is the capacitance coupling to the steel substrate at high frequencies.

4.1.2 Physical Properties

Size and Tolerance The size of the most commonly used alumina substrates (Table 4.2) range from 1/8 by 1/8 in^2 (0.32 by 0.32 cm^2) or $\frac{1}{8}$ in. (0.32 cm) in diameter to about 2 by 2 in^2 (5.1 by 5.1 cm^2) or 2 in. (5.1 cm) in diameter. Substrate in the larger sizes can be purchased with breaklines already embossed within the substrate for easy separation of multiple screened circuits [54]. The thickness of the alumina substrate varies between 0.010 in. (0.025 cm) and 0.080 in. (0.2 cm) (±0.002 in. or ±0.0051 cm). The most popular single-function substrate sizes are 1 by 1 by 0.024 in. (2.54 by 2.54 by 0.061 cm) and 1 by 2 by 0.025 in. (2.54 by 5.1 by 0.06 cm). Linear dimension tolerances are usually ±1%, or better available on special order.

Surface Finish Since the resistance of the screened and fired resistor depends on film thickness, resistor paste must be screened to a uniform height to maintain resistor tolerance. An uneven substrate surface may result in cracks and stress on the printed film, which with aging will end up in discontinuous film. Adhesion of the printed film also depends on the surface finish of substrate. The substrate surface requirements for thick-film deposition are not as stringent as those for thin-film deposition. Surface finish is defined as the centerline average (CLA), the arithmetical average of measured deviations in a surface profile from an imaginary mean centerline located between the peaks and valleys [55]. Mathematically, it can be written as

$$\mathrm{CLA} = \frac{A + B + C + D}{L}$$

TABLE 4.2 Standard Thickness and Common Sizes of Alumina Substrates Used in Thick-Film Hybrid Circuits (CoorsTek ADS-96R)

Standard Thickness (mm)	Common Size (mm^2)	Dielectric Constant	Dissipation Factor	Thermal Conductivity $[kcal/(s \cdot cm \cdot °C)^{-1}]$	Average Grain Size (μm)	Coefficient of Linear Expansion (ppm/°C)
0.254–3.048	88.9	9.5	0.00101 kHz to 0.00041 MHz	0.062, 0.048, and 0.028 at 20, 100, and 400°C	4–7	6.4, 7.2, 7.9, and 8.2 at 25–200, 25–500, 25–800, and 25–1000°C

Source: CoorsTek, Golden, Colorado.

where A, B, C, and D represent areas under or above the centerline and L is the assigned length of the substrate.

It has been observed that circuit losses increase quite rapidly with increasing CLA. CLA units are usually in./in. (or cm/cm). Typical thick-film substrates should have surface finishes between 20 and 40 in./in. (50.8 to 101.6 cm/cm).

Camber or Flatness The trend toward smaller geometry to afford more dies per wafer and the advances in photolithography have resulted in an ever-growing demand for wafer flatness of less than 4 mils/in. (40 μm/cm) of the substrate length [55–57]. *Camber* is defined as the amount of overall warpage present in a substrate. Camber can result in major problems in film screening, photo processing, and circuit stability. A nonflat substrate surface may result in an uneven film surface. This will result in uneven film exposure during photolithography. A nonflat surface will also develop stress in screened film. Such stress may change the resistor value in an unpredictable way and can cause microcracking and even discontinuity in a circuit.

Holes, Flats, and Lands Although substrates are most frequently used in rectangular form, they can be specified in disk shapes to fit can-type packages. Holes, identifications, alignment keys, score lines for multiple substrates, and the like are generally made when substrates are in the *green* state. It is very difficult to cut holes or slots after the substrate material is fired. Figure 4.2 shows an alumina substrate with laser-drilled holes [58]. The CO_2 laser is often used for precision drilling and machining of ceramic substrates. This process is capable of drilling a small hole about 0.005 in. (0.013 cm) with a feature tolerance of ±0.002 [58]. Typical hole dimensions vary from 0.020 to 0.030 in. (0.051 to 0.076 cm) in diameter with standard tolerances of ±10%.

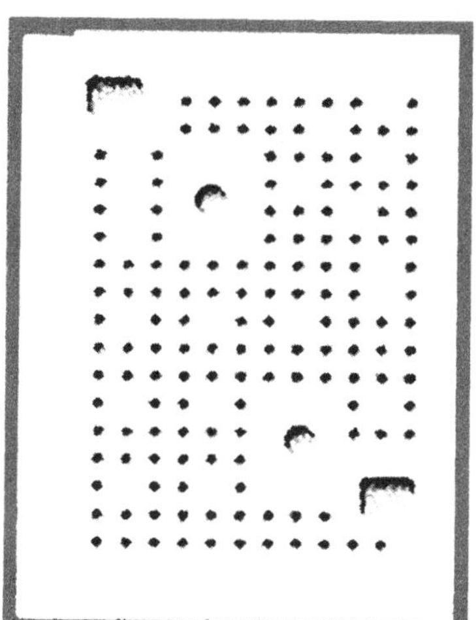

Figure 4.2 Laser-drilled holes on alumina substrate [58]. (Courtesy of Accu-Tech Laser Processing, San Marcos, California.)

4.1.3 Substrate Fabrication

Alumina powder with particle sizes ranging from 0.1 to 2.0 mils (3 to 50 μm) in diameter is blended with various additives. These additives vary from time to time and are proprietary. The most common additives are clay (silica), talc (magnesia), and calcia (calcium carbonate). The magnesia and calcia act as promoters, and silica acts as a glass-forming agent and binds the materials together. The powders are ball or roll milled for 12 to 16 h with plasticizers, organic binders, lubricants, and solvents, to ensure mixing. For dry press fabrication the slurry is spray-dried to obtain a uniform particle size.

Dry Powder Processing Dry powder processing is used when the required thickness of the substrate is higher than 0.04 in. (0.1 cm). The powder alumina is packed into an abrasion resistance die (silicon carbide) under high pressure (12,000 to 20,000 psi). Shrinkage during sintering is about 16 to 20%, and the die is designed oversize to compensate for this. Before sintering, holes are made in the substrate; they should not be located too close to an outside edge. In contrast to dry powder pressing, the powder can be subjected to isostatic pressure. Under isostatic pressure the powder, with additives, is packed in a bag and loaded inside a chamber. The chamber is first evacuated and filled with viscous liquid. Pressures ranging from 5000 to 10,000 psi are applied to the liquid, which is transmitted uniformly though the viscous liquid to the powder. The advantage of this method is that it permits fabrication of substrates with a relatively high length/width ratio [59]. Figure 4.3 is a flowchart for typical ceramic substrate fabrication processes, and Table 4.2 lists some of the physical characteristics of alumina substrate available commercially.

Sheet Casting The sheet casting method is adopted where the required thickness of the substrate is less that 0.04 in. (0.1 cm). In this method a slurry is made out of alumina powder mixed with additives,* organic binders, plasticizers, and solvents. The slurry is allowed to flow out on a long, smooth Mylar belt. The Mylar belt and slurry are allowed to move under a metal doctor blade positioned above the film. This blade actually controls the film thickness. The material is air-dried for 20 to 24 h to remove 25 to 50% of the solvent. The air-dried film is usually referred to as being in the green state because of its color. Holes and slots are punched, at this time. Scribed lines approximately 0.005 to 0.010 in. (0.013 to 0.25 cm) thick can also be made in the green sheet. The scribed lines facilitate breaking up the main substrate into individual parts after screening. The individual substrates are stacked on flat kiln plates and fired. Firing is done in two steps.

*Silica (SiO_2), magnesia (MgO), soda (Na_2O), lime (CaO), potash (K_2O), zirconia (ZrO_2), and ferric oxide (Fe_2O_3).

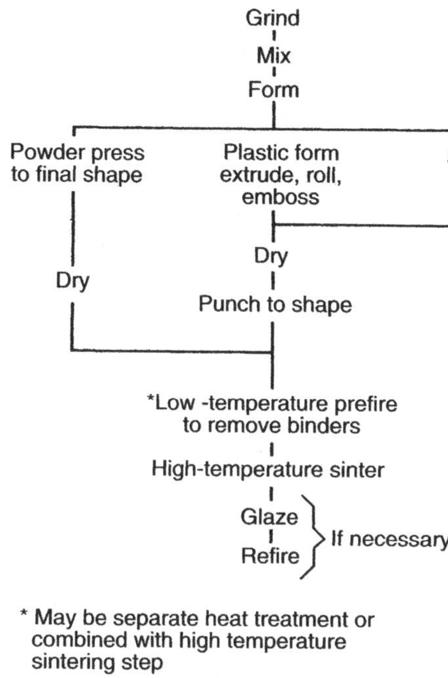

Figure 4.3 Ceramic substrate forming processes [59]. (Reproduced with permission of The McGraw-Hill Companies.)

Firing

Prefire The substrate fabricated by either of two processes is subjected to prefiring between 400 and 550°C to burn off the organic solvents, lubricants, and plasticizers.

Sintering Sintering of the substrate is followed by prefiring between 1500 and 1700°C to consolidate the fine particles by recrystallization [60]. This process helps the larger grains to grow at the expense of the smaller ones [61,62]. This results in a ceramic body of considerable mechanical strength. Excessive grain growth is responsible for the undesirable roughness exhibited by many *as-fired* ceramics. Recently, it has been possible to inhibit grain growth during firing and to produce dense high-purity alumina with markedly improved surface and mechanical and electrical properties. During sintering, ceramics shrink as much as 18 to 25% in linear dimensions or up to 40% in volume because of loss of binder and particle coalescence.

4.1.4 Electrical Properties

An important parameter in the selection of an insulating substrate material is its dielectric loss. When an electric field acts on any matter, the latter dissipates a certain quantity of electric energy in the form of heat. This dissipation of electric power in the specimen is a measure of electric loss [62,63]. The loss of energy will be higher if the frequency of the applied voltage is higher. The dissipation factor is a function of the frequency of the applied voltage, temperature, and dielectric material.

At room temperature and below 10^{10} Hz, vibration and deformation losses in ceramics are small. The two other important sources of dielectric loss in ceramics are ion migration and space-charge polarization [17]. Ion migration is responsible for electric breakdown, which develops as a result of the interaction of free charged particles (electrons and ions) accelerated by an electric field. The theory of electric breakdown of dielectric materials is extremely complicated and the values of electric strength of dielectrics are calculated on the basis of the structures and various parameters of the materials. For Al_2O_3, BeO, and AlN, the electric breakdown voltages are well above the requirements for normal hybrid circuits. The second type of breakdown is *electrothermal*. At a given electric field intensity the dielectric loss depends on the conductivity of the material. If alternating voltage is employed, the loss factor is dependent on $\varepsilon \tan \delta$ of the material, where ε is the permittivity and $\tan \delta$ is the loss tangent [64]. If the coefficient of thermal conductivity is high and the electrical conductivity and $\varepsilon \tan \delta$ values are insensitive to temperature, thermal breakdown within the material will develop only with difficulty, and the dielectric material will operate for a long time under comparatively high voltage without damage. Typical dielectric strength for alumina 0.025 in. (0.06 cm) thick is 600 V_{rms}/mil (24 kV_{rms}/mm), which is far off the requirements for a normal hybrid circuit. In the fabrication of functional film assemblies, substrates with a high dielectric constant (DIC = K) are frequently desired to achieve a high capacitance per unit area [65]. Titanates with a K value of 1700, for example, have been used as substrate for thin-film amplifiers. On such a substrate, interelectrode capacitance becomes significant as the distance between conductance paths decreases [18,19].

4.1.5 Multilayer Technology Using LTCC

The driving force for higher system performance, together with increased circuit density at minimal cost, is multilayer circuit technology. In the thick-film process, multilayer circuit technology is a sequential process where multilayers are built up layer by layer by consecutive printing and firing of conductive and insulating layers. The process uses high-temperature cofired technology, which is not cost-effective. An obvious trend or need in multilayer circuits is to have the equipment perform more functions faster and at lower cost. Low-temperature cofired ceramic substrate is an alternative approach in ceramic technology that can meet the electrical and physical requirements of advanced packaging [66,67]. It is also expected that the LTCC approach will be cost-effective in a high-volume manufacturing process [68–70]. The development of LTCC technology has successfully incorporated both external and internal (buried-in) circuit resistors within the LTCC structure. The concepts of external and buried-in thick-film LTCC resistors are logical progressions toward higher circuit density and savings in top-level real estate. Many thick-film and ceramic manufacturers have embarked on programs to develop materials suitable for LTCC technology [71–73].

The commercially available low-temperature cofired ceramic substrate is a formulation of a mixture of glass and ceramic fillers. The standard LTCC process is very similar to high-temperature alumina cofired processes [74]. However, compared to high-temperature (ca. 1500°C) sintering of alumina, cofired LTCC composites are sintered at about 850°C in less time (1–2 h) in air than in hydrogen [69]. Through-holes or via holes are formed by punch-and-die methods on a green sheet blanked to the size required. The via holes are filled with conductive materials and the circuitry is screen printed onto each sheet. The metallized sheets are stacked in the proper order and then laminated together between flat plates at low temperature (70°C) at a pressure of 2000 to 4000 psi. The laminated body is trimmed to the shape desired and fired once to burn out the organics. This helps to sinter the inorganics. LTCC and thick-film materials show great potential for increased circuit density, better packaging and performance, and a wide range of applications [71]. LTCC substrates are able to integrate microwave structures for antenna filters [75].

LTCC is rapidly becoming the technology of choice for RF components and modules for portable wireless applications [76]. High Q or low loss is an important feature of LTCC technology that contributes to enhance performance. Lower loss mean reduced power consumption and extended battery life, critical for portable applications. As a matter of fact, LTCC technology opens the door to a wider range of wireless packaging innovations, including next-generation portable wireless and gigahertz clock rate applications. Table 4.3 presents a comparative analysis of some of the properties of the LTCC and alumina substrates, and Figure 4.4 shows a flow diagram for LTCC processing.

TABLE 4.3 Properties of LTCC and Alumina

Property	LTCC	90–96% Alumina
CTE (ppm/°C)	7.9	6.5–8.0
Density (g/cm^3)	2.9	3.5–3.8
Camber (μm/cm)	1–4	1–2
Surface smoothness (μ·cm)	22.1	36.8
Thermal conductivity [kcal/(s·cm·°C)$^{-1}$ × 10^{-4}]	71–118	300–600
Flexural strength (dyn/cm^2 × 10^7)	152	275–345
Thickness/layer after firing (μm)	89–254	12.7–25.4
Insulation resistance (Ω at 100 V)	>10^{12}	>10^{12}
Dielectric constant at 1 MHz	7.1	9.0–10.0
Dissipation factor (%)	0.3	0.3

Source: From [83]; reproduced with permission of The McGraw-Hill Companies.

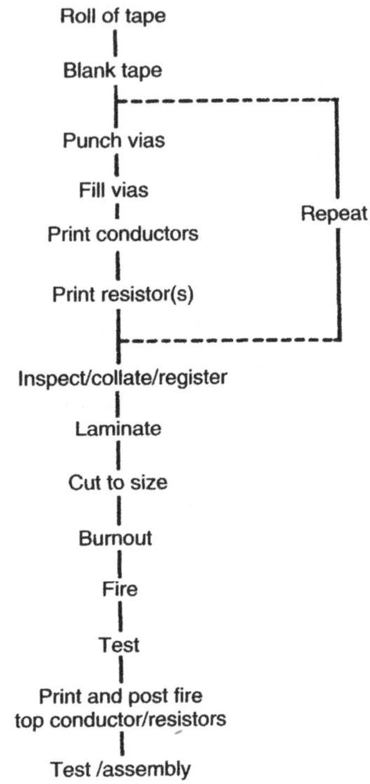

Figure 4.4 Flow diagram for LTCC processing [72].

Green Dielectric Tape Green dielectric tape is cast on a Mylar carrier and treated with a release agent to facilitate stripping. Proprietary coating technology produces high-quality green tape with minimal defects and thickness control within 2% of the nominal thickness [0.004 in. (0.01 cm)] across the width of the roll. The tape contains a glass-phase composition and particle size distribution engineered for viscosity and shrinkage control. Refractory fillers are selected to match thermal expansion of alumina and to minimize the dielectric constant. The firing profile is a modified thick-film profile run at a reduced belt speed to yield a peak temperature 850°C, dwell time at peak of 15 minutes, and total cycle time of approximately 90 minutes. To reduce signal propagation delays, low-dielectric-constant LTCC substrate materials together with low-impedance conductors have been used successfully by Baba and co-workers [77].

4.2 THICK-FILM CONDUCTORS

4.2.1 Conduction in Metals

Electrical conduction in metals is governed by Ohm's law, and the current density along the conductor can be written as [78]

$$J = \sigma\varepsilon = -nev \tag{4.1}$$

where J is the current density; σ the conductivity; ε the electric field applied, and n the number of free electrons available per unit volume, each of charge e and mobility v. According to Ohm's law, the electron velocity will remain constant for any particular field and can be explained in terms of collisions of the electrons with the lattice. If the average time between collisions is τ, the number of collisions per second is $1/\tau$. Following the equation of motion, we can write

$$v_{Dx} = \frac{-e\tau\varepsilon_x}{m}\left[1 - \exp\left(\frac{-t}{\tau}\right)\right] \tag{4.2}$$

and the current density

$$J = n(-e)v_{Dx} = \frac{ne^2\tau\varepsilon_x}{m}\left[1 - \exp\left(\frac{-t}{\tau}\right)\right] \tag{4.3}$$

The free-electron model has been used successfully to explain many aspects of electronic conduction in metals, particularly when Fermi statistics is employed [79,80]. The electrons near the Fermi level are governed by

$$T_F = \tau_F v_F \tag{4.4}$$

where v_F is the velocity of an electron with Fermi energy and can be approximated as

$$v_F = \left(\frac{2E_{F0}e}{m}\right)^{1/2} \tag{4.5}$$

For copper, $E_{F0} = 7.0$ eV, $\tau_F = 10^{-10}$ s, and $v_F = 10^6$ m/s and the conductivity $(\sigma = ne\mu = ne^2\tau_F/m) = 6 \times 10^7$ S/m.

In a monovalent metal, each atom contributes one valence electron to the conduction band. The concentration of n conduction electrons is the number of metal atoms per unit volume, or the Avogadro number divided by the molar volume. The molar volume is the molecular weight divided by the density [81]. In a good conducting material the essential requirement is a partially filled energy band.

4.2.2 Conductor Materials

Thick-film conductive compositions must satisfy a number of requirements to be viable in high-volume microcircuit applications [82]. Not only must they offer low-cost alternatives to competing technologies but must also have excellent performance under a wide range of processing conditions. Thick-film pastes [83] have three ingredients: the *binder*, glass frit; the *vehicle*, organic solvents and plasticizer; and the *host material*, the metal. The ink formulation is a blend of finely divided metal and glass powder that is suspended in organic vehicle much like paint thinner and helps define the printing characteristics of the ink. The vehicle usually contributes about 15 to 25% by weight of the conductor paste. Typical solvents used include alcohol, pine oil, and *d*-terpineol. The binder for thick-film ink consists of low-melting-point glasses, which hold the metal particles in contact and bond the film to the substrate. Viscosity, surface tension, chemical reactivity, and coefficient of thermal expansion (CTE) of the conductive paste are controlled mostly by the composition of the glass. So the manufacturer should be concerned about the selection of the glass in formulating the paste. The temperature flow of the glass during processing must be controlled properly to wet or coat the metal particles sufficiently to hold them and to wet the substrate.

The conductive metal particles are small and less than 0.00002 mil (5 Å) in size and typically cover 50 to 75% by weight of the conductive paste. The particle size, size distribution, and particle shape of the metal have significant influence on the electrical and physical properties of the final film conductor. Conductor materials include precious metals such as silver and gold as well as alloy of either silver–palladium, gold–platinum, or gold–palladium [84–86]. The conductor ink is selected on the basis of cost, compatibility with resistor and dielectric pastes, packaging concepts, and individual requirements. An important property of the thick-film conductor paste is adhesion of the film to the substrate [87,88].

The adhesion of the fired film is tested from the peel strength of the conducting film by applying a perpendicular force to the film. The measure of this perpendicular force to separate the film from the substrate divided by the pad width defines the *peel strength*. The magnitude of this force should be on the order of 75 to 100 lb/in. (13.4 to 17.86 kg/cm) for a well-adhering conductor paste. The adhesive strength between the conductor inks and the ceramic substrate depends on various factors, two of which are most important: the glassy phase of the glass frit composite inks, and the amount of additive (e.g., CuO or a similar material which promotes interaction during firing between the ink and the ceramic substrate).

Conductor sheet resistance is an important criterion for selecting a paste. Table 4.4 presents comparative data of the sheet resistivity and the processing techniques of the most commonly used conductor pastes in hybrid circuits. In almost all applications sheet resistance below 0.05 Ω/sq and low contact resistance at interfaces with metals and other conductors are preferred. The paste must also exhibit chemical stability under circuit operating conditions and should be free of phase changes when subjected to high- and low-temperature cycles (125 to −50°C).

Firing temperature plays an important role in establishing good bonding strength [89]. Careful examination of the conductor films by a scanning electron microscope after each step of firing processes has shown that *glass bleedout*

TABLE 4.4 Thick-Film Conductor Paste Characteristics

Type of Metal or Metal in the Paste (Thick-Film Ink)	Sheet Resistivity (Ω/sq) at 20–25 μm Thickness	Printing and Drying at 125–130°C for 10–15 min	Firing Schedule Peak for 10 min (°C)	Thinner
Silver	0.003–0.005	200–300 mesh	600–800	Bertyl Cellosolve
Gold	0.003–0.01	200–325 mesh	850–1000	Proprietary
Palladium–gold	0.04–0.07	200–325 mesh	850–1000	Pine oil
Palladium–silver	0.04–0.06	200–325 mesh	850–950	Proprietary
Platinum–gold	0.08	200 mesh	750–950	Proprietary

is generally more pronounced as the number of 850°C firing steps increases and the firing time decreases. Firing sensitivity studies are common for thick-film resistor [87] because conductors are generally assumed to function under a wide range of processing conditions and have received little attention, although thermodynamic equilibrium is not achievable under the conditions used to fire thick-film compositions. The firing process is determined largely by kinetics, which means that heating and cooling rates might contribute significantly. Recent trends toward rapid furnace firing profiles, including infrared furnace firing, might be expected to cause performance variations in conductors [90].

In high-frequency applications, skin effect (described in Chapter 11) must be accounted for, and gold or silver conductors should be used. Use of silver should be restricted because of the metal electromigration problem [72]. Solder-dip techniques normally used to reduce resistivity at low frequencies is not effective at very high frequencies [91].

Silver Silver is the less expensive conductor. It is easy to solder and can be bonded with high strength to the substrate. It can be fired simultaneously (cofired) with resistor compositions. However, silver has a tendency to migrate under the influence of dc fields [92–95]. Silver is also inferior to palladium–silver in solder wetability (leach resistance) and in susceptibility to tarnishing in air. When considering skin effect, silver is seen to be a relatively excellent conductor at high frequencies. A low-temperature cofired green tape [67,94] has been used successfully with silver as conductor material.

Gold Gold paste shows low sheet resistance and outstanding bondability by die and thermocompression bonding methods. As a result, thick-film gold as a conductor is widely used in chip and wired circuits. However, gold film affords poor adhesion to the substrate and it is difficult to solder to gold conductors because of the solubility of gold in solder materials. Moreover, typical gold paste degrades its properties, especially solderability, when subjected to multiple firing. The multiple fired film shows a phase on the surface which is rich in elements usually present in glass frit. Another disadvantage with gold is cost and price fluctuation. Very recently, a gold conductor paste has been reported [94] to produce fired circuits with 1.5-mil (0.04-mm) line with 2 mils (0.05 mm) of space and is compatible with both gold and aluminum wire bonding. The conductor paste has good adhesion to 96 and 99% alumina, beryllia, and LTCC tape. Recommended firing temperature is between 600 and 850°C. Like silver, gold performs excellently in high-frequency applications.

Binary Metal Alloy System Metals are used to form ohmic contact and/or to form interconnections between the devices. So it is expected that the contacts should have (1) low resistivity ($\rho = 1/\sigma$) or should offer the least resistance to the current flow, (2) strong adhesion to the surface it is applied, and (3) strong mechanical/electrical stability under harsh conditions. The difficulty with the single-metal metallization system is that it suffers from poor adhesion and or electromigration of the metal. To minimize electromigration and to promote adhesion, a single-metal metallization system is alloyed with a second or even a third metal. According to the Hume–Rothery rule [95], to form a substitutional solid solution between two composite systems, each of the metals taking part in the metallization system should have similar electronegetivity, valence, crystal structure, and atomic number. All the time it is observed that minimization of electromigration or promotion of adhesion costs increased resistivity of the alloy system. The increased resistivity that arises from solid solution may be expressed as

$$\rho_x = y_x x(1 - x)$$

where x is the atom fraction of the solute and $1 - x$ is the atom fraction of solid solvent. The solution resistivity coefficient y_x is specific for each binary metal system. The solid solution resistivity ρ_x originates from the same mechanism that varies the resistivity with temperature [96].

Platinum–Gold To overcome some of the difficulties encountered with gold paste, platinum–gold conducting pastes have been formulated [97]. This paste is used extensively in thick-film applications and has major advantages over gold. It has good solderability and retains reasonable wire bonding characteristics. However, the presence of platinum increases the degree of solderability at the expense of conductivity [98]. A 25-μm (ca. 1-mil) paste is fired between 700 and 1400°C for 10 to 20 minutes and exhibits sheet resistance between 20 and 100 Ω/sq. Platinum–gold paste is more expensive than gold paste, but adhesion of Pt–Au paste to alumina substrate is similar to that of gold paste.

Palladium–Gold Palladium–gold conductors have become very popular because they perform essentially as well as platinum–gold while delivering a considerable savings in cost [99]. Moreover, this paste exhibits excellent adhesion to alumina and beryllia substrates. The presence of palladium increases the sheet resistance of the paste but promotes the solderability and wire bonding capability.

Palladium–Silver Palladium–silver [100] paste has the highest use as a thick-film conductor in hybrid microcircuit. The material is less expensive than gold and shows higher sheet resistance than silver. The paste yields excellent solderability with tin–lead, tin–silver, or tin–lead–silver solders. Palladium–silver compositions adhere excellently to alumina and satisfactorily to beryllia. The material has a slight silver migration problem, however; the presence of palladium reduces the metal migration. The migrated silver diffuses inside resistors and Pd–Ag termination and decreases their resistance values. During firing in air, formation of palladium oxide affects the solderability of the terminals, and the presence of glass films on the surface creates a problem due to oozing from the pores between the metal grains. Abnormally high current noise, hot spots in the depletion regions, and high drift in resistor values are all thought to be adverse effects of bad termination. The substitution of gold for silver in the conductor paste alleviates the diffusion of gold but results in poor wire bondability. Pd–Ag silver conductors with less than 1% Pd show excellent leach resistance and solderability, and solderability remains above 95% even after 10 or 11 leach cycles [101] when the Ag/Pd ratio reaches 3:1.

Copper As the size of the hybrid microcircuit decreases, higher-density multiple circuits [102], fine lines, and the reduction of spaces between lines are essential. The low cost and high conductivity (next to that of silver and superior to aluminum) of copper, together with its high solder leach resistance and low migration tendency, can meet all the requirements of high-density multilayer circuits. Silver, silver–palladium, and gold can meet some of the problems; however, the high cost of these materials makes their practical use difficult. As a result, the use of copper thick-film technology is gaining momentum and wide acceptance. Copper-based thick-film inks have several important advantages over noble metal systems in multilevel high-density circuitry on ceramic base boards [103]. The disadvantage with copper-bearing materials is that they oxidized very easily and need an inert atmosphere for firing. Recent advancement in furnace design has eased the firing of copper-bearing materials [104,105]. Copper conductors have high density, adhesion, good solderability, and works well in the metallization of alumina, beryllia, and stealite substrates. Copper-conductive thick-film paste is made up of copper powder, glass frit, and reactive bonding agents, which are generally oxides of bismuth, lead, and cadmium. Copper conductor–compatible thick-film resistor and dielectric pastes are also available. The resistor materials contain many conductive materials, such as Pb, Sn, Mo, and Ta, together with their oxides. The sheet resistance of a 1-mil (25-μm)-thick film can be varied from 100 Ω/sq to 1 MΩ/sq by varying the percentage composition of the oxide. The typical TCR value of the resistor is ±150 ppm/°C, and the stability is less than 0.5%. Nitrogen firable dielectrics compatible with copper metallization show high density with an excellent CTE matching property and have an excellent dissipation factor (<2.5%), insulation resistance (>10^{12}), leakage current (<10 μA/cm^2), and stability. These dielectric materials are excellent for high-frequency applications. A recent study shows that copper film can exhibit a higher electromigration rate at the surface and interfaces than in the bulk film. However, copper paste doped with cadmium and zinc offers significantly improved electromigration with a slight increase in resistivity [106]. It has a wide firing range (850 to 1050°C). Copper has very good metallurgical properties together with low electrical resistivity and is suitable for high-speed packaging applications [107].

The principal back-end-of-line technology advances in the 1990s have influenced the progress of the microelectronic industry so dramatically that the 1990s will be remembered as the interconnect decade. This impact is due to (1) chemical mechanical planarization; (2) use of copper metal for fine lines and vias, which provides better electrical conduction and adhesion and prevents electromigration; (3) dual-damascene patterning; and (4) low-K dielectric property [108]. Materials engineering developments in ceramic-to-copper bonding are provide metallized substrates. These substrates have superior properties for certain hybrid applications. As a matter of fact, pure copper metallization on ceramics performs better when used in hybrid microcircuitry [109].

Some observers feel that a production-ready system compatible with copper is already available. This includes copper conductors, resistors, dielectric crossover, resistor overglaze, ultraviolet protection, overcoat, and solder paste. As a matter of fact, it can hardly be denied that copper technology has largely determined the shaping of modern hybrid circuit materials, and the advantages and drawbacks of this development will be judged in the near future [110].

4.2.3 Conductor Pastes

When selecting thick-film conductive pastes, the following performance characteristics should be studied.

1. *Conductivity.* This should be high enough to minimize voltage drops, heat generation, and interference with other circuit functions.
2. *Adhesion.* Attached wire leads and discrete components must not come loose during assembly or use.
3. *Bondability.* There must exist the capability of being soldered to or bonded to by thermal compression, ultrasonic, or eutectic die techniques.
4. *Clarity.* The paste should be able to produce fine lines without runoff, smearing, or scalloping.
5. *Compatibility.* The paste should be compatible with the processing and use of thick-film resistors and dielectric compositions.

6. *Shelf life.* There should be long shelf life.
7. *Stability.* The paste should remain stable during processing.

4.3 THICK-FILM RESISTORS

The relative elements of thick-film networks require the greatest precision and stability. Electrical performance is the primary consideration in choosing a resistor. As the state of the art improves, producers of thick-film systems are developing materials with TCR characteristics of 25 ppm/°C (−55 to +125°C) with close-tracking capabilities. Improvements in resistance drift under high loads and temperatures, current noise, power-handling capabilities, and voltage coefficient of resistance are continually being made, so that users are testing thick-film systems now in situations where only thin films were a candidate in the past.

Thick-film resistor inks, such as conductor inks, contain a suspension of metallic and glass powders within an organic vehicle. Originally, carbon was used as the conducting material. The first successful and commercially available thick-film resistor pastes were based on compositions of palladium, palladium oxide, and silver. The disadvantage with this system was that the ink was very sensitive to the firing profile and susceptible to reduction by a hydrogen atmosphere. To assure resistor stability, an overglaze coat is necessary. So the composition of the new ink [101] is formulated from ruthenium, iridium, and rhenium oxide. Ruthenium oxide is very stable and does not change its chemical properties when fired at an elevated temperature (850°C). Formulations using ruthenium oxide and lead glass give pastes with a wide range of sheet resistivity, good stability, relatively low TCR, and low electrical noise. It has been observed that after firing, ruthenium oxide (RuO_2) sometimes forms a glass–RuO_2 network by coalescence of glass grains and agglomeration of RuO_2 particles [111,112]. This reduces the resistivity of the fired resistor. Ruthenium compounds are employed as the conductive phase in the majority of commercial thick-film resistors. The small ionic radius 2.5×10^{-6} mil (0.62 Å) of Ru^{4+} allows the occupation of octahedral sites in a variety of stable oxide lattices [113]. The electrical conductivity of ruthenium compounds arises from the overlap of radial extensions of the 4*d* orbital in adjacent Ru^{4+} atoms. The glass phase is responsible for the physical integrity of the film. The glass binds the conductive phase together and bonds the film to the substrate while providing a stable interface to both. The glass is the most chemically active continuous phase in the composite, and it dictates the CTE of the resistor material. Commercially available resistor pastes are blended to a value on an individual basis considering the screening and firing equipment used. Typical characteristics of DuPont Birox 1600 series commercial thick-film resistor paste are shown in Table 4.5. Careful control must be maintained over material aspects of the screening cycle to ensure that thick film inks are thoroughly mixed and

TABLE 4.5 Typical Characteristics of a Commercial Thick-Film Resistor Paste

Type	Sheet Resistance (Ω/sq)	TCR (ppm/°C)	Noise (dB)	Drift (dB)	VCR (ppm/V·cm)	Conductor Termination	Firing Temperature (°C)
DuPont Birox	1–10 M	0–100	−5 to −25	0.1–0.2 at 1000 h	<−7 at 10 kΩ/sq	Pd/Ag 9061 or 9308	850 for 1600 series

have a uniformly constant viscosity. This is important since the amount of paste deposited determines the final resistance of the screened and fired resistor. Resistor shape will also determine final screened and fired values.

4.3.1 Physical Properties

Noise Spontaneous fluctuations in current, voltage, and temperature of the system under test are referred to as *noise* [114]. Noise is an important problem in science and engineering, since it sets lower limits on the accuracy of any measurement and on the strength of signals that can be processed electronically. In thick-film resistors, two components of current noise have been observed: One is generated in the resistor material itself and the other is generated at the resistor/conductor material interface. The latter is often called the *contact noise* of the resistor. Contact noise voltage is known to be proportional to the square of the direct current flowing through the contact [101,103]. There is a close relationship between the contact noise and the contact geometry, since the contact noise is a function of the direct current distributed through the contact area. Thus smaller effective contact areas have higher contact resistance and higher contact noise. Therefore, measurements of contact noise can be used as a tool to determine the quality of the contact. Experimental observations show that the contact noise of a thick-film resistor has $1/f$ characteristics and is proportional to the dc biasing resistance. The primary variables affecting resistor noise for a given composition are sheet resistance and resistor volume.

Another type of noise very frequently observed in trimmed thick-film resistor is *burst noise* (also called *popcorn noise*), and the magnitude of this noise is larger than $1/f$ noise fluctuations [115,116]. When burst noise exists in a resistor, the frequency spectrum includes burst, shot, current, and thermal noise and is found to follow a k/f^{α} relation. The average value of α is found to have a value of 1.0 regardless of the integration time. The burst noise spectra observed are pronounced for 30 Hz $< f <$ 10 kHz. Below 30 Hz, burst noise seems to level off and is dominated by conventional current $(1/f)$ noise. Because of its spectrum, burst noise is of special interest to the designers of telecommunications and data processing systems. Experimental data show that the burst noise frequency spectrum is dependent on (1) bias voltage and (2) annealing temperature. Excess burst noise can be reduced through use of thick-film resistor ink with lower sheet resistance [117,118].

Tolerance Tolerance of a thick-film resistor is determined by furnace profile, ambient during firing, interaction of the surface with foreign gases and materials, and the composition of the substrate and its interaction during firing. The other variables that are important are composition of the thick-film paste, uniformity of print thickness, screen mesh uniformity, screen emulsion and tension, and the physical location of the resistor. Screened and fired resistor can be produced within 5 to 10% tolerance.

Stability Another important aspect of thick-film resistor is the stability of the resistor at various temperatures (−50 to +125°C) and humidities under load and/or no-load conditions. Thermal expansion and chemical changes within the film alter the value of the resistor by more than 2% after several thousand hours of loading. But some good thick-film resistors have shown stability better that 0.2% even after 1000 h under load. Over a wide resistance range, increasing resistor temperature without increasing power dissipation increases resistance, whereas increasing resistor power dissipation under constant temperature decreases resistance. Fortunately, these effects tend to compensate mutually. It has been observed that thick-film paste having sheet resistance less than 1 kΩ/sq suffers a major drift problem.

Termination Anomalies Thick-film resistor is printed on thick-film terminals. The compositions of the terminal material and that of the resistor are different. The metallurgical and electrical data show a large deviation of the resistor value from the value predicted. This deviation is due to diffusion of silver from the resistor into the terminals. The deviation is greater in a silver-based ruthenium oxide thick-film resistor system. Terminations containing silver, such as palladium silver, exhibit the smallest deviation and are preferable. The reaction described above has numerous adverse effects on the performance of thick-film resistors, such as (1) departure from resistor values predicted by scaling, (2) abnormally high and geometry-dependent temperature coefficient of resistance, (3) abnormally high drift in resistance during the life of the resistor, (4) abnormally high noise figure, (5) hot spots in the depletion regions, and (6) catastrophic failure under gross overload.

Power Rating Prediction and Evaluation As the power densities and operating temperatures of thick-film networks continue to escalate, the problem of dissipating maximum power in a resistor of fixed size is being encountered more and more frequently. Consider as an example the design of a 1.25-W resistor to be operated in a 70°C ambient on a heat sink with a 0.015°C/Watt·cm^2 thermal sheet resistivity. The resistance of the resistor is to be 120 Ω, and it is to be printed with 100-Ω/sq paste. The resistor is to be laser trimmed with a kerf width of 0.025 cm. The aspect ratio should be 1:1. The resistor is to be qualified under MIL-R-55182, which allows 0.5% drift in the 70°C power rating test. This 0.5% includes drift from all sources.

Hot Spot A hot spot is defined as the small area that is unable to dissipate the proper amount of heat that it is rated to dissipate. As a result, a part of the film resistor becomes hotter than the surrounding portion of the film. The concept of hot-spot analysis is simply that the drift of the trimmed resistor is determined by the drift expected in the region of the resistor, which operates at the highest temperature. This is usually in the region of maximum power density.

4.3.2 Resistor Characteristics

1. *Temperature coefficient of resistance (TCR).* Thick-film resistor in hybrid circuits should have a TCR value of about 100 ppm/°C in the temperature range from −55 to 125°C.
2. *Resistance tolerance.* The resistance tolerance for an untrimmed thick-film resistor should be ±10%.
3. *Long-term stability.* At 125°C after 5000 h of operation the drift for thick-film resistor should not be more than ±0.25%, and the tracking between resistors should be better than 0.1%.
4. *Power dissipation.* The rated dissipation power for a thick-film resistor is typically 14 W/cm^2 at 125°C. This assumes that the total power dissipation on the substrate should not exceed 0.4 to 0.8 W/cm^2.

4.4 DIELECTRIC INKS AND PASTES

One method for lowering costs in hybrid microcircuits is to eliminate some add-on components by screening on capacitors and crossovers. Screen-printable thick-film dielectric materials have been available commercially [119,120] for a long time for encapsulation, crossovers (multilayer circuitry), and capacitor uses. Table 4.6 represents the typical characteristic of a dielectric paste. The satisfactory use of some dielectric materials is affected because of their poor abrasion resistance, low breakdown voltage and quality factor (Q), and high sensitivity to humid conditions [121,122]. The dielectric materials used for crossovers are usually materials with low dielectric constant (low K), whereas high- and medium-K materials (from 5 to 1500) are used for capacitors. Sometimes the K value is varied by adding various amounts of barium titanate. Ferroelectric materials also show high K values. The temperature coefficients of these capacitors with ferroelectric materials as dielectrics are found within 5000 ppm/°C [123]. Several low-permittivity, nonferroelectric dielectric materials (low K) are based on magnesium titanate, zinc titanate, titanium oxide, and calcium titanate [124]. The dielectric constant for these materials ranges from 15 to 150 with temperature coefficients of 200 ppm/°C. By changing the percentage weight of the titanates of either magnesium, zinc, or calcium or titanium oxide, the temperature coefficients of these materials can be made negative, zero, or positive.

Dielectrics often are deposited to provide crossovers and multilayer circuits. These dielectrics have low K and usually give less than 2 pF per crossover. Two types of dielectrics are commonly used for this purpose: glass and crystallizable glass. Glass dielectrics will resoften during subsequent refiring. Crystallizable glass dielectrics, however, do not exhibit this property. The material partially crystallizes during firing and does not permit the *swimming* of conductors even when it is refired at 1000°C. Large-area multilayer structures suffer from ther-

TABLE 4.6 Typical Characteristics of Capacitor Dielectric Paste Materials

Type	Dielectric Constant	Temperature Coefficient of Capacitance (ppm/°C)	Insulation Resistance (Ω/sq)	PF/cm	Dielectric Strength (V/μm)	Stability
DuPont series	10–250	200–500	10–5 × 10	787–19,685	20–30	0.2% CDP

mal expansion mismatch between substrate and dielectric materials, and there is an urgent need for a dielectric composition compatible with 96% alumina substrates. In addition to matching the substrate expansion coefficient, the dielectric must meet several visual and electrical standards. These include dissipation factor ($\tan\delta$), dielectric constant, and compatibility with precious metal conductors. Recently, a new commercially available material is reported by Suess and Skurski [94] that can produce a 3-mil (76-μm) via in a 1.6- to 1.7-mil (40- to 45-μm) fired layer and is compatible with new and conventional gold thick film.

Dielectrics for screened thick-film capacitors should fulfill some of the following criteria:

1. The film should be free of pinholes.
2. The film material should flow reasonably well.
3. The film should show high Q values, especially when it will be used in the high-frequency tuned circuits. Passing high current through capacitors at high frequencies requires high-Q components to minimize losses and avoid heating. High-Q components are also needed in filters to reduce insertion loss. But Q is not as important a factor for coupling and bypass applications as it is for narrowband and power applications [125].

4.4.1 Low-*K* Dielectric Materials

Ceramic thick-film multilayer technology is particularly suited to resolve many of the problems that existed in the high-frequency microcircuits. Good fine lines and via resolution are achievable with screen printing of cermet pastes. There has been much improvement with the introduction of laminate materials. Compared to ceramic pastes [dielectric constant (DIC) between 8 and 12], these laminate materials show DIC values between 4 and 5 [126]. The 7240D series capacitor inks from EMCA-Remax are used in portable transceivers and have DIC values between 20 and 100, with df values between 0.1 and 0.8 [127]. Low-*K* material properties are shown in Table 4.7.

4.4.2 High-*K* Dielectric Materials

The functional high-*K* dielectric pastes are generally inorganic compounds such as titanates, niobates, titanium oxides, and ferroelectric materials in matching

TABLE 4.7 Low-*K*-Material Properties

Properties	Low *K*	Planarized
K at 1 MHz	4.2	4.8
df at 1 MHz	0.006	0.0022
IR at 100 V (in.)	10	10
Dielectric strength	1010	1100

binders and solvents. Unfortunately, most high-K dielectric pastes require a high (1300 to 1400°C) sintering temperature to produce nonporous film [128]. However, the firing temperature of the dielectric pastes can be lowered by different formulations (below 900°C), and the porosity of the film can be reduced by adding glass to its compositions. The addition of glass to the composition also promotes the adherence of the film but at the expense of a drastic reduction in the dielectric constant (DIC = K). However, dielectric pastes having moderately high dielectric constants (25 to 500) can be processed at relatively low temperature (ca. 850°C), which is the typical thick-film processing temperature.

4.5 THICK-FILM INDUCTORS

Thick-film inductors are becoming popular in thick-film hybrid circuits because of recent developments of ferrite pastes (inks) and ever-growing applications of thick-film technology to RF frequencies [129]. Typically, the term *thick-film inductor* is used to refer to inductors fabricated mainly by a printed conductor coil with or without some form of deposited magnetic material (ferrite). This is to be distinguished from hybrid discrete inductors, which are simply miniature wound coils in the form of a chip attachable to thick-film hybrid circuits [130].

Thick-film inductors are generally air/ceramic cores and have been used in RF circuits and matching networks [131], TV filters [132], power amplifiers [133], and power electronic applications [134,135]. However, several limitations imposed by both material availability and the technology available for thick-film screen printing have limited the use of thick-film screen-printed inductors. Some of these limitations are limited ferrite layer achievable in a single print, large physical size, lack of tunability, relatively low permeability of the pastes, and lack of accurate knowledge about the ferrite materials.

Inductors generally used in thick-film hybrid circuits have two distinct configurations, spiral type and ferrite-core inductors [133,135]. In the case of spiral configuration, inductance values depend on the number of turns, and large inductance values can easily be obtained by printing many turns (50 to 60) [136]. One should be careful with a spiral inductor with a large number of turns, as it increases the parasitic capacitance of the coil. The Q factor for these spiral inductors at any frequency f is related to the inductance L and series resistance r of the inductor and follows a mathematical relation as

$$Q = \frac{2\pi f L}{r} \tag{4.6}$$

The inductor value in a ferrite-core inductor depends on the geometry and the ferrite-core material [137]. To realize a large inductor value in ferrite-core inductors, special ferrite-core materials have been developed by many manufacturers. With a special ferrite-core material with high permeability, a large inductor value can be fabricated with a small geometry, which is important for large, dense multilayer hybrid circuits. In thick-film hybrid circuits, screened inductors are always avoided. An active filter, a gyrator [138], *RC* instead of *LC* or *RC* frequency tailoring, or a chip inductor is used instead [139,140].

REFERENCES

1. J. A. King, J. Freer, and R. Woodard, *Materials Handbook for Hybrid Microelectronics*, Artech House, Norwood, MA, 1988.
2. M. R. Haskard, Paste and substrate system, in *Thick-Film Hybrids Manufacture and Design*, Prentice Hall, Upper Saddle River, NJ, 1988, Chap. 3.
3. D. W. Hamer and J. V. Biggers, *Thick-Film Hybrid Microcircuit Technology*, Wiley, New York, 1972.
4. C. P. Wong, Materials for electronic packaging, in *Electronic Packaging and Interconnection Handbook*, 3rd ed., C. A. Harper, ed., McGraw-Hill, New York, 2000.
5. A. Christou, *Electromigration and Electronic Device Degradation*, Wiley, New York, 1993.
6. L. H. Van Vlack, *Materials Engineering*, Addison-Wesley, Reading, MA, 1982, p. 446.
7. J. M. Ziman, *Principles of the Theory of Solids*, Cambridge University Press, Cambridge, 1964, pp. 110–114.
8. W. Jones and N. H. March, *Theoretical Solid-State Physics*, Vol. 2, Dover Publications, Mineola, NY, 1973 pp. 939 and 988.
9. D. S. Campbel, Mechanical properties of solids, in *Handbook of Thin-Film Technology*, 2nd ed., L. I. Maissel and R. Glang, eds., McGraw-Hill, New York, 1983, Chapter 12, p. 6.
10. C. A. Harper and R. N. Sampson, eds., *Electronic Materials and Processing Handbook*, 2nd ed., McGraw-Hill, New York, 1994, Chaps. 2 and 7.
11. R. Brown, Thin-film substrates, in *Handbook of Thin-Film Technology*, 2nd ed., L. I. Maissel and R. Glang, eds., McGraw-Hill, New York, 1983, Chapter 6, p. 29.
12. L. F. Miller, *Thick-Film Technology and Chip Joining*, Gordon & Breach, New York, 1972.
13. News, Thick-film materials enter home video picture, *Adv. Packag.*, May/June 1994, p. 6.
14. *Encyclopedia of Science and Technology*, Vol. 4, 8th ed., McGraw-Hill, New York, 1997.
15. J. J. Licari and L. R. Enlow, *Hybrid Microcircuit Technology Handbook*, Noyes Publications, Park Ridge, NJ, 1998, Chap. 2, p. 32.
16. J. C. Anderson, *Dielectrics*, Chapman & Hall, London, 1964.
17. B. Tareev, *Physics of Dielectric Materials*, Mir Publishing, Moscow, 1975, Chaps. 1 to 3.
18. H. R. Kaiser and P. S. Castro, *Proc. IRE*, Vol. 50, 1962, p. 2142.
19. W. W. Happ, *Electron. Ind.*, Vol. E4-7, June 1962.

20. R. H. Alderson and F. Ashworth, *Br. J. Appl. Phys.*, Vol. 8, 1957, p. 205.
21. S. Greiner, The substrate spit, *Adv. Packag.*, Nov./Dec. 1995, p. 36.
22. W. D. Kingery, *Introduction to Ceramics*, Wiley, New York, 1960.
23. J. E. Hove and W. C. Riley, eds., *Ceramics for Advanced Technologies*, Wiley, New York, 1965, p. 105.
24. W. S. Young and S. H. Knickerbocker, *Ceramic Materials for Electronics*, Marcel Dekker, New York, 1986.
25. J. E. Hove and W. C. Riley, eds., *Ceramics for Advanced Technology*, Wiley, New York, 1965, p. 29.
26. L. H. Van Vlack, *Materials for Engineering: Concept and Applications*, Addison-Wesley, Reading, MA, 1982, Chap. 12.
27. K. Okazaki, *Ceramics Engineering for Dielectrics*, Gakkensha Publishing, Tokyo, 1969.
28. L. L. Hench and D. B. Dove, *Physics of Electronic Ceramics*, Part A, Marcel Dekker, New York, 1972.
29. R. S. Jensen, Ceramic substrate for hybrid application, in *Hybrid Microelectronics Handbook*, 2nd ed., J. E. Sergent and C. Harper, eds., McGraw-Hill, New York, 1995.
30. L. L. Hench and J. K. West, *Principles of Electronic Ceramics*, Wiley, New York, 1990, Chap. 6.
31. R. D. Jones, Substrate materials and properties, in *Hybrid Circuit Design and Manufacture*, Marcel Dekker, New York, 1982, p. 9.
32. H. B. Bakoglu, *Circuits, Interconnections, and Packaging for VLSI*, Addison-Wesley, Reading, MA, Chap. 3; R. C. Sundahl and E. J. Sedora, Effects of alumina substrate surface defects on thin-film interconnect patterns, *Proc. IEEE*, Vol. 59, No. 10, 1971, pp. 1462–1467.
33. C. A. Harper and R. N. Sampson, eds., *Electronic Materials and Processes Handbook*, 2nd ed., McGraw-Hill, New York, 1994, p. 3.15.
34. R. Brown, Thin film substrate, in *Handbook of Thin-Film Technology*, 2nd ed., L. I. Maissel and R. Glang, eds., McGraw-Hill, New York, 1983, p. 6-4; A. E. Riad and F. Barlow III, *Properties of Thin-Film Materials*, McGraw-Hill, New York, 1998.
35. D. W. Richardson, *Modern Ceramic Engineers*, Marcel Dekker, New York, 1992.
36. M. M. Schwartz, *Handbook of Structural Ceramics*, McGraw-Hill, New York, 1992; H. Brian and B. Wellman, Beryllia packages for PC board heat dissipation, *Electron. manuf.*, Vol. 32, No. 2, Mar. 1986, pp. 51–53.
37. W. H. Kohl, *Handbook of Materials and Techniques for Vacuum Devices*, Reinhold, New York, 1967; J. J. Licari and L. R. Enlow, *Hybrid Microcircuit Technology Handbook*, Chap. 2, p. 45.
38. B. V. Hiremath, Ceramic thick and thin films, *Ceram. trans.*, Vol. 11, 1980.
39. M. B. Powers, *Potential Beryllium Exposure While Processing Beryllia Ceramics for Electronic Applications*, Ceramics Division, Brush Wellman Engineered Materials, Cleveland, OH, 1975.
40. D. M. Mattox, Ceramics, glasses, and diamond, in *Electronic Materials and Processes Handbook*, 2nd ed., C. A. Harper and R. M. Sampson, eds., McGraw-Hill, New York, 1994.

41. J. Isigane, T. Kubota, S. Sekihara, and K. Fujimura, Technological advances of thick-film resistors for AlN substrate with conductive composition, *Proc. Int. Symp. Microelectron.*, 1988, p. 349; M. Johnson, The performance of electronic devices mounted on aluminum nitride, *Proc. EEC*, 1989, pp. 562–565.
42. A. A. Mohammed and S. J. Corbett, Thin-film metallization and performance of a power hybrid module on aluminum nitride substrate, *ISHM Proc.*, Anaheim, CA, Oct. 1985, pp. 218–220; H. Hiroyuki, K. Tohru, K. Yasuhiro, N. Shozo, S. Yuzo, T. Hideo, and V. Kazuaki, Development of high thermal conductive aluminum nitride substrate by green sheet, *Proc. EEC*, 1986, pp. 412–418.
43. E. S. Dettmer, H. K. Charles, Jr., S. J. Mobley, and B. M. Romenesko, Hybrid design and processing using aluminum nitride substrate, *ISHM Proc.*, Seattle, WA, 1988, p. 545.
44. E. S. Dettmer and H. K. Charles, Jr., AlN and SiC substrate processing and characteristics, in *Advances in Ceramics*, Vol. 31, American Ceramic Society, Westerville, OH, 1989, pp. 87–106.
45. C. Schardt and W. Kinzy Jones, Co-fired AlN pad grid MCM-C/D for hybrid digital application, Int. J. Microelectron., Vol. 17, No. 3, 1994.
46. R. Lashway, AlN thermal management, *Adv. Packag.*, May/June 1994, p. 24; J. Dalcin, U.S. government contract DE-AC04-76-DP000613.
47. J. J. Licari, Thin and thick film, aluminum nitride, in *Electronic Materials Handbook*, C. A. Harper and R. M. Sampson, eds., McGraw-Hill, New York, 1994, Sec. 8.2.4, p. 8.13.
48. R. Brown, Thin-film substrates, in *Handbook of Thin-Film Technology*, 2nd ed., L. I. Maissel and R. Glang, eds., McGraw-Hill, New York, 1983, Chap. 6, p. 6-10.
49. N. D. Kingery, H. K. Bowen, and D. R. Uhlman, *Introduction to Ceramics*, Wiley, New York, 1976, Chap. III. pp. 308–309.
50. D. E. Harrison and C. J. Moratis, Steatite, in *Electronic Materials Process Handbook*, C. A. Harper, ed., McGraw-Hill, New York, 1970.
51. T. R. Allington and R. E. Cote, *Characterization of Thick-Film Compositions for Porcelain Steel Substrates*, Electronic Materials Division, DuPont, Wilmington, DE, 1978; M. Norio, Circuit assembly innovation by insulated metal substrate technology, *Proc. IEMT*, 1987, pp. 2–9.
52. I. H. Van Vlack, Nonconducting circuit materials, in *Materials for Engineering*, Addison-Wesley, Reading, MA, 1982, Chap. 19, pp. 525–530.
53. Special issue on porcelain enamel steel substrates, *RCA Rev.*, Vol. 42, No. 2, June 1978.
54. A. Kosloff, *Screen Printing Electronic Circuits*, 2nd ed., Signs of the Times Publishing, Cincinnati, OH, 1980.
55. R. D. Jones, Substrate materials and properties, in *Hybrid Circuit Design and Manufacture*, Marcel Dekker, New York, 1982, Chap. 2.
56. R. A. Rikoski, *Hybrid Microelectronic Circuits: Thick Film*, Wiley, New York, 1982, p. 8.
57. D. C. Guidica, Wafer flatness: an overview of measurement considerations and equipment correlation, *Solid State Technol.*, Sept. 1989, p. 56.
58. Bulletin, Accu-Tech Laser Processing, Inc., San Marcos, CA, 1994.

59. L. Berrin and Y. S. Kim, *Preparation of Alumina Substrate by an Isostatic Pressure Process*, American Ceramic Society, Westerville, OH, 1970; R. Brown, Thin-film substrates, in *Handbook of Thin-Film Technology*, 2nd ed., L. I. Maissel and R. Glang, eds., McGraw-Hill, New York, 1983, p. 6.6.
60. M. M. Schwartz, *Handbook of Structural Ceramics*, McGraw-Hill, New York, 1992; W. D. Kingery, *Introduction to Ceramics*, Wiley, New York, 1960.
61. D. W. Hammer and J. V. Biggers, *The Fundamentals of Thick-Film Hybrid Technology*, State of the Arts, Inc., State College, PA, 1980.
62. C. Kittel, Dielectric properties, in *Introduction to Solid State Physics*, Wiley, New York, 1966, Chap. 12.
63. L. Solymar and D. Walsh, Dielectrics, in *Lectures on Electrical Properties of Materials*, 4th ed., Oxford University Press, Oxford, 1988, Chap. 10.
64. J. M. Herbert, Properties of dielectrics, in *Ceramic Dielectrics and Capacitors*, Gordon & Breach, New York, 1985.
65. W. D. Fuller, *NEC Proc.*, Vol. 17, 1961, p. 32; A. E. Feuersanger, Preparation and properties of high permittivity thin film dielectrics, in *Thin-Film Dielectrics*, F. Vranty, ed., Electrochemical Society, Pennington, NJ, 1969, pp. 209–236.
66. A. L. Eustice, S. J. Horowitz, J. J. Stewart, A. R. Travis, and H. T. Sawhill, Low-temperature co-firable ceramics: a new approach for electronic packaging, *Proc. 36th Electron. Comput. Conf.*, Seattle, WA, 1988, pp. 37–47.
67. J. J. Licari and L. R. Enlow, *Hybrid Microcircuit Technology Handbook*, Noyes Publications, Park Ridge, NJ, 1998, Chap. 3, p. 36.
68. C. Harper and R. M. Sampson, eds., *Electronic Materials and Processes Handbook*, McGraw-Hill, New York, 1994, p. 1.88.
69. M. F. Bender, F. K. Patterson, E. A. Kemp, and J. E. Gantzhorn, Low-temperature cofired ceramic system: a cost-effective solution for multilayer packaging, *Int. Symp. Microelectron.*, 1988, pp. 12–19.
70. J. J. Licari and L. R. Enlow, Co-fired ceramic tape substrate, in *Hybrid Microcircuit Technology Handbook*, Noyes Publications, Park Ridge, NJ, 1998.
71. R. G. Pond, C. J. Sabo, W. A. Vitriol, and R. L. Brown, Processing and reliability of resistors incorporated with low-temperature co-fired ceramic structures, *ISHM Proc.*, Atlanta, GA, Oct. 1986, p. 461.
72. H. T. Sawhill, A. L. Eustic, S. J. Horowitz, J. Gar-El, and A. R. Travis, Low-temperature co-firable ceramics with co-fired resistors, *ISHM Proc.*, Atlanta, GA, Oct. 1986, p. 473.
73. Y. Shemada, K. Utsumi, M. Suzuki, H. Takamizawa, N. Mitsuru, and Y. Seihen, Low-temperature multilayer glass ceramic substrate, *IEEE Proc.*, 1983, pp. 314–319.
74. J. E. Sergent, The hybrid microelectronics and MCM technology, in *Electronic Packaging and Interconnection Handbook*, 3rd ed., C. A. Harper, ed., McGraw-Hill, New York, 2000, Chap. 9.
75. Horizons, *DuPont Thick Film News*, global ed., No. 14, Dec. 1999.
76. D. I. Ames, M. T. Dirks, R. R. Draudt, S. J. Horowitz, and C. R. S. Needes, Opening the door to wireless innovations, *Adv. Packag.*, Mar. 2000.
77. Y. Baba, H. Ochi, and S. Segawa, *IEEE Trans. Comput. Packag. Manuf. Technol.*, Pt. A, Vol. 8, No. 1, Mar. 1993, pp. 170–173.

78. J. Allison, *Electronic Engineering Semiconductors and Devices*, McGraw-Hill, New York, 1990.

79. R. Rief, *Fundamentals of Statistical and Thermal Physics*, McGraw-Hill, New York, 1965, pp. 350–393.

80. J. M. Ziman, Static properties of solids, in *Principles of the Theory of Solids*, Cambridge University Press, Cambridge, 1964, Chap. 4, p. 116.

81. C. Kittel, Free-electron Fermi gas I, in *Introduction to Solid-State Physics*, 3rd ed., Wiley, New York, 1966, Chap. 7, p. 217.

82. T. R. Arlington, H. E. Schmidt, L. H. Slack, and B. E. Taylor, A low-cost thick-film multilayer hybrid system, *Proc. 34th Electron. Comput. Conf.*, 1984, pp. 314–323.

83. J. J. Licari, Thick-film paste materials, in *Electronic Materials and Processes Handbook*, C. A. Harper and R. M. Sampson, eds., McGraw-Hill, New York, 1994, Sec. 8.4.3, p. 8.77.

84. J. R. Rairden, Thick and thin films for electronic application: materials process review, *Solid State Technol.*, Vol. 13, No. 1, Jan. 1970, pp. 38–39.

85. M. F. Romano, *Hybrid Microelectron. Rev.*, July 1970, p. 3; T. C. Reissing, An overview of today's thick-film technology, *Proc. IEEE*, Vol. 59, No. 10, Oct. 1971, pp. 1448–1454.

86. A. W. Postlethwaite, Hybrid thick-film printed components: materials and process, *Proc. IEEE Workshop Thick Film Hybrid IC Technol.*, Mar. 1968, p. 5-2-9.

87. P. E. Bacher and J. S. Murday, Factors affecting thick-film adhesion, *Proc. Int. Symp. Microelectron.*, Vancouver, BC, Canada, 1986.

88. J. J. Licari and L. Enlow, Paste materials, in *Hybrid Microelectronics Handbook*, 2nd ed., Noyes Publications, Park Ridge, NJ, 1998, p. 130.

89. W. A. Crossland and L. Itailes, Thick film conductor adhesion reliability, *Proc. Int. Symp. Microelectron.*, Nov. 1970, pp. 3.1–3.13; J. E. McDonald and J. G. Eberhert, Adhesion of aluminum oxide metal systems, *Trans. Met. Soc. AIME*, Vol. 233, Mar. 1965, pp. 512–516.

90. D. L. Hankey, A. S. Shaikh, S. Vasudevan, and C. S. Khadilkar, Thick-film materials and process, in *Hybrid Microelectronics Handbook*, 2nd ed., J. E. Sergent and C. A. Harper, eds., McGraw-Hill, New York, 1995, Chap. 3.

91. R. J. Klein Wassink, *Soldering in Electronics*, 2nd ed., Electrochemical Society, Pennington, NJ, 1989.

92. H. B. Huntington and A. R. Grone, Current-induced marker motion in gold wires, *J. Phys. Chem. Solids*, Vol. 20, 1961, pp. 76–87.

93. D. E. Reimer, Materials selection and design guidelines for migration-resistant thick-film circuits with silver-bearing conductors, *Proc. 31st Electron. Comput. Conf.*, 1981, pp. 287–292.

94. T. R. Suess and M. A. Skurski, Thick-film changes its image, *Adv. Packag.*, Sept./Oct. 1995, p. 54.

95. J. M. Ziman, Static properties of solids, in *Principles of the Theory of Solids*, Cambridge University Press, Cambridge, 1962, p. 114.

96. L. H. Van Vlack, Metallic materials, and Metallic conductors, in *Materials for Engineering*, Addison-Wesley, Reading, MA, 1982, Chaps. 2 and 17.

97. G. W. A. Dummer, Sliding controls wear and noise phenomena, in *Metals for Conducting and Resistive Functions*, Hayden Book Co., New York, 1970, Chap. 15, pp. 245–246.
98. I. M. Twerski, Characterization of a solderable gold thick films capable of withstanding multiple firing, *Proc. ISHM*, Minneapolis, MN, 1987, p. 56.
99. R. A. Rikoski, Screened and fired films, in *Hybrid Microelectronics Circuits: Thick Film*, Wiley, New York, 1980, p. 30.
100. R. D. Jones, Thick-film materials, in *Hybrid Circuit Design and Manufacture*, Marcel Dekker, New York, 1983, Chap. 3, p. 24.
101. S. Tong and C. E. Falleta, Silver palladium conductors with excellent leach resistance and solderability but containing less than 1% palladium, *Proc. ISHM*, Minneapolis, MN, 1987, p. 8.
102. R. C. Headley and R. E. Cote, Tracking of resistance and TCR in thick-film resistors: distinction, definition, and derivations, *Proc. ISHM*, Los Angeles, 1979.
103. J. H. Wood, Base material conductor pastes: a new challenge for the precious metal system, *Proc. Int. Symp. Microelectron.*, 1976, p. 257.
104. S. E. Carmassi and C. M. Stedom, Effective furnace design for large volume copper firing, *Hybrid Circuit Technol.*, Vol. 6, 1987, pp. 23–27.
105. W. F. Yext, E. A. Hayduk, Jr., and C. K. Fisher, Process for manufacturing thick-film electrical components, U.S. patent 4,622,240, Nov. 11, 1986.
106. J. A. Cunningham, Improving copper interconnects: a search for dopiness, *Semicond. Int.*, Vol. 23, No. 4, Apr. 2000, p. 98.
107. P. S. Palanisamy and D. H. R. Sarma, Thermodynamics of processing copper thick-film system in reactive atmosphere, *Proc. Int. Symp. Hybrid Microelectron.*, 1986, p. 848.
108. I. V. Verner, Future direction in copper technology, in *Copper: Fundamental Mechanism for Microelectronics*, S. P. Murarka, I. V. Verner, and R. J. Gittmann, eds., Wiley, New York, 2000, Chap. 10, p. 313.
109. T. C. Evans, Pure copper metallization technology, in *Hybrid Microelectronics Handbook*, 2nd ed., J. E. Sergent and C. A. Harper, eds., McGraw-Hill, New York, 1995, Chap. 5.
110. G. Joseph, *Copper: Its Trade, Manufacture, Use, and Environmental Status*, ASM, Washington, DC, 1999.
111. K. Tizuka and T. Yamaguchi, Mechanism of RuO_2–glass interaction and microstructure development in thick-film resistors, *Proc. Int. Symp. Microelectron.*, 1988, p. 345.
112. J. A. Loughran and R. A. Sigsbee, Transmission anomalies in thick-film resistors, *Proc. Int. Symp. Microelectron.*, 1969, pp. 271–280.
113. R. C. Sutterlin, G. O. Dayton, J. Adair, and J. V. Biggers, Resistor interaction in integrated ceramic packages, *Proc. Int. Symp. Microelectron.*, 1988, p. 358.
114. A. van der Ziel, *Noise in Solid-State Devices and Circuits*, Wiley, New York, 1986.
115. J. G. Cotte and T. M. Chen, Characteristics, sources, and minimization of thick-film resistors burst noise, *Proc. ISHM*, Minneapolis, MN, 1987, p. 837.
116. A. Bell, *Electrical Noise*, Van Nostrand, New York, 1960, pp. 258–261.

117. W. H. Card and P. K. Chaudhuri, Characteristics of burst noise, *Proc. IEEE*, Vol. 53, 1965, pp. 652–653.

118. A. Norell, M. Garierson, and T. M. Chen, Characterization of burst noise in thick-film resistors, *Proc. ISHM*, Los Angeles, 1979, pp. 14–18.

119. R. Reddy, L. C. Burton, and F. W. Stephenson, A study of high dielectric materials in conjunction with a multilayer thick-film system, *Proc. Int. Symp. Microelectron.*, 1988, p. 335.

120. P. J. Harp, *Dielectrics*, Butterworth, London, 1972.

121. J. B. Birks, *Modern Dielectric Materials*, Heywood, London, 1960.

122. L. Solymar and D. Walsh, Dielectrics, in *Lectures on Electrical Properties of Materials*, 4th ed., Oxford University Press, Oxford, 1988, Chap. 10, p. 262.

123. B. Tareev, *Physics of Dielectric Materials*, Mir Publishing, Moscow, 1975, p. 136.

124. L. Hozer, *Semiconductors Ceramics*, Ellis Horwood, Warsaw, Poland, 1994, Chap. 4; W. P. Mason, Piezo and ferroelectric devices, in *Molecular Engineering*, A. R. Von Hippel, ed., Wiley, New York, 1959, pp. 292–308.

125. S. J. Stein, C. Huang, and L. Ugol, Thick film capacitors with high *Q* factors, *Proc. Int. Symp. Hybrid Microelectron.*, 1979.

126. G. A. Breed, Capacitor behavior at radio frequencies, *RF Des.*, Jan. 1996, pp. 55–61.

127. D. Kellerman, Digital Equipment Corporation, U.S. patents 4,781,968, 4,865,875, and 5,178,934, Microelectronic devices and methods of manufacturing system.

128. J. Harer, V. Brown, D. Nabatian, and S. Shahbazi, RF characterization of thick-film materials for portable transceivers, *Proc. ISHM*, 1996, pp. 462–470.

129. J. J. Licari, Thick and thin film, in *Electronic Materials and Processing Handbook*, C. A. Harper and R. N. Sampson, eds., McGraw-Hill, New York, 1993, p. 8.89.

130. C. A. Harper, Passive electronic component handbook, in *Transformer and Inductive Devices*, L. E. Wilson, ed., McGraw-Hill, New York, 1997, Chap. 3.

131. A. E. Riad and J. C. Toscano, Wideband characterisation and modeling of thick-film inductors, *Proc. ISHM*, Minneapolis, MN, 1987, p. 73.

132. P. Barnwell, *Electron. Components*, Vol. 12, No. 5, 1970, pp. 681–684; Vol. 12, No. 11, 1971, pp. 749–758; Vol. 13, No. 15, 1972, pp. 723–724.

133. J. Corkhill and M. Martin, *Electron. Eng. J.*, Vol. 43, No. 519, 1971, pp. 65–69.

134. R. E. Chaddock, *Radio Electron. Eng.*, Vol. 44, No. 8, Aug. 1974, pp. 414–420.

135. R. L. Remke and G. A. Burdick, Spiral inductors for hybrid and microwave applications, *Proc. 1974 Electron. Comput. Conf.*, 1974, pp. 152–161.

136. W. Olshewski, The hybrid compatible transformer, *IEEE Trans. Comput. Hybrid Manuf. Technol.*, Vol. 2, No. 4, Dec. 1979, pp. 487–489.

137. J. F. Shackelford, *Introduction to Materials Science for Engineering*, Macmillan, New York, 1988; B. Lax and K. J. Button, *Microwave Ferrites and Ferromagnetics*, McGraw-Hill, New York, 1962.

138. P. M. Pozar, Theory and design of ferromagnetic components, in *Microwave Engineering*, Addison-Wesley, Reading, MA, 1990, Chap. 10, p. 570.

139. N. K. Bose, Film distributed networks, *Microelectronics*, Vol. 5, No. 3, 1974, p. 30-7.

140. K. E. G. Pih, Thick-film distributed notch filters, *Microelectronics*, Vol. 9, No. 1, 1978, pp. 18–21.

RECOMMENDED READING

Birks, J. B., *Modern Dielectric Materials*, Heywood, London, 1960.

Dingle, J. W. D. ed., *Oxide and Oxide of Metals*, Vol. I, Marcel Dekker, New York, 1972.

Dummer, G. W. A., *Materials for Conductive and Resistive Functions*, Hayden Book Co., New York, 1970.

Hamer, D. W., and J. V. Biggers, *Thick-Film Hybrid Microcircuit Technology*, Wiley, New York, 1972.

Harper, C. A., and R. M. Sampson, eds., *Electronic Materials and Processes Handbook*, 2nd ed., McGraw-Hill, New York, 1993.

Haskard, M. R., *Thick-Film Hybrids Manufacture and Design*, Prentice Hall, Upper Saddle River, NJ, 1988.

Hench, L. L., and D. B. Dove, *Physics of Dielectric Ceramics*, Marcel Dekker, New York, 1972.

Holmes, P. J., and R. G. Loasby, *Handbook of Thick-Film Technology*, Electrochemical Society, Ayre, Scotland, 1976.

Jain, T. C., *Properties of Electrical Engineering Materials*, Harper & Row, New York, 1967.

Johnson, R. W., *ISHM Modular Series in Hybrid Microelectronics*, ISHM, Reston, VA, 1991.

Jones, R. D., *Hybrid Circuit Design and Manufacture*, Marcel Dekker, New York, 1981.

Kosloff, A., *Screen Printing Electronic Circuits*, 2nd ed., Signs of the Times Publishing, Cincinnati, OH, 1980.

Licari, J. J., and L. R. Enlow, *Hybrid Microcircuit Technology Handbook*, Noyes Publications, Park Ridge, NJ, 1998.

Miller, L. F., *Thick-Film Technology and Chip Joining*, Gordon & Breach, New York, 1972.

Moran, P., *Hybrid Microelectronic Technology*, Gordon & Breach, New York, 1984.

Murarka, S. P., I. V. Verner, and R. J. Gutmann, *Copper: Fundamental Mechanism for Microelectronics Applications*, Wiley, New York, 2000.

Ozazaki, K., *Ceramics Engineering for Dielectrics*, Gakkensha, Tokyo, 1969.

Rikoski, R. A., *Hybrid Microelectronics Circuits: Thick Film*, Wiley, New York, 1980.

Sergent, J. E., and C. A. Harper, eds., *Hybrid Microelectronics Handbook*, 2nd ed., McGraw-Hill, New York, 1995.

Tareev, B., *Physics of Dielectric Materials*, Mir Publishing Moscow, 1975.

Topper, M. L., *Thick-Film Microelectronic Circuit Fabrication: Design and Application*, Van Nostrand Reinhold, New York, 1971.

Van Vlack, L. H., *Materials for Engineering*, Addison-Wesley, Reading, MA, 1982.

5

THICK-FILM DEPOSITION TECHNIQUES

The thick-film hybrid microcircuit technology is becoming one of the most popular and cost-effective technologies [1,2]. It is being used in almost every sphere of life. It is a relatively old technology which requires a considerably large area compared to monolithic circuits and is suitable for manufacturing thick-film hybrid microcircuits for high-power applications. Thick-film hybrid circuit can be applied to high frequencies (up to 1 GHz) with high circuit yield [3,4]. Thick-film printing applications evolve rapidly in areas that are still considered to be state of the art. The process uses well-established screen-printing techniques, and the screen printer should be familiar with some of the necessary processes employed in microcircuitry such as artwork preparation, process camera work [5], step-and-repeat work, varied photographic reproduction processes, and screen-printing equipment employed in printing, drying, and firing [6].

No other process can match the economics of screen printing for the deposition of large, accurate quantities of material with well-defined dimensions needed in the fabrication of thick film [7]. Due to accurate repeatability, screen printing can be used for printing any material, on any substrate, in any size or thickness, and in almost any shape. As a matter of fact, screen printing is an additive process and can be used for mass production of thick films. Using silk or stainless-steel fabric as volumetric containers that are filled by hydraulic pressure in the roll of ink (slurry) forms deposits before the squeegee. This process leads to uniform deposition of the ink on the substrate. The substrate, along with the deposited slurry (ink), is removed and dried in air followed by

Handbook of Thick- and Thin-Film Hybrid Microelectronics, By Tapan K. Gupta
ISBN 0-471-27229-9

oven drying. The dried film is then fired. The entire process can be manual, semiautomatic, or fully mechanized [8,9].

An enlarged drawing (10 to 20 times that of the original circuit) or design is made from the schematic consisting of graphical symbols representing electronic components and interconnecting lines between conductors, resistors, capacitors, and sometimes inductors. The enlarged drawing makes it possible to better control errors in line width and spaces. A master is made from the circuit layout. The master is reduced to the required size, which later serves as a negative. From the negative a positive is made on either a stainless-steel or silk mesh. This mesh is framed in a proper size frame and serves as a *screen* [10].

The best that can be achieved routinely in production with screen printing is still 5-mil (125-μm) lines with 5-mil (125-μm) spaces and 10-mil (250-μm) vias. By evaluating selected mesh and wire sizes, it has been observed that 2-mil (50-μm) and 3-mil (75-μm) lines in the screen pattern can easily be resolved with good definition using a 400-mesh screen with 0.75-mil (19-μm) wire and 0.4-mil (10-μm) emulsion [11]. A smooth emulsion is required for high-resolution printing to ensure optimum emulsion gasketing.

5.1 THICK-FILM PROCESSING

The actual processing of thick film starts after selecting a suitable substrate and ink. It is also important to choose an ink that will disperse thoroughly through the screen [12]. The typical process sequences are (Figure 5.1): (1) print and dry conductor and bottom capacitor electrodes; (2) fire conductors; (3) print and dry capacitor or crossover dielectrics; (4) print and dry crossover conductors and top electrodes; (5) cofire dielectric (conductor combinations); (6) print, dry, and fire resistor compositions; (7) trim resistors; (8) encapsulate resistors (print, dry, and fire glass encapsulants); (9) solder conductive patterns; (10) attach active devices, leads, discrete components, etc.; (11) encapsulate or otherwise package circuit; and (12) test.

In thick-film processing, material (ink) ingredients play major roles in circuit fabrication and contribute several variables in the firing [13]. The conductivity of the paste becomes important when fabricating two different values of resistors. The resistance path of the conductor can cause measurement errors when holding tight tolerances for low-ohmic-value resistors. The resistor–conductor interface also plays an important role in the final resistor product. Physically, the interface must be uniform if power dissipation problems are to be avoided. The other variables to consider with these are the cost, electrical properties, and degree of difficulty to obtain reproducible results [14]. On the subject of cost, the palladium–silver system is by far the most inexpensive, the ruthenium oxide more expensive, and the iridium oxide system the most expensive. In terms of processing of these pastes (inks), silver palladium is the most difficult paste to process. This is due to the fact that equilibrium between palladium oxide and palladium alloy is unstable and depends critically on the heating and cooling

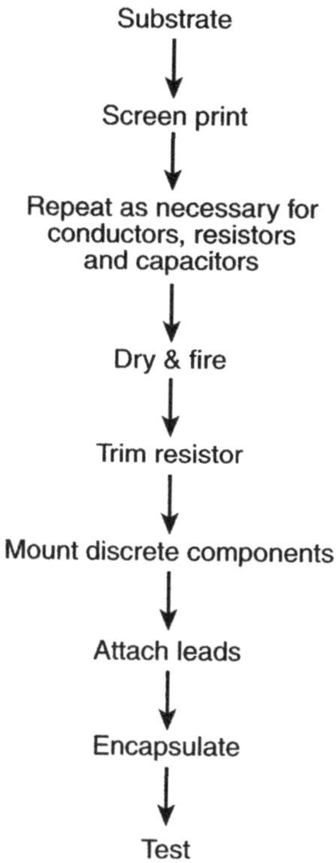

Figure 5.1 Typical thick-film process sequence.

rate. Interconnect resistance and capacitance have become increasingly dominant in packaging density, reliability, and manufacturing cost of hybrid circuits.

5.2 SCREEN PRINTING

The screen-printing technology is one of the most important steps in the manufacture of thick-film hybrid microcircuit. There are several parameters [15,16], which are considered important in the screen printing. Most of these parameters are logical consequences arising from the requirements established for the screen itself [17]. Some of the requirements are (1) processing temperature; (2) relative humidity; (3) screen tension; (4) mesh count; (5) coated emulsion; (6) squeegee speed, attack angle, pressure, stability, and flexibility; and (7) screen-to-substrate spacing. Figure 5.2 shows the release of the squeegee from the screen after the paste is released through the opening. Figure 5.3 shows the screen distance off the substrate before the squeegee at the start and end of printing. The distances are given in mils (1 mil is equivalent to 0.001 in. or approximately 25.4 μm).

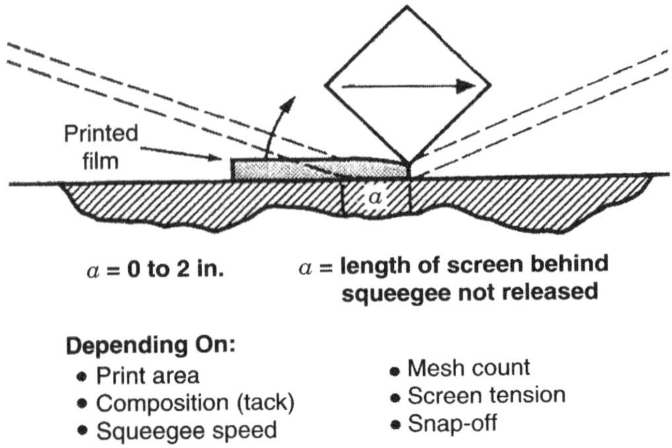

Figure 5.2 Screen release (peel). (From [11].)

Almost all thick-film hybrid circuits are produced using stainless-steel mesh. The screen printer consists of (1) a platform on which the substrate is held, usually a vacuum chuck; (2) a structure for holding the screen; and (3) a squeegee blade, which can be passed over the screen surface, forcing the inks through its opening onto the substrate. A cast aluminum frame is generally used to mount the stainless steel mesh. The proper image pattern is transferred to the screen by an ultraviolet exposure-developing technique following photolithography [5,18].

The ink is the most complex item of the three basic components of screen printing: the screen, the squeegee, and the ink [19]. The transfer of ink [20,21]

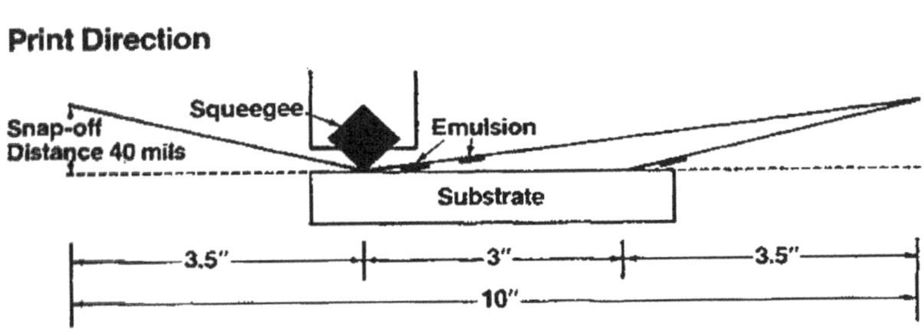

	Start of Print	End of Print
Distance before Squeegee	Distance of Screen off Substrate	Distance of Screen off Substrate
500 mils	3.1 mils	5.7 mils
200	1.2	2.2
100	0.6	1.1
81	0.5	0.8
50	0.3	0.6
44		0.5

Figure 5.3 Screen distance off substrate before squeegee [11].

through the screen to the substrate is a complex phenomenon and relates directly to the rheological or flow property of the ink. Rheology requirements for screen inks have been a subject of study for more than 15 years [22,23]. One factor in the rheology of the ink is the viscosity, which applies a resistance to the shearing force. The behavior that describes the inks used when subjected to a shear force [11] such as the squeegee provides to the ink during the printing process is thixotropic or pseudoplastic [24]. It means that as the squeegee speed is increased, the ink becomes more fluid (the viscosity drops). That characteristic is more desirable for easy flow of the ink through the openings of the screen. It also helps to wet the substrate. Pseudoplastic behavior supposedly improves ink printability.

Under ideal processing conditions, the volume of ink deposited by a screen is given by

$$V_{\text{min}} = AF_t$$

where A is the percent open area of the screen and F_t is the thickness of the fabrics. This equation does not take account of the shape of fabric threads. If the threads are considered to be circular of diameter D and the mesh count is taken to be M_c, the equation is transformed to

$$V_{\text{max}} = F_t - \frac{\pi M_c D^2}{2} \tag{5.1}$$

However, empirical data and some research evidence suggest that the actual values measured during printing are closer to V_{min} than to V_{max}. The theoretical deviation shows that the ink roll in the front of the squeegee represents a hydraulic pump, which converts movement of a boundary and transmits shear forces into the viscous fluid. As a result, the fluid layers shift against each other and established a relative motion between the two layers.

Printing resistor ink requires special attention [25]. The electrical resistance of a resistor depends on the geometry of the resistor (i.e., length, width, and thickness). If the length/width ratio is kept constant, the only variable in the process is then the thickness of the printed resistor. The primary print thickness control comes from the screen. Minor changes in print thickness are possible by adjusting the squeegee downstop (the mechanism that allows one to set the stopping position of the squeegee up from its downmost position in the *print* mode), pressure, speed, or screen breakaway. Most of these adjustments result in increased print thickness. Applying too much emulsion to the screen or applying tape to the bottom of the screen to repair a hole or cut in the screen will also affect the resistor thickness. Printing a resistor to close to a crossover will also make the resistor run thick.

Successful ink removal from the screen mesh is generally due to the adhesion between the ink and the substrate. The pull of the adhesive forces between substrate surface and ink on one side and between mesh surface and ink on the other side are assumed to stress the ink during snap-off until the ink yields under the applied tension and *breaks* in the vicinity of the mesh. The creation of vacuum in the ink during snap-off, while helping with the pattern transfer of ink from screen to substrate, causes a problem with small pattern features surrounded by ink (e.g., via holes). Ink gets drawn underneath the screen emulsion, which defines the pattern feature on the screen during snap-off. This destroys edge resolution on sequential prints unless the underside of the screen is wiped clean with dry absorbent material before the next point stroke. The most critical parameter during setup of a screen printer is the snap-off distance. To determine the maximum snap-off distance permissible within the elastic limit of the screen fabric requires knowledge of the yield point of the material and of the tension existing in the screen fabric prestretched on a frame. If the fabric is prestretched to an elongation of 0.1%, the snap-off distance can be set for an additional stretching of 0.2% during the printing stroke.

5.2.1 Screen Printer

Screen There are two types of screens available in the market [26]: silk screen and stainless steel screen. Silk screen printing process has proved to be dependable and economical for the low-cost production of reliable circuits. In stainless-steel screen, stainless-steel cloth is used instead of silk or nylon cloth. The screen is framed with uniform tension. Wooden frame is generally used for the silk screen, whereas cast aluminum frame is used in stainless steel mesh. To achieve better resolution and repeatability, automatic screen printing system is recommended. The screen sits in a larger holder with a sloppy fit. Using a light chase, the pattern to be printed on the substrate is positioned accurately on pins. The substrate normally sits on a vacuum chuck connected to a rotary pumping system. The substrate holder allows fine and coarse movement of the substrate. By adjusting the positions of both the substrate and the screen, the patterns to be printed on the substrate are aligned. A pneumatic (or hydraulic) system is used to operate the squeegee and the flood blade and to move the substrate in and out.

The stainless-steel screens are usually available in 5 by 5 in^2 (12.7 by 12.7 cm^2) and 8 by 10 in^2 (20.32 by 25.4 cm^2) sizes and mounted mechanically in a cast aluminum frame with stretched mesh and deliver resolution up to 0.004 in. (0.01 cm) with 0.004-in. (0.01-cm) spacing. Tension control is very important and can be monitored by applying a given load to the center of the screen and measuring the deflection [27]. The deflection of the screen is a function of the mesh type and screen size. The screen can be purchased with photosensitive emulsion applied to any thickness specified, but many firms choose to apply their own emulsion to be able to choose the thickness to suit particular situation. The minimum thickness of the emulsion required to get sufficient gasketing is about 8 μm. The emulsion holds the screen wires off the substrate near the edge of the emulsion, and as the screen deforms, the wires touch the substrate at a distance away from the emulsion [28–30]. The emulsion can be

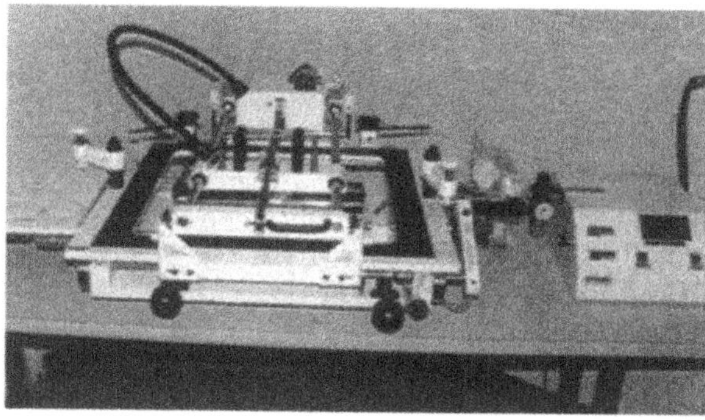

Figure 5.4 Screen printer. (Courtesy of VR Industries, Warwick, Rhode Island.)

direct, indirect, or direct–indirect, which is also called a hybrid emulsion. An indirect emulsion differs from direct and hybrid emulsions in the method of application of the stencil to the screen. Figure 5.4 shows an automated screen printer.

Direct-emulsion stainless-steel screen is most commonly used. Direct emulsions are applied in liquid form directly to the screen wires. In the indirect-emulsion method, the print is first developed on the sheet of a Mylar or some type of release paper carrying the emulsion and then transferred on the screen mesh [31]. The direct–indirect emulsion method is a combination of the direct and indirect emulsion systems. The advantage of this method is the good adhesion of the stencil to the screen with higher resolution of the image. The emulsions are made of a base material, polyvinyl alcohol, and some sensitizers [32].

The emulsion coating is dried out completely in a room with yellow lights and heated with very moderate heat. Special light sources (ultraviolet, from calibrated mercury lamps or carbon arcs) are required to expose the emulsion [33]. Energy of the radiated source and the exposure time are adjusted to reproduce the original pattern dimensions on the screen. The unexposed pattern is washed out with warm water [27,34]. All thick-film pastes, whether they are conductive, resistive, or dielectric, possess very fine particles suspended in organic vehicle. The viscosity of the ink can range from 20,000 to 250,000 cp. The homogeneity in the emulsion is strictly maintained. The vehicle vapor pressure should not be too high and must provide lubrication for the squeegee during screening. A moderate degree of thixotrophy is most desirable; otherwise, the paste is apt to loose screening definition [35]. Screens can be purchased with a variety of mesh counts (wires per inch) and with different wire diameters. Different screens are used for different inks, depending on the thickness of the ink to be deposited. As a rule of thumb, the mesh count is approximately 2.5 to 5 times larger than the average particle size of the paste material. The wire diameter, mesh count, and percentage of open area for stainless-steel screens are outlined in Table 5.1. Table 5.2 gives the size of the

TABLE 5.1 Stainless-Steel Wire Mesh

Mesh Count	Nominal Wire Diameter in.	Nominal Wire Diameter cm	Mesh Opening in.	Mesh Opening cm	Open Area (%)	Weave Thickness Range in.	Weave Thickness Range cm
80	0.0020	0.0050	0.0105	0.02	70.5	0.0036–0.0046	0.0091–0.0117
80	0.0055	0.014	0.0070	0.018	31.4	0.0110–0.0125	0.028–0.031
120	0.0026	0.0066	0.0057	0.0127	47.3	0.0052–0.0058	0.0132–0.0147
150	0.0026	0.0066	0.0041	0.0184	37.2	0.0051–0.0057	0.013–0.0127
180	0.0018	0.0045	0.0038	0.0096	45.7	0.0037–0.0043	0.008–0.011
200	0.0016	0.0040	0.0034	0.0086	46.2	0.0032–0.0038	0.0081–0.0096
200	0.0021	0.0053	0.0029	0.0073	33.6	0.0041–0.0046	0.0104–0.0117
230	0.0014	0.0035	0.0029	0.0073	45.9	0.0028–0.0034	0.0071–0.0086
230	0.0011	0.0028	0.0039	0.0099	54.0	0.0023–0.0027	0.0584–0.0068
250	0.0016	0.0040	0.0024	0.0061	36.0	0.0034–0.0038	0.0086–0.0096
280	0.0012	0.0025	0.0024	0.0061	44.1	0.0026–0.0032	0.0066–0.0081
325	0.0009	0.0023	0.0022	0.0056	50.1	0.0020–0.0025	0.0050–0.00635
325	0.0011	0.0028	0.0020	0.0051	41.3	0.0023–0.0028	0.0058–0.0071

Source: Microcircuit Engineering Corporation, Mount Holly, New Jersey. (Reprinted with permission.)

cast aluminum frame used to frame the stainless-steel screen, and Table 5.3 gives the tension recommended for stainless-steel screens.

Squeegee The action of the squeegee that pushes the ink and produces the pattern on the substrate in the printing operation carries the thick-film paste (ink) in front of it. The paste is carried across the screen and fills the pattern openings. Squeegee material is a medium-hard rubber or neoprene. The most popular material for squeegee is polyurethane [36]. The material is chosen for wear resistance and chemical inertness. The attack angle of the squeegee should be sufficiently acute to promote full loading of the screen opening during printing. An attack angle of 45° gives excellent rigidity and force on the paste [37]. The effect of the squeegee speed is critical and one should be careful about not operating at speeds high enough to impart large momentum to the paste resulting in air entrapment, or low enough to produce long contact times between the screen and the substrate during printing. This can cause some loss of line definition. At very high speeds, the squeegee can actually *hydroplane* over the ink on the screen. At this point a thin film of ink is left on the top of the screen and the film above the openings will be pulled through with the ink in the openings. This produces columns of ink taller than the thickness of the screen. The applied squeegee pressure and the amount of stress developed on the mesh wire are dependent on the size of the squeegee used in screen printing [38]. A new commercial screen printer is available which includes a floating and rotating squeegee head [39].

TABLE 5.2 Aluminum Frames

I.D.		O.D.		Flatness		Frame Thickness	
in.	cm	in.	cm	in.	cm	in.	cm
5 × 5	12.7 × 12.7	6.625 × 6.625	16.82 × 16.82	0.010	0.254	0.750 ± 0.010	1.9 ± 0.025
8 × 10	20.32 × 25.4	9.625 × 11.625	24.44 × 24.44	0.010	0.254	0.750 ± 0.010	1.9 ± 0.025
12 × 12	30.48 × 30.48	14.00 × 14.00	35.56 × 35.56	0.003	0.0762	1.0 ± 0.002	2.5 ± 0.025
15 × 15	38.1 × 38.1	17.00 × 17.00	43.18 × 43.18	0.003	0.0762	1.0 ± 0.001	2.5 ± 0.025

Source: Microcircuit Engineering Corporation, Mount Holly, New Jersey. (Reprinted with permission.)

TABLE 5.3 Tensions Recommended for Stainless-Steel Screens

Mesh Count	Deflection		8 × 10 in. frame	
	mils	μm	mils	μm
185	35	889	45	1143
200	38	965	48	1219
235	40	1016	50	1270
325	45	1143	55	1397

Source: From [27].

5.2.2 Drying and Firing

Drying The screened substrate is first dried in an oven before firing. Several types of ovens have been developed to implement drying process, including gravity convection and forced-air mechanical convection. The first one is used for general-purpose applications. Heat flows directly from the oven bottom up through the oven chamber and exits through vents in the top. An automatic hydraulic or bimetal thermostat controls the temperature of the oven. For precise control of the temperature, faster drying times, and for larger loads, the second type of oven is recommended. In this oven heated air moves up the oven chamber sides, where it exits through the openings in the sidewall plenums. The air is then circulated down through the center of the chamber, reheated, and recirculated. The oven has a primary hydraulic-control thermostat and a secondary independent safety thermostat for automatic oven temperature protection. Temperature control is better than ±0.5°. The operating temperature of the oven is between 200 and 250°C. The prime function of the drying process is to remove most of the organic solvents of the paste and to prevent rapid evaporation during the actual firing process, which could cause physical defects (e.g., voids and blister). The drying process is somewhat critical in that the rate and the temperature (100 to 150°C) used must allow the solvent to evaporate completely. The boiling temperature of the solvent is quite high, so that little is lost during the screening process. However, the solvent must not be allowed to boil while drying. Excessive drying may cause film shrinkage and peel back.

Firing After drying, the particles of various materials in the film are bound together with a plasticlike material called ethyl cellulose. This bonds the film to the substrate and holds the particle together until the substrate is fired. Firing is done in furnaces, kilns, lehrs, or furnace belt conveyors. Firing schedules affect the degree of adhesion of the films to the substrate. However, the firing schedules for conductors and dielectrics are not considered as critical as for resistor compositions. With increased interest in thick-film manufacturing, the selection of the furnace and firing process are considered very critical in establishing uniform high-yield production. A furnace with three or four zones of control allows flexibility over the temperature profile and controls over small increments that help to achieve adhesion and solderability of the films. In the furnace preheat zone, the solvents are thoroughly evaporated. Removal of the solvent vapors and prevention of their entry into high-heat zones are maintained with a positive flow of air or inert gas. A baffle between the preheat and high heat zones prevents further migration of vapors into high heat zones. In open furnaces (no baffle), a vent at the furnace entry aids this removal [40].

The flow of air should be of magnitude and direction such that clean fresh air is drawn through the main firing sections to flush the initial heating zones of contaminating organic materials. Draft velocities of 2 to 5 in./s (5 to 12.5 cm/s) give good results. A higher draft can disrupt temperature control. Commercial kilns are provided with conveyer belt. For a given profile, control should be held within 2°C/min for the rising segment of the profile and about the same for the fall rate from the peak temperature [37]. The typical firing profile is usually about 60 in. (152 cm) overall with 10 to 12 min at the peak temperature in the hot zone. For best results, the peak firing temperature should be held within ± 1°C for a selected value.

For a critical firing profile it is not the static temperature profile of the furnace that is important but the time and temperature profile of the parts. This is determined not only by the temperature throughout the furnace, but also by the speed of the conveyor belt. Fluctuations of $\pm 5\%$ in the belt speed could change the soak time from 4.75 to 5.25 minutes. It is generally recommended that for a critical firing profile, belt speed control should be within $\pm 2\%$ or better [41]. Commercial kilns are usually constructed with four or more zones of controllable temperature, exhaust, conveyor belt, and atmospheric control. This belt varies in sizes from about 4 to 12 in. (10 to 30.5 cm) in length, depending on the production volume required. The belt is usually made of nichrome or stainless steel [42]. The need for a contamination-free air supply with additional entry for dry nitrogen, atmospheric control, interior gas exhaust, and no entry for the contaminated gas inside the firing zones is undoubtedly a first priority for hybrid manufacturers [43]. Such a continuous furnace (TFM 241-3B5-192N84) is designed and fabricated by BTU Engineering Corp., Billerica, Massachusetts, and is shown in Figure 5.5. The furnace has eight zones, with a 6-ft (183-cm) burnout zone, two air rake entries, a 10-ft (305-cm) zone with one air entry, a 3-ft (92-cm) cooling zone, and a 7-ft (213-cm) water-cooling zone.

During the firing process the nonvolatile portion of the organic vehicle decomposes and the glass binder softens to wet the ceramic base and bond the metal particles together in a glass matrix. Firing is done in a well-ventilated section of the firing zone of the furnace. The effect of multiple refiring is shown to degrade the adhesion of the film to the alumina substrate as a result of a high percentage of conductor–ceramic interface [44]. The firing process of resistors is complex. The organic binders are burned out, the metallic elements oxidized or reduced, and the sintered glass materials provide adherence to the resistor film. The furnace used to produce the fired resistors should be examined in the light of many variables, which can influence the properties of the resistor. The obvious variables are time, temperature, and composition of the furnace atmo-

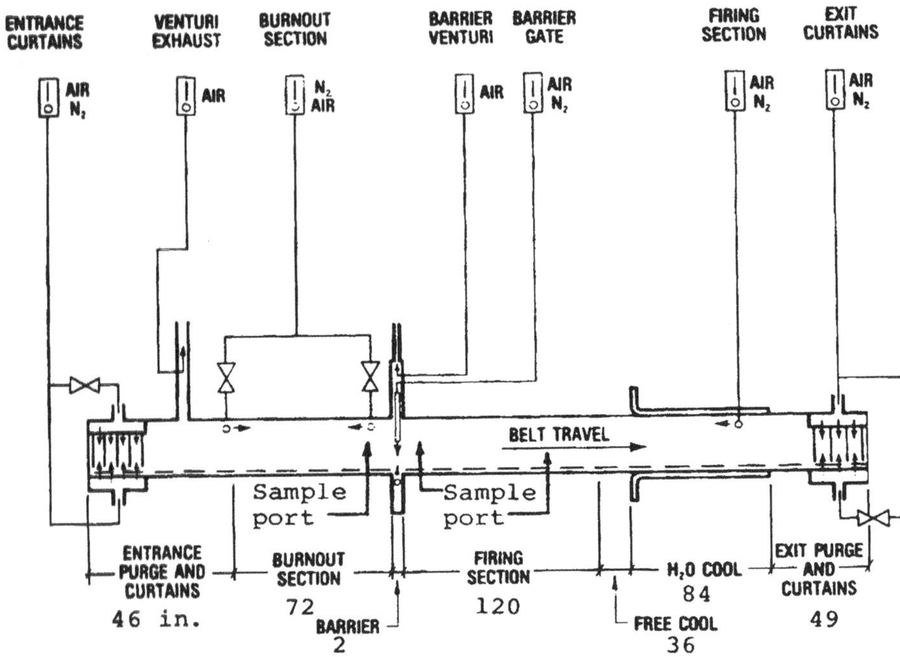

Figure 5.5 Schematic of a furnace used for thick-film processing. (Courtesy of BTU International, North Billerica, Massachusetts.)

sphere [45,46] and its pressure, positive or negative. To control atmospheric variables, two types of furnace are used: the forced air and the inclined plane (muffler inclined about 2° from the horizontal, with the entrance end high). Using a quartz tube can eliminate contamination from the muffler.

The firing cycle for the resistors is very critical. Due to the sensitivity of resistor materials to the temperature profile and the difficulty of maintaining a stable profile, it is preferable not to change a single kiln to the profile required. Dielectric and conducting films can share a kiln provided that production loading is such that adequate profile transition time is available. Resistor pastes with different sheet resistances may need different profiles as well as different kilns. Typical firing temperatures for the various thick-film materials are tabulated in Table 5.4.

Recently, BTU Engineering has reported a new version of an industrial furnace (Fast Fire II) that does not have a muffler (shown in Figures 5.6 and 5.7). The furnace is suitable for thick-film processing and can provide an air–nitrogen atmosphere during firing. It is claimed that the furnace has a uniform cross-belt temperature and can control the belt speed. This mufflerless furnace allows the furnace to heat and cool very rapidly, and the furnace can be turned off when not in use, to save energy and money. This furnace can reach 900°C from the ambient in less than 10 minutes. The furnace belt drive system has

TABLE 5.4 Firing Temperature for Thick-Film Materials

Material	Firing Temperature (°C)
Conductor	850–900
Dielectric	850–900
Resistors	700–850
Dielectric encapsulation	550–750

been reengineered to provide longer life for the belt. The Fast Fire II furnace drive pulleys are rubber-legged and machined for concentricity and uniform diameter. As a result, the clamping force between pulley, belt, and snub roller is uniform across the width of the belt. The belt wrap is a full 270°, to reduce the clamping force requirement.

The atmosphere within the furnace is well controlled and provided with a venturi exhaust system with a chamber-wide plenum for removal of gases. Removal of the evolved vapors and prevention of their entry into the hot zone is accomplished by maintaining a positive airflow through the preheat section toward the furnace entry. Data on a belt system for thick-film firing are given in Table 5.5.

Nitrogen Firable Furnace Aluminum has very good metallurgical properties, but its electrical resistivity is higher than that of silver or copper, making it less

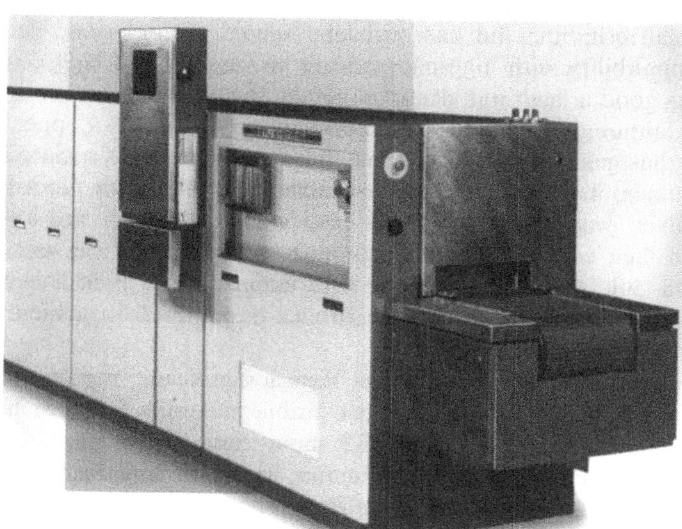

Figure 5.6 BTU high-temperature Fast Fire II furnace. (Courtesy of BTU International, North Billerica, Massachusetts.)

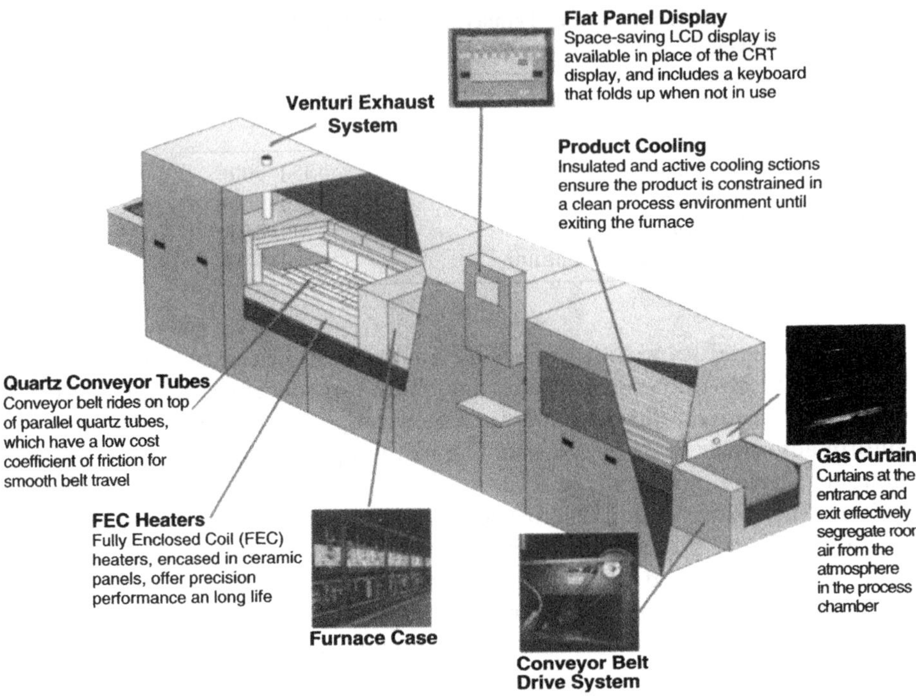

Figure 5.7 Mufflerless design feature of the Fast Fire II furnace. (Courtesy of BTU International, North Billerica, Massachusetts.)

desirable for high-speed packaging applications. Silver has very low thermal and electrical resitivities but has problems related to corrosion, electromigration, incompatibility with high-temperature processing, and high cost. Copper is almost as good a heat and electrical conductor as silver, yet costs much less than silver, although it has the same corrosion problem [47]. Copper thick-film technology has gained wide acceptance due to advantages associated with low sheet resistance, high solder leach resistance, low migration tendencies compared to silver (yet its conductivity is next to that of silver), and low cost. As a matter of fact, copper interconnects have revolutionized the semiconductor industry [48–50]. A major drawback with copper-based hybrid technology is that a special atmosphere inside the furnace is required to achieve optimum properties of the fired film [51].

Over the past few years there has been a significant improvement in the characteristics achievable with nitrogen firable materials for both hybrid and multilayer applications. Test data [52] show that with a good, controllable nitrogen furnace, copper-based conductors, resistors, and dielectric materials can be fired. The performance of these films is comparable with those of regular air-fired conductor, resistor, and dielectric films. As a matter of fact, copper holds promise as a significant material in interconnects [53] and as a dielectric material for dense circuit design [54].

TABLE 5.5 Belt Furnace for Thick-Film Firing[a]

Model[b]	Belt System in.	Belt System mm	Clearance in.	Clearance mm	Number of Zones	Heating Zone Length in.	Heating Zone Length cm	Active Cooling Zone Length in.	Active Cooling Zone Length cm
TFF9-2-9-108A24	9	229	2	50.8	9/8/7/11	108	274	12	30.5
						90	228	24	61
						126	321		
TFF14-2-9-108A24	14	356	2	50.8	9/8/7/11/13	108	274	12	30.5
						90	228	24	61
						126	321	30	76.2
						156	396		
TFF18-2-9-108A24	18	458	2	50.8	9/7/11/14/13	108	274	12	30.5
						126	321	24	61
						156	396	30	76.2
						216	548		
						234	594		
TFF5-2-13-156A48	25	635	2	50.8	13/14	156	396	30	76.2
						216	548		
						234	594		
TFF36-2-14-216A60	36	914	2	50.8	14/13	216	548	30	76.2
						234	594		

[a] The ambient is air for all models.

[b] All models are manufactured by BTU International, Billerica, Massachusetts.

Infrared Furnace For conventional belt furnaces, belt speed, the environmental conditions inside the furnace, and the firing zone temperatures all contribute to the optimization of film processing. In *IR* heating the processing parameters depend primarily on the absorption of heat and the dimension of the film, whereas the substrate is almost transparent to the *IR* radiation [55]. As the firing temperature is sensitive to the material absorption, the sheet resistance of a resistor drops dramatically with increased resistor area [29]. However, *IR* furnaces are gaining acceptance by copper thick-film manufacturers for greater speed, flexibility, and increased productivity [56–58].

REFERENCES

1. M. L. Topfer, *Thick-Film Microelectronics Fabrication, Design, and Applications*, Van Nostrand Reinhold, New York, 1971.
2. D. W. Hamer and J. V. Biggers, *Thick-Film Hybrid Microcircuit Technology*, Wiley, New York, 1972.

3. *The Thick-Film Microcircuit Handbook*, DuPont Electrical Products, DuPont, Wilmington, DE, 1970; J. Gitsman, *Thin-Film Microwave Circuits*, South Australia Institute of Technology, Adelaide, SA, Australia, 1974.
4. J. Hardy, *High-Frequency Circuits*, Reston Publishing, Reston, VA, 1979.
5. R. Glang and L. V. Gregor, Photolithography: generation of patterns in thin film, in *Handbook of Thin-Film Technology*, 2nd ed., L. I. Maissel and R. Glang, eds., McGraw-Hill, New York, 1983; D. McPhail, Screen printing is a science not an art, *Solder. Surface Mount Technol.*, No. 23, June 1996, p. 25.
6. J. J. Licari and L. R. Enlow, Thick-film processes, in *Hybrid Microcircuit Technology Handbook*, Noyes Publications, Park Ridge, NJ, 1998, Chap. 4, p. 104.
7. B. M. Austin, Thick-film screen printing, *Solid State Technol.*, June 1969, p. 53.
8. A. Kosloff, *Screen Printing Electronic Circuits*, Signs of the Times Publishing, Cincinnati, OH, 1980, Chap. 10, p. 89.
9. D. C. Hughes, Jr., *Screen Printing Microcircuits*, Press Co. Somerville, NJ, 1967.
10. G. V. Planer and L. S. Phillips, *Thick-Film Circuits*, Crane, Russak, New York, 1972; J. A. Gaglani, Obtaining fine-line geometries in today's hybrids, *Proc. ISHM*, 1986, pp. 819–825.
11. J. Bacher, *High-Resolution Thick-Film Printing*, ISHM, Reston, VA, 1986, p. 576.
12. A. Kosloff, Screen printing, in *Screen Printing Electronic Circuits*, Signs of the Times Publishing, Cincinnati, OH, 1980, Chap. 1.
13. M. R. Haskard, *Thick-Film Hybrids: Manufacture and Design*, Prentice Hall, Upper Saddle River, NJ, 1988, Chaps. 3 and 4.
14. P. J. Holmes and R. G. Loasby, *Handbook of Thick-Film Technology*, Electrochemical Publishing, Ayre, Scotland, 1976.
15. J. J. Licari and L. R. Enlow, Thick-film process, in *Hybrid Microcircuit Technology Handbook*, 2nd ed., Noyes Publishing, Park Ridge, NJ, 1998, Chap. 1.1, p. 104.
16. D. L. Hankey, A. S. Shaikh, S. Vasudevan, and C. S. Khadilkar, Thick-film materials and processes, in *Hybrid Microelectronics Handbook*, 2nd ed., J. E. Sergent and C. A. Harper, eds., McGraw-Hill, New York, 1995.
17. R. A. Rikoski, Screens and mask circuit production, in *Hybrid Microelectronic Circuits: Thick Film*, Wiley, New York, 1980, Chap. 6, p. 160.
18. L. F. Thompson, C. G. Wilson, and J. M. Frechet, *Materials for Microlithography*, ACS Symposium Series, American Chemical Society, Washington, DC, 1984.
19. D. E. Reimer, The shear and flow experience of ink during screen printing, *Proc. ISHM*, Minneapolis, MN, 1987, p. 335.
20. T. Freska, Theoretical model for multilevel screen printing of thick-film compositions, *Proc. ISHM*, Minneapolis, MN, 1987, p. 314.
21. H. Ranchi, B. Hunter, and P. K. Ajmera, A model for deposition of thick films by the screen printing technique, *Proc. ISHM*, Minneapolis, MN, 1978.
22. G. M. Johnson, The importance of rheology in the screen printing ink, *Proc. ISHM*, Minneapolis, MN, 1987.
23. C. C. Miller, The viscous flow of disperse suspensions in tubes, in *Rheology of Disperse Systems*, C. C. Miller, Pergamon Press, New York, 1959.
24. G. F. Kinney, *Engineering Properties and Applications of Plastics*, Wiley, New York, 1957.

25. T. C. Reissing, An overview of today's thick-film technology, *Proc. IEEE*, Vol. 59, No. 10, Oct. 1971; D. L. Hankey, A. S. Shaik, S. Vasudevan, and C. S. Khadilkar, Thick-film materials and processes, in *Hybrid Microelectronics Handbook*, 2nd ed., J. E. Sergent and C. A. Harper, eds., McGraw-Hill, New York, 1995.
26. A. Kosloff, Screen fabrics for electronic circuits, in *Screen Printing Electronic Circuits*, Signs of the Times Publishing, Cincinnati OH, 1980, Chap. 11, p. 101.
27. F. Franconvile and S. G. Stalnecker, Screen: essential tool for thick-film printing, *Proc. Int. Symp. Hybrid Microelectron.*, 1974, pp. 154–163.
28. R. J. Bacher, Thick-film processing for high yields, 1: screen printing, *Insulation Circuits*, July 1982, pp. 21–25.
29. J. R. Larry, R. M. Rosenberg, and R. O. Uhler, Thick-film technology: an introduction to the materials, *IEEE Trans. Comput. Hybrid Manuf. Technol.*, Vol. 3, 1980, p. 211.
30. C. C. Wonding and G. D. Haitt, *Polymeric Materials*, McGraw-Hill, New York, 1961.
31. C. Missele, Screens and screen printing, in *ISHM Modular Series in Hybrid Microelectronics*, R. W. Johnson, ed., ISHM, Reston, VA, 1991.
32. F. W. Billmeyer, *Textbook of Polymer Science*, Interscience, New York, 1962.
33. W. S. Deforest, *Photoresist Materials and Processes*, McGraw-Hill, New York, 1975.
34. T. M. Place, Practical production of thick-film conductors and resistors, *Proc. Int. Symp. Hybrid Microelectron.*, Palo Alto, CA, 1967, p. 13.
35. H. N. Nester, T. E. Salzer, and J. M. Woulbroum, Thick-film production techniques, *Proc. ISHM*, Palo Alto, CA, 1967, p. 3.
36. E. Graber, A. Klingsbers, and P. M. Siegel, *Kirk–Othmer Concise Encyclopedia of Chemical Technology*, Wiley, New York, 1985, pp. 400, 1211–1212.
37. R. C. Headley, Reproducibility of electrical properties of thick film resistors, *Proc. First Tech. Thick Film Symp.*, Palo Alto, CA, 1967, p. 58; M. R. Haskard, Paste and substrate materials, in *Thick-Film Hybrids*, Prentice Hall, Upper Saddle River, NJ, 1988, Chap. 3, pp. 31–37.
38. D. O. Brown, Screen printing: an integrated system, *Proc. Int. Symp. Microelectron.*, 1986, p. 582.
39. Juki Automation Systems, Morrisville, NC, Screen printing: The KS-1700 printer, *Surface Mount Technol.*, Vol. 15, No. 2, Feb. 2001, p. 131.
40. *Thick-Film Technology* (MPT 133), microelectronics processing technician course curriculum, ISHM, Reston, VA, 1987, p. 55.
41. D. J. Spigarelli, Selection of furnace equipment for thick-film firing, *Proc. First Tech. Symp. Thick Film*, Palo Alto, CA, 1967.
42. D. W. Hamer and J. V. Biggers, *The Fundamentals of Thick-Film Hybrid Technology*, State of the Arts, State College, PA, 1980.
43. S. Banerjee, Furnaces and ovens, in *Heat-Treating Furnaces*, Business Communication Co., Norwalk, CT, 2000.
44. D. R. Johnson, The effect of multiple refiring on the adhesion of thick-film conductors, *Proc. Int. Symp. Microelectron.*, 1977, p. 1; A. Singh, Techniques of

adjusting thick- and thin-film resistors in hybrid circuits, *Microelectron. Reliab.*, Vol. 15, 1979, pp. 123–129.
45. C. R. Delott, Control of furnace atmosphere, *Proc. Int. Symp.*, 1985, pp. 592–594; M. R. Haskard, Trimming processes, in *Thick-Film Hybrid Manufacturing*, Prentice Hall, Upper Saddle River, NJ, 1988, Chap. 4, pp. 45–49.
46. L. C. Hoffman, Precision glaze resistors, *Am. Ceram. Soc. Bul.*, Vol. 42, No. 9, Sept. 15, 1963.
47. G. W. A. Dummer, Properties of conductive materials, in *Materials for Conductive and Resistive Function*, Hayden Book Co., New York, 1970, Chap. 2.
48. S. P. Murarka, I. V. Verner, and R. J. Gitmann, *Copper: Fundamental Mechanisms for Microelectronic Applications*, Wiley, New York, 2000.
49. G. Joseph, *Copper: Its Trade, Manufacture, Use, and Environmental Status*, ASM, Washington, DC, 1999.
50. A. Elshabini and F. Burlow, Copper (properties of thin film materials), in *Thin-Film Technology Handbook*, A. E. Riad and F. Barlow, eds., McGraw-Hill, New York, 1998.
51. M. Colgan, C. Morath, G. Tas, and M. Grief, Copper damascene CMP metrology, *Solid State Technol.*, Feb. 2001.
52. P. S. Palanisamy and D. H. R. Sarma, Thermodynamics of processing copper thick-film systems in reactive atmosphere, *Proc. Int. Symp. Microelectron.*, 1986, p. 848.
53. P. Sayers and W. Petsis, Characteristics of nitrogen firable materials for hybrids and multilayers, *Proc. Int. Symp. Microelectron.*, 1987, p. 11.
54. J. Cong and C. K. Koh, *IEEE Trans. VLSI*, Dec. 1994.
55. X. W. Lin and D. P. Paramanik, Future interconnect technologies and copper metallization, *Solid State Technol.*, Oct. 1998, p. 63; M. K. Rao and C. W. Loch, Investigation on infrared-fired thick-film resistor layers, *Proc. ISHM*, 1985, pp. 552–561.
56. J. L. Resutek and R. E. Cote, Infrared firing of thick-film compositions, *Solid State Technol.*, June 1981.
57. W. Smetana, Parameters affecting the integrated firing process of thick-film resistors, *Proc. Int. Symp. Microelectron.*, 1985, p. 574.
58. D. R. Taschler, P. E. Herman, and W. F. Yext, Furnace atmosphere effects on copper thick film-conductors fired in an IR furnace, *Proc. Int. Symp. Microelectron.*, 1986, p. 859.

RECOMMENDED READING

DeForest, W. S., *Photoresist Materials and Process*, McGraw-Hill, New York, 1975.

Kosloff, A., *Photographic Screen Printing*, Signs of the Times Publishing, Cincinnati, OH, 1979.

Kosloff, A., *Screen Printing Electronic Circuits*, Signs of the Times Publishing, Cincinnati, OH, 1980.

Kosloff, A., *Screen Printing Photographic Techniques*, Signs of the Times Publishing, Cincinnati, OH, 1978.

Miller, C. C., ed., *The Viscous Flow of Disperse Suspension in Tubes*, Pergamon Press, New York, 1959.

Thompson, L. F., C. G. Wilson, and J. M. Frechet, *Materials for Microlithography*, ACS Symposium Series, American Chemical Society, Washington, DC, 1984.

U.S. Department of Commerce, *New Advances in Printed Circuits*, National Bureau of Standards Publication 192, U.S. Government Printing Office, Washington, DC, 1948.

6

THIN-FILM FUNDAMENTALS

Fulfillment of all the requirements of a hybrid microcircuit is not possible using thick-film alone. Thus thin-film technology has become an essential part of the hybrid microcircuit program despite the fact that thin-film technology is much more expensive than its thick-film counterpart [1,2]. The interconnection density (interconnection length per unit area available) for packaging technology should be at least one order of magnitude lower than that at the least expensive level to achieve higher speed and circuit density [3,4]. On the other hand, high-speed, high-frequency circuits for reliable performance require passive circuit elements with high sheet resistance, a low temperature coefficient of resistance (TCR), and a circuit stability greater than 0.01% with a minimum consumption of real estate. As a matter of fact, thin-film hybrid provides an alternative way to achieve high-density system-level interconnection [5,6]. Thus, acquisition of thin-film technology is an extremely useful addition to the capability of a thick-film hybrid manufacturing house, because it provides a way to obtain high packaging density as well as resistors with characteristics of stability and TCR far superior to what can be achieved using thick-film technology [7,8].

A *thin film*, in our context, is generally defined as a film whose surface is less than 0.04 mil (1.0 μm) thick. However, it is difficult to draw a line between thick and thin films on the basis of their thickness. The rapid development of hybrid microcircuits has placed thin film in the forefront, especially in high-frequency and high-density circuits [9–11]. A particular advantage of the thin-film process is high reproducibility besides high quality, stability, reliability, and flexibility. Moreover, thin-film passive and active devices can be fabricated

Handbook of Thick- and Thin-Film Hybrid Microelectronics, By Tapan K. Gupta
ISBN 0-471-27229-9

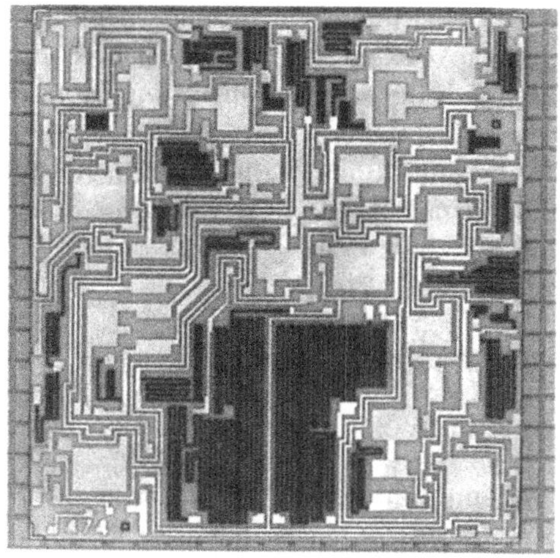

Figure 6.1 Substrate carrying thin-film metallized pads and resistors. (Courtesy of Satcon Electronics, Marlboro, Massachusetts.)

in one uninterrupted vacuum cycle. Resistive heating of the material under vacuum primarily does thin-film deposition. However, the variety of materials and processes required to meet the total needs of the industry adequately have necessitated the development of several deposition technologies other than resistive heating vacuum evaporation [also called physical vapor deposition (PVD)]: sputtering, ion-beam deposition (IBD), chemical vapor deposition (CVD), and epitaxy (including hot wall epitaxy), especially molecular-beam epitaxy (MBE) [12]. Figure 6.1 shows a high-density packaging unit on a ceramic substrate. Most of the circuit elements are from thin-film hybrid circuit technology.

One of the important physical characteristics of the vacuum-deposited thin films is smaller grain size compared to that of bulk materials. However, the grain size is mostly the function of deposition conditions and the annealing temperature [13]. The major problem with thin film is adhesion. Adhesion is dependent on the surface condition [14–16] of the substrate (e.g., roughness, cleanliness, and adsorption energy of the deposit and/or the higher number of nucleation centers in the early growth stage of the film). However, compared to the thick-film process, thin-film processing is expensive. Films of a desired pattern can be deposited over the substrate either through a mask or by photolithography followed by either dry or wet etching [17,18]. In the latter case, submicron-level resolution can be achieved. The conducting materials are metals, with high conductivity, and the metals that have a higher adherence to substrate are generally preferred. Aluminum, silver, gold, palladium, and chromium are the primary choice, according to circuit requirements. In some cases, bimetal, even trimetal, systems are being used. The resistor material, typically nickel chromium (nichrome) or tantalum nitride, is deposited initially, followed by the deposition of adhesion and/or barrier metals such as nickel, chromium, titanium, or palladium. The primary conductor material, such as gold, is the last material to be deposited. Historically, thin-film technology is applied to those areas requiring precision line dimensions, tight tolerance, and matched temperature-coefficient resistors and capacitors with wire-bonded or gold-bearing solder interconnections. Perhaps only thin films can produce the small, low-power, high-current-density devices required for high-density hybrid circuits.

6.1 THIN-FILM SUBSTRATES

6.1.1 Substrate Materials

Alumina Substrates Aluminum oxide (99.5%) ceramic substrate with a surface finish typically 1 to 5 μin./in. is generally used for thin-film hybrid circuits [19,20]. These substrates are presently used in microwave circuits, resistor networks, analog-to-digital (A/D) or digital-to-analog (D/A) converters, and various other analog circuits [21]. The commercially available 99.5% alumina substrates for thin-film application exhibit a centerline average (CLA) surface finish of approximately 2 to 4 μin. (500 to 1000 Å), and the peak-to-valley height of the surface profile may have an average value as high as 40 μin. (1 μm). The average grain size of the substrate is 1 μm with occasional grains as large as 0.08 to 0.12 mil (2 to 3 μm). It is demonstrated that an alumina substrate with a polished surface finish of 0.25 μin. (62.5 Å) or better can be manufactured with virtually no surface defects. These alumina substrates are often used in thin-film circuit fabrication because of their high thermal conductivity, good chemical stability, and resistance to thermal and mechanical shocks. To improve surface roughness (<1 μm/in.), alumina substrates are often glazed. Figure 6.2 shows the picture of a typical alumina substrate with laser-drilled holes used in thin-film hybrid circuits.

The common choice for substrate material for thin-film deposition is 99.6% alumina [23]. Substrate dimensions vary from 1 by 1 in. (2.54 by 2.54 cm) to 3

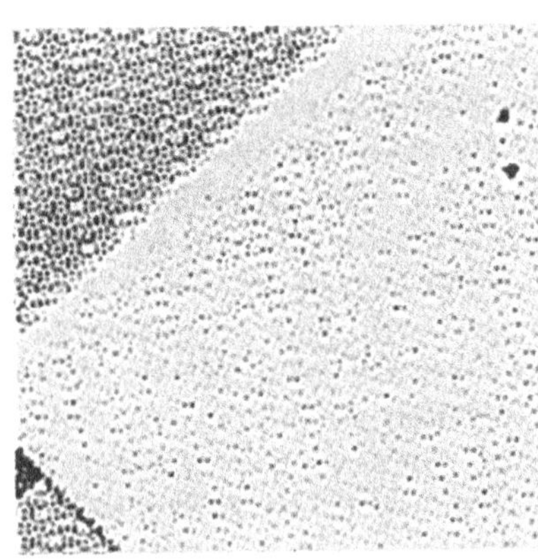

Figure 6.2 Alumina substrate for a thin-film hybrid circuit with laser holes [22]. (Courtesy of Accu-Tech Laser Processing, San Marcos, California.)

by 3 in. (76 by 76 cm), with the thickness varying from 0.010 to 0.025 in. (0.025 to 0.063 cm) in increments of 0.015 in. (0.038 cm). Ideally, the substrates should meet the following criteria: (1) good mechanical strength, (2) atomically smooth surface, (3) inertness to chemicals used in processing, (4) high thermal conductivity, (5) high volume and surface resistivity, (6) coefficient of thermal expansion similar to the nested films, (7) high mechanical strength, (8) high thermal shock resistance, (9) zero porosity, (10) surface waviness less exceed 1 mil (ca. 25.4 μm), and (11) and low manufacturing cost [24].

Commercial thin-film alumina substrates are manufactured from calcined alumina [25]. The calcined powder is ground for a period of 48 to 120 h to produce very fine submicrometer particles. However, the grinding and milling processes are performed so as to leave some coarse particles, which act as nuclei of exaggerated grain growth. The narrow powder size produced by grinding and milling is utilized to produce a green ceramic tape. The green tape is dried, cut to the size required, and sintered to about 1500°C with $\frac{1}{2}$ to 1 h of soaking time [26]. The dense substrates are then characterized for surface finish with Tally Surf and a 1 μm stylus. The dielectric losses of these commercially available substrates are also significantly higher in comparison to the loss of the polished substrates with uniform microstructure. Thin metal films can adhere well to these substrates and resistors screened onto these substrates have shown good aging characteristics.

Single-Crystal Substrates Single-crystal substrates such as silicon, germanium, and sapphire (alumina) are used for deposition of thin films. Hagge [27] has identified 26 companies and universities working with silicon as a substrate for multilayer thin-film hybrid circuit. Sapphire is also used widely in thin-film hybrid circuits. These substrates are generally cut from large boules. Most crystals, such as silicon and germanium, are grown by Czochralski method [28]. Another method of crystal growth is flame fusion (Verneuli technique), which is developed for the preparation of artificial gems such as sapphire. In this method, alumina powder is dropped through an oxyhydrogen flame directed at a refractory pillar. As more powder is added and the pillar is lowered, the sapphire crystal grows. This method has been able to grow boules up to 1.4 in. (3.55 cm) in diameter and 6.0 in. (15.2 cm) in length. Laboratory-grown diamond substrate is also used for thin-film deposition [29]. Diamond [30] has an excellent thermal conductivity, low thermal expansion coefficient, high electrical resistivity, and mechanical strength. However, the cost of laboratory-grown diamond is very high compared to that of alumina and silicon substrates.

6.2 PHYSICAL CHARACTERISTICS

6.2.1 Characteristics of Substrates [31,32]

Surface Texture The thin-film technology is interested primarily in surface features in the micrometer and submicrometer ranges. Drawn glasses and glazed ceramics have smooth surfaces that fulfill the requirements of thin-film hybrid microcircuits. Sintered ceramic materials (particularly, Al_2O_3), however, generally provide unsatisfactory surfaces for thin-film deposition. Polishing does not improve the surface significantly, presumably because there is little cohesion between the grain boundaries. To overcome the difficulty, coating with a thin glassy layer of very fine grained ceramics is performed.

Surface Flatness Flatness of the surface is another important criterion for thin-film applications, especially for photoresist work. It is required to facilitate close contact with photomasks and the surface of the substrate. A microfinish surface is needed for very thin film deposition. Uneven surface of the substrate introduces unwanted stress and results in microcracks and sometimes, in discontinuous film.

Porosity Porosity of thin-film substrate is referred here to the *volume property* of the substrate. The presence of pores can accumulate dust and cleaning solution during cleaning. During vacuum deposition, foreign materials from these pores are released and cause serious problem. The size and distribution of these pores may vary and depend on the conditions of surface preparation.

Volume Resistivity In addition to providing mechanical stability for the film, substrate material for the hybrid circuit should not contribute to electrical conduction through the film. So an ideal substrate should have infinite volume resistivity. Glazed alumina substrates generally provide high-volume resistivity. In general, ceramic oxides (including Al_2O_3) have large bandgaps and are very good insulators at room temperature (300 K). Conduction by mobile ions is usually low except at very high temperatures.

Dielectric Properties In the fabrication of functional thin-film assemblies, substrates with high dielectric constants are frequently desired to achieve high capacitance per unit area [33,34]. Below 10^{10} Hz, vibrational and deformation losses at room temperature are low with ceramics, including alumina. The two important sources of dielectric loss in ceramics are ion migration and space-charge polarization. Glasses and crystalline solids with a cubic lattice structure expand uniformly, whereas alumina has a trigonal structure and different expansion coefficients parallel and perpendicular to the *c*-axis. This behavior, called *anisotropy*, manifests itself in nonisotropic stress.

Thermal Expansion The effect of thermal expansion mismatch between a substrate and the nesting films results in stress on the film. In certain cases, the forces due to the stress may be high enough to exceed the elastic limit of the film, which can help to develop microcracks, even discontinuity, in the film. For normal deposition temperatures (50 to a few hundred degrees Celsius), the stress in the metal film is typically 10^8 to 10^{10} dyn/cm^2 and tensile. The refractory metals occupy the upper region; the soft ones occupy the lower end of the stress spectrum [35–37]. Matching of thermal expansion coefficients between the nesting films, the surface-mounted chips, the bonding, and the substrate is most important because the thin-film substrate carries device chips and deposited films made from different materials. In fact, thin-film hybrid circuits carry silicon transistors, diodes, ICs, and resistor or capacitor chips in addition to thin films of different materials, and differences in thermal expansion between these devices result in considerable shear stress at the interface between the chips, the films, and the substrate.

Thermal Conductivity The temperature of thin-film materials increases in proportion to the electric power applied. At elevated temperatures, thin films become susceptible to recrystallization and oxidation, which may lead to resistance changes or even destruction of the film. Therefore, temperature control is an important design consideration for thin-film circuits. At elevated temperatures substrates might even react with the film. In contrast to glasses, thermal conductivity of ceramics decreases with temperature more widely (from 0.2 to 0.6 cal/cm per second per degree) than do other materials used for thin-film deposition.

6.2.2 Characteristics of Metals

Electrical Characteristics

Conductivity Conductivity, which is the reciprocal of resistivity (ρ), is the most important factor in determining the quality of a metal as a conductor. The conductivity of a conductor

$$\sigma = \frac{1}{\rho} = \frac{ne^2\tau}{m} = ne\mu \tag{6.1}$$

where n is electron density of charge $-e$, μ the electron mobility, m the mass of the electron, and τ the mean free time. From equation (6.1) we can see that the two quantities that determine the conductivity of a metal are the electron density n and the mean free time τ, which is dependent on temperature and the defect structure of the metal.

Temperature Dependency The temperature dependency of τ is connected with the probability of a phonon–electron interaction (i.e., a *collision* or *scattering event*). If the probability is small, a long time will elapse before a collision occurs and the mean free time will be long. The probability of a collision occurring in time δt is $\delta t/\tau$. The probability of a phonon–electron interaction occurring will be proportional to the product of the numbers of electrons and phonons present. Now the number of electrons is constant and independent of temperature, but the number of phonons is dependent on temperature.

$$n_{\text{phonon}} = \frac{1}{[\exp(h\nu/kT)] - 1} \tag{6.2}$$

where k is Boltzmann's constant, T is the temperature in absolute degrees, h is Planck's constant, and ν is the lattice wave frequency. Equation (6.2) can be approximated and we can write [38]

$$n_{\text{electrons}} n_{\text{phonons}} = n_{\text{electrons}} \frac{kT}{h\nu} \tag{6.3}$$

and the probability

$$p = \frac{1}{T} = CT \tag{6.4}$$

where T is the temperature in absolute degrees, and C is a constant. At ordinary temperatures $h\nu$ is very small ($<4 \times 10^{-4}$ eV), and the value of kT at room temperature (300 K) is 0.025 eV.

When measurements are made at low temperatures, the number of phonons becomes very small and electron–phonon collisions become less important than those with crystal defects such as impurities or grain boundaries [39,40]. So at low temperatures conductivity tends to a finite value. This suggests that at ordinary temperatures it might be possible to decrease the conductivity of a metal at the same time that TCR value can be changed by introducing some defects. This is achieved in some disordered solid solutions [41].

Electromigration Electromigration involves the movement of atoms of conducting material as a result of momentum exchange between mobile carriers and the atomic lattice [42,43]. The moving electrons of the conducting material collide with the atoms and push them toward the positively biased electrode. As a result, metal piles up near the electrode and is depleted from other parts of the conductor, especially from the regions near the intersection of grain boundaries in the polycrystalline metal film. This mass transport of the metal occurs when the metal strip making an ohmic contact is stressed to high current densities (higher than 10^5 A/cm^2). Huntington and Grone [43] estimated this atomic flux:

$$F = \frac{ND(Z^* q \rho j)}{kT} \tag{6.5}$$

where N is the density of metal atoms, D is the diffusion coefficient of vacancies, k is Boltzmann's constant, T is the absolute temperature, Z^*q is an effective charge of the metal atoms, and ρ is the resistivity of the metal film. The diffusion coefficient $D = D_0 \exp(-E_a/KT)$, where E_a is the activation energy. Flux F depends, then, through D, ρ, and Z^* on the temperature, composition, and structure of the metal film. As a result, the gradients in the interconnect temperature, composition, or changes in its structure are responsible for void formation [44–46].

Mechanical Characteristics

Adhesion Adhesion is the mechanical strength joining two different objects or materials and is a fundamental requirement of most deposited film systems. The *apparent adhesion* is determined by applying external mechanical stress, and failure may occur in the interfacial region or nearby material. Failure of adhesion (or *deadhesion*) can be due to mechanical failure by fracture or deformation, chemical causes such as corrosion or dissolution, or diffusion processes such as the diffusion of material to or away from the film–substrate interface [35,47]. The wide range (0.1 to 10 eV) of adhesion energy between the film and the substrate surface leads us to postulate two different mechanisms. For low energy, weak van der Waals forces are supposed to bind adsorbed atoms, a physisorbtion of the atoms. The second mechanism that gives rise to higher binding energies between the adsorbed atoms and the substrate surface (above 0.4 eV) results from chemisorption of the atoms. This is due either to formation of an intermediate layer (oxides) or to a higher number of nucleation centers in the early stage of the growth of the film. As a matter of fact, nucleation density of the film atoms deposited is an early indication of good or poor adhesion. A high nucleation density indicates strong chemical interaction of the deposited atoms with the substrate surface and results in good adhesion. There are many film adhesion tests. Most of them, except the contact angle test, are destructive methods [16]. The destructive tests are the tape test, stud-pull test, wire bond and ball bond pull test, shear or topple test, scratch test and scratch pattern, indentation test, and indentation pattern [46] test.

Stress The stress in the films deposited develops during the initial growth of the film when the thickness of the film is very thin. However, there are certain depositions and preparation conditions such that films of zero stress may be prepared. This stress arises generally due to the differences in the thermal coefficient of expansion between the film and the substrate. However, this stress is seen to subside as the film grows. High-modulus materials such as chromium, tungsten, and compound materials can add to this initial thermal stress. If the magnitude of these stresses is high, it might result in blistering (if it is compressive) or microcracking (if it is tensile). If the compressive stresses are isotropic, the blistering will be in the form of *worm tracks*, whereas isotropic tensile stresses ended up in the form of *dried mudflats* [47]. It is, however, observed that the compressive stresses are minimized when deposition takes place under high vacuum (higher than 10^{-6} torr).

6.3 THIN-FILM CONDUCTORS

6.3.1 Conductor Materials and Properties

Silver Silver shows the highest conductivity. The conductivity requirements for thin-film hybrid circuits are met adequately with silver [48]. Besides conductivity, another basic requirement for a usable conducting film is that it should adhere well to the substrate when subject to a variety of changes in nature and level of such environmental factors as temperature, humidity, electrical stress, and the chemical nature of the ambient. Adherence of silver to alumina or glass is better than gold, but it forms sulfide tarnish film quite readily. It reacts with oxygen and forms silver oxides. Silver also shows metal migration when subjected to high dc current. To minimize the silver migration, an alloy of silver and palladium is sometimes recommended for metallization. It also has poor solder leach resistance.

Gold Gold has a number of desirable properties and has gained widespread use. From a practical point of view, it has not appeared possible to get direct adherence of gold to almost any substrate, and most efforts have been aimed at using gold with some intermediate layer or layers to improve adherence [16,49,50]. These include chromium, nichrome, titanium, tungsten, and molybdenum [51,52]. The principal disadvantage of titanium–gold metal composite is based on its diffusion effects, which can be minimized successfully by introducing a sandwich layer of platinum between the titanium and gold [53]. Another disadvantage of gold is its price fluctuation in the world market.

Copper Copper is the second-highest conductive material in the row of materials used for interconnecting material in the thin-film hybrid circuit. The melting point of copper is higher than that of silver and aluminum. Adhesion of the film on alumina substrate is good, and it remains adherent when subjected to a variety of changes in nature. Copper has been used in almost all alloy materials, and copper interconnects offer the promise of 30% faster devices with fewer metal levels and lower-cost production. A worldwide effort has been made to develop copper damascene processes for making smaller, faster, and more sophisticated circuits [54]. Chemical mechanical planarization (CMP) is used to polish away the excess copper, leaving only the interconnect lines. The dependence of the CMP process on feature size has a strong relation to the pad viscoelastic (time-dependent) deformation model [55]. Now this transition from tungstan ILD CMP to copper CMP is not simply a change in the metal but also a change in the process technology involved. So the success of this interconnection technology does not rest on copper metal alone but also on several other things (e.g., step height, copper dishing, polishing, dielectric erosion, and device geometry), including successful operation of the copper surface topology. The barrier metal layer has a negligible effect on the measured sheet resistivity of wide lines, but it can be a significant portion of the total cross-

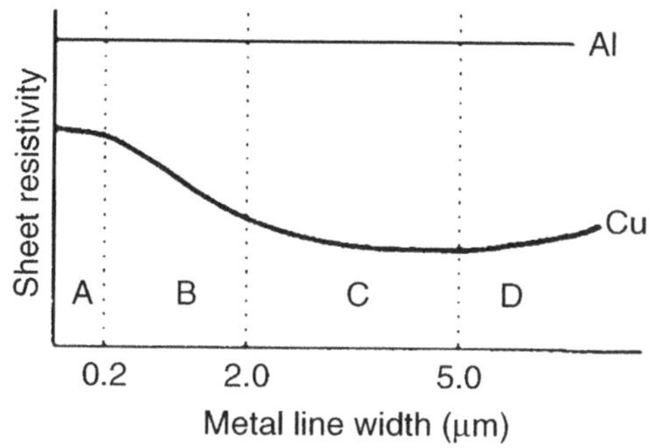

Figure 6.3 Variation of sheet resistivity with line width [56]. (Courtesy of Keithley Industries, Cleveland, Ohio; reprinted from the March 2001 edition of *Solid State Technology*. Copyright 2001 by PennWell Publishing.)

sectional area on narrow lines. Thus the measured sheet resistivity will show an apparent increase with smaller line widths, as shown in Figure 6.3 (section C compared to section D). Section D of Figure 6.3 shows an increase in sheet resistivity with a larger line width. This is due to the CMP dishing effect [56,57].

Aluminum Aluminum is by far the most commonly used metal in thin-film circuits today [58]. It fulfills almost all the requirements for thin-film conductor. Adhesion of the metal to alumina is good. It can handle current densities of more than 6.25×10^5 A/in. (10^5 A/cm^2) at about 150°C. Aluminum films used for interconnections adhere well to insulating as well as semiconductor surfaces. The melting temperature of aluminum is low (659.7°C) and can be deposited by a relatively low temperature process. The metal is highly conductive and the resistivity in thin-film form (ca. 3.0 μΩ·cm) is not very different from its bulk resistivity (2.65 μΩ·cm) [59]. Deposited aluminum film can easily be patterned with high resolution. Aluminum bonds well with gold and is used for ultrasonic bonding. However, a gold–aluminum binary system is generally avoided at the lead–chip interface. Aluminum metal on silicon or silicide, however, leads to metallurgical interactions leading to serious instabilities [60] and a diffusion barrier between two layers (either Al–Si or Al–silicides) is recommended. For Al–Si, Pt, Pd, Co, Ni, and Ti and for refractory silicides of Mo, Ta, Ti, and W, TiN, Ti–W, and TiC are recommended [60]. Aged at 350°C in air for 100 h, metal aluminum shows an 11% increase in sheet resistance [61]. Properties of some conducting materials used in thin-film hybrid circuits are given in Table 6.1.

Table 6.2 shows that the effective dielectric constant is related to changes in technology node. A hybrid scheme with mixed aluminum and copper metallization will be adopted in the near future. Figure 6.4 shows a typical dielectric etching process for subtractive aluminum, and Figure 6.5 shows a via–first copper dual damascene process [63,64]. The damascene process consists of (1) etching a trench 4000 to 5000 Å deep with an aspect ratio approaching 6 : 1, (2)

TABLE 6.1 Properties of Commonly Used Conducting Materials

Material	Resistivity (μΩ·cm)	TCR (ppm) per °C	Melting Point (°C)	Tensile Strength lb/in^2	Tensile Strength MPa	Coefficient of Linear Expansion (ppm/K)	Density (g/mL)
Aluminum	2.65	0.00429	660	35,000	241.15	24	2.45
Copper	1.67	0.0068	1083	37,000	254.93	17	8.96
Silver	1.59	0.0041	960.8	42,000	289.38	18.9	10.5
Gold	2.35	0.004	1063	20,000	137.8	14.2	19.3
Chromium	12.9	0.003	1900	—	—	—	7.1
Indium	5.3	0.00392	156.4	—	—	—	7.3
Nickel	6.85	0.0069	1453	70,000	482.3	14.0	8.9
Palladium	10.8	0.00377	1552	—	—	—	12.0
Platinum	10.6	0.00393	1769	50,000	344.5	8.9	21.45
Tantalum	12.45	0.00383	3000	—	—	—	16.6
Tungsten	5.65	0.0051	3380	6,000,000	41,340	4.0	19.3
Titanium	42	—	1680	—	—	—	4.0
Molybdenum	5.0	0.00435	2620	—	—	5.0	10.2
Manganese	185	—	1250	—	—	—	7.4

Source: Handbook of Chemistry and Physics, 55th ed., CRC Press, Cleveland, OH, 1974–75.

TABLE 6.2 Interconnect Requirements for Logic Devices

	Subtractive Aluminum		Hybrid or Copper Damascene			
Year of first product shipment	1997	1999	2002	2005	2008	2011
Technology node (nm)	250	180	130	100	70	50
Number of metal levels	6	6–7	7–8	8–9	9	9–10
Effective dielectric constant	4	4–3.5	3.5–2.7	2.2–1.6	1.5	<1.5

Source: From [62].

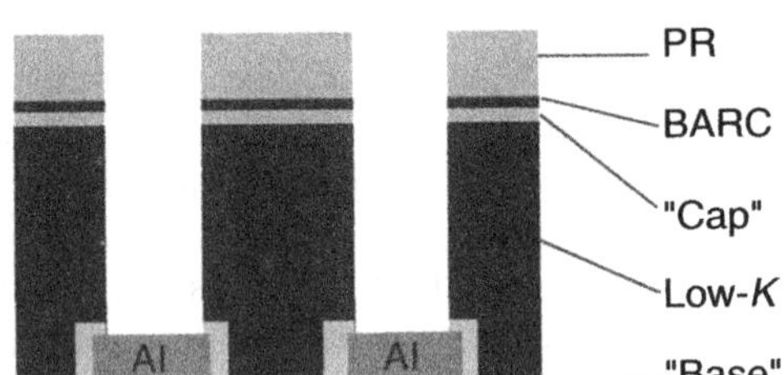

Figure 6.4 Dielectric etch process for subtractive aluminum. PR, photoresist; BARC, barrier antireflective coating [64]. (Courtesy of Trikon Technologies, Newport, Wales, UK.)

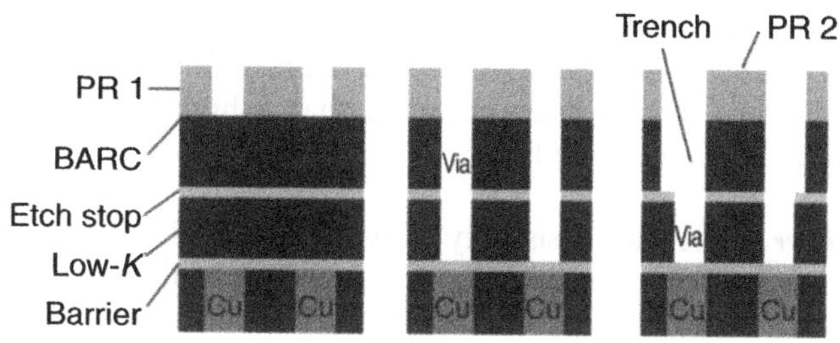

Figure 6.5 Dielectric process for via–first copper damascene [64]. (Courtesy of Trikon Technologies, Newport, Wales, UK.)

electrochemical plating of copper, and (3) polishing off the copper overfill with chemical mechanical polishing. A tungsten plug connects silicon to metal using a single damascene process. Dual-damascene processes developed at Motorola and IBM take advantage of the high conductivity of copper interconnects. Indeed, the change in metallization from tungsten–aluminum to copper has reduced line resistance. Table 6.3 illustrates the cost, adhesion, and environmental stability, chemical resistance, and electromigration of the common metals used as ohmic contacts in thin film.

Titanium Titanium is a very good conductor which shows sheet resistances from 1 to 10^9 Ω/sq for films between 4000 and 400 Å thick. The TCR values for these films range from −1000 to 1000 ppm/°C [65]. A TCR value near zero is observed for a film with a sheet resistance value of about 50 Ω/sq. Load life

TABLE 6.3 Thin-Film Conductor Material

Type	Material Cost	Manufacturing Cost	Environmental Stability	Adhesion to Alumina	Chemical Resistance Stability	Electromigration
Gold	Very high	High	Excellent	Bad	Good	Low
Silver	Moderate	Moderate	Moderate	Moderate	Moderate	Moderate
Aluminum	Low	Moderate	High oxidation rate, poor stability at elevated temperatures	Moderate	Poor	High
Copper	Low	Moderate	High oxidation rate in the presence of moisture, poor oxidation stability	Moderate	Poor	Lowest

tests on 100-Ω/sq film shows a stability of 0.5%, and a change in sheet resistance is noticed after 1000 h at 65°C [66]. Titanium films can be deposited by vacuum evaporation. It has been observed that a high deposition rate minimizes the contamination within films.

Palladium–Silver The bulk resistivity of palladium is almost sevenfold that of silver. Despite the low conductivity value of palladium, it is sometimes mixed with silver to minimize metal migration of silver with a trade-off between conductivity and metal migration [67,68]. Adherence of palladium silver to alumina is good. In the conductor the ratio of palladium to silver is often maintained in the ratio 60:40 and shows a resistivity value between 42 and 44 μΩ·cm [69].

Titanium–Palladium–Gold–Titanium Sometimes more than two metal systems are used, to have an ohmic contact from the terminals. One of the metals is generally chosen from the good conductive materials. Unfortunately, most good conductors have one or more of the properties of poor adherence to the substrate, metal migration, and poor bondability. To compensate, other metals are combined with the first metal to promote adhesion and bondability and to minimize metal migration [70]. So the presence of a second and third metal are necessary to compensate for the other properties of a contact pad. But unfortunately, there is always a trade-off between the conductivity, adhesion, and metal migration and bondability. The complex Ti–Pd–Au–Ti metal system provides excellent bondability, migration resisting characteristic, and adhesion strength at the cost of the conduction property of a single gold metal system [71]. Table 6.4 shows the behavior of two-, three-, and four-conductor metallurgy.

6.4 THIN-FILM RESISTORS

Thin films have been used as resistive materials for at least 50 years. The cracked carbon resistor, which was the first thin-film resistor, has been in production for over 40 years. When used for the fabrication of discrete resistors, thin films offer improved performance and reliability then those of composite resistors and lower cost for comparable performance. Most film resistor requirements can be met with films having sheet resistance 10 to 1000 Ω/sq. Resistors below 10 Ω are rarely used, whereas resistors with values of several megohms can be realized. Besides sheet resistance, thin-film resistors should possess a low TCR value and should be sufficiently stable during their operating life. The electrical properties of a thin-film resistor differ from those of the bulk material from which it is deposited. The change is due to structural differences and thickness, especially when the film thickness is comparable with the mean free path of the electron. The properties of the film are also influenced greatly by film deposition conditions, substrate temperature, and the environ-

TABLE 6.4 Behavior of Two-, Three-, and Four-Conductor Metallurgy

Composite System	Adhesion	Aging Effects	Thickness	Current Noise
Ti–Au	Starts peeling after 10 h at 150 to 250°C	Aging at 150°C poor	100 Å Ti, 1000 Å Au	Degradation of current index with aging
NiCr–Au	Better than the Ti–Au system	Better than Ti–Au; after 300 h, degrades rapidly	500 Å Ni–Cr, 10,000 Å Au	−20 dB; does not degrade with aging
NiCr–Pd–Au	No peeling at 150°C for 1000 h; at 250°C starts to peel after 600 h	Better than NiCr–Au; aging sheet resistance is worse than NiCr–Au	500 Å NiCr, 10,000 Å Au, 3000 Å Pd	−20 dB at 150°C; at 250°C, noise index increases
NiCr–Cu–Pd–Au	Peeling after only 5 h at 250°C	As good as NiCr–Pd–Au	500 Å NiCr, 3000 Å Cu, 3000 Å Pd, 10,000 Å Au	−20 dB at 150°C; at 250°C, noise index increases
Ti–Pd–Au	Peeling after 1000 h at either 150 or 250°C	As good as NiCr–Pd–Au	750 Å Ti, 3000 Å Pd, 10,000 Å Au	−20 dB at 150°C; Pd diffuses into gold

mental conditions during deposition. The resistivity of a thin film is a function of several parameters: temperature [72], purity [73] of the material deposited (e.g., density of the film deposited is always found to be different from the bulk material), thickness, crystalline structure [74,75], and defect structure [76] of the film.

6.4.1 Resistor Properties

Contact Resistance Contact resistance can be defined as the resistance offered to the path of the current flow between two solid materials. The contact resistance can be formed between two conductors, between a semiconductor and a conductor, or between a conductor and an insulator. Contact resistance is inversely proportional to the area of contact and can be expressed mathematically as

$$R_c = \frac{\rho_c}{A} \tag{6.6}$$

where ρ_c is the specific contact resistivity and A is the area of contact.

The primary cause for higher contact resistance might be due to the non-uniform surface contact between the total areas of two adjoining solids. The other reason might be due to the potential differences between the two contact solids (which is called the *contact potential*) [77]. In the case of metals when two metals of different work functions are brought into contact, electrons will flow from the higher-energy states to a lower-energy state. As a result, excess of negative charge accumulation will set up an electric field with a polarity, which will hinder further flow of electrons.

When contact is made between a semiconductor and a conductor with a lower workfunction, electrons will flow from the metal to the semiconductor until equilibrium is established. But if there are enough empty surface states, the electrons will first occupy those without much affecting the height of the potential barrier. So the potential barrier will stay and will form a rectifying junction. In the case of an insulator (sufficiently thick, so that no tunneling of electrons is possible) and metal, there will be no flow of electrons, and there will be no flow of current through the insulator.

The contact resistance of a thin-film resistor is an important factor, especially on lower-valued components. The contact resistance is made up of two parts: the resistance of the thin film underneath the contact through which current must flow, and the interface resistance resulting from the alloy mixture or compound formed between the resistor surface and the metal pad contact. Making the contact larger can minimize both of them. But for a high-density circuit that is not a good choice. Other things that might happen at the interface are the formation of metal oxide and high contact resistance.

Conductor materials serve as ohmic contacts to resistors and as an ohmic path to the outer world. The contribution of these conductor materials as interfacial materials for resistors has been modeled by Naguib [78]. The model assumes the interface to be characterized by a uniform interfacial conductance per unit area measured in S/mm^2. At elevated temperatures some interface materials degrade and sometimes lose completely their adherence to substrates. To prevent adherence and preserve the conductance value, many different composites [79] have been formulated for termination materials for thin-film resistors. They are composed of at least two layers, the top layer usually gold for thermocompression bonding, and as the bottom layer either titanium, chromium, or nichrome for adequate adhesion.

Unfortunately, the difficulties with the metals mentioned above and their alloys are (1) variation in contact resistance with time, (2) contact noise [80], (3) tunnel effect [81], (4) coherent action, (5) constriction resistance [65], (6) contact resistance [82], (7) arc discharge, and (8) adherence [35]. The contact resistance of a pair of contacts in air increases with time due to their affinity for ever-increasing oxide film. These metals and their alloys also sometimes form sulfide and chloride and tarnish with time. Contact noise is allied to the problem of contact resistance. This phenomenon is due to external vibrations setting up variable contact pressure and, in turn, variable contact resistance. Variable contact resistance may reach values of several thousand ohms. Contact noise is lowest with low-resistivity contact materials, such as rhodium, gold, silver, and platinum and their alloys.

When two clean metallic films are not in good contact with each other, *constriction* resistance is pronounced and the resistor value can be expressed mathematically as [65]

$$R = \frac{\rho}{2na} \tag{6.7}$$

where ρ is the resistivity of contact material, n the number of discrete contact areas, and a the mean radius of the area.

Stability One of the most important electrical characteristics of resistive films is their long-term stability. This is generally measured as a percentage change in resistance when the resistor is kept at an elevated temperature, usually 100 to 150°C under electrical load at full-rated voltage for 1000 to 2000 h. The electrical load may be continuous or cycled: for example, $1\frac{1}{2}$ h on and $\frac{1}{2}$ h off. During this operation, resistor values might change due to several factors: (1) mismatch in linear expansion between the film and the substrate (2) migration of alkali ions from an alkali-containing substrate, (3) chemical reactions between film and the surrounding atmosphere, and (4) structural change in the film-drying operation. The stability of a thin-film resistor might be affected by the amount of scattering of conduction electrons at the film surface. The resistor material might contain impurities or imperfections, which may lead to instability of the resistor as a whole. The presence of porosity in the film will also contribute to instability of the film resistor. Some metallic films, such as chromium and nichrome, form stratified films, which might cause instability in the resistor film. The stability of resistors is also measured under shelf-life conditions, in which case the resistance change is measured after several years, during which resistors are kept under normal room conditions without load. Sputtered film in the presence of nitrogen as dopant has shown long-term stability with variable sheet resistance [83,84].

Reliability Reliability is the probability that an item will perform a required function under stated conditions for a stated period of time [85,86]. The classic definition of reliability is the probability of a product performing without failure a specified function under given conditions for a specified period of time [87]. The reliability figure of merit is presented in Table 6.5.

Inherent in the establishment of reliability requirements is the need to estimate or predict reliability in advance of a manufacturing product. This prediction is a continuing process which takes place at several stages of the progression, from design through use. Table 6.6 shows the stages of reliability analysis and prediction, and Table 6.7 shows the unit breakdown time period with a total part failure rate.

TABLE 6.5 Reliability Figure of Merit

Figure of Merit	Meaning
Mean time between failures	Mean time between successive failure of a repairable product
Failure rate	Number of failures per unit time
Mean time to failure	Mean time to failure of a nonrepairable product or mean time to first failure of a repairable product
Mean life	Mean value of life (life may be related to major overhaul, wear-out, etc.)
Mean time to first failure	Mean time to first failure of a repairable product
Mean time between maintenance	Mean time between a specified type of maintenance action
Longevity	Wear-out time for a product
Availability	Operating time expressed as a percentage of operating and repair time
System effectiveness	Extent to which a product achieves the requirements of the user
Probability of success	Same as reliability (but often used for one-shot or non-time-oriented products)
B10	Life during which 10% of the population would have failed
B50	Median life, or life during which 50% of the population would have failed
Repairs/100	Number of repairs per 100 operating hours

Source: From [86]; reproduced with permission of The McGraw-Hill Companies.

Mean time between failures (MTBF), the mean (average) time between successive failures of a production, is represented as

$$\text{MTBF} = \frac{1}{\text{failure rate}} = \frac{1}{\sum n\lambda} = \frac{1}{0.00004} = 25{,}000 \text{ h} \tag{6.8}$$

This definition assumes that the product in question can be repaired and placed back in operation after each failure. These conditions are not always met. If the failure rate is constant, the probability that a product will operate without failure for a time equal to or greater than its MTBF is only 37%. MTBF is not the same as operating life, service life, or replacement time. An increase in an MTBF does not result in a proportional increase in reliability (the probability of survival). When the failure rate is constant, the probability of survival (reliability) can be written as

$$\text{PS} = Re^{-t/\mu} = e^{-t\lambda} \tag{6.9}$$

where R is the probability of failure-free operation for a time period equal to or

TABLE 6.6 Stages of Reliability Analysis and Predictions

	Step 1: Start of Design	Step 2: During Detailed Design	Step 3: At Final Design	Step 4: From System Tests	From Customer Use
Basis	Prediction based on approximate part counts and part failure rates from previous product use, little knowledge of stress levels, redundancy, etc.	Prediction based on quantities and types of parts, redundancies, stress levels, etc.	Prediction based on types and quantities of parts failure rates for expected stress levels, redundancies, external environments, special maintenance practices, special effects of system complexity, cycling effects, etc.	Measurement based on the results of the tests of the complete system; approximate reliability indexes are calculated from the number of failures and operating time	Same as step 4 except calculations are based on customer use data
Primary uses	(a) Evaluate feasibility of meeting a proposed numerical requirement (b) Help in establishing a reliability goal for design	(a) Evaluate overall reliability (b) Define problem areas	(a) Evaluate reliability (b) Define problem areas	(a) Evaluate overall reliability (b) Define problem areas	(a) Measure achieved reliability (b) Define problem areas (c) Obtain data for future designs

Source: From [88]; reproduced with permission of The McGraw-Hill Companies.

TABLE 6.7 Unit Breakdown

Circuit Component	Number Used, n	Failure Rate per Part λ (%/1000 h)	Total Part Failure Rate $n\lambda$ (%/1000 h)
Thin-film resistor	100	0.04	4.0

greater than t, $e = 2.718$, t is a specified period of failure-free operation, μ is the mean time between failures, and λ is the failure rate (the reciprocal of μ) [89].

6.4.2 Resistor Materials

Nickel–Chromium Nickel–chromium [90] has so far proved to be one of the best metal alloys for resistor material [91]. The most popular method of deposition of Ni–Cr films is vacuum deposition [92]. The major problem associated with Ni–Cr film relates to the considerable differences in vapor pressure between Ni and Cr and the high reactivity of Ni with many reducible materials. Because of the high vapor pressure of chromium, the stoichiometric compositions of Ni–Cr vary with deposition time. Another problem in the deposition of Ni films is the partial oxidation of chromium films during deposition [93]. In addition, since films normally receive subsequent stabilization treatment, changes in resistance due to oxidation will be a function of amount of chromium at the surface of the film. The properties of Ni–Cr films depend strongly on the film composition, which can vary over wide limits depending on the deposition method employed. The TCR value of Ni–Cr can go from negative to zero to positive values depending on the percentage of chromium present in the matrix of nickel and chromium. On the other hand, the explanation suggested for increasingly positive TCR of 80%–20% Ni–Cr composition at a high ohms per square value is that the films have an islandlike structure of nickel grains in a matrix of chromium oxide. TCR values of evaporated films varied from +350 ppm/°C for films around 3 Ω/sq to −300 ppm/°C for films around 3000 Ω/sq, whereas sputtered films had TCR values of 150 ppm/°C in the range 5 to 1200 Ω/sq [94]. The problem of composition control of Ni–Cr films can be solved either by a flash evaporation technique or by sputtering rather than vacuum evaporation. The films deposited either on glass or alumina are very stable, and the sheet resistivities can be varied from 100 to 300 Ω/sq with a thickness of about 0.008 mil (200 Å). According to some researchers, Ni–Cr resistor is better than tantalum nitride resistors except in moist atmosphere [95], and they do not show any posttrim drift [96].

Tantalum Tantalum belongs to a group of metals known as *valve metals* which form tough, self-protective oxides, either through heat treatment in oxygen or through anodic oxidation. β-Tantalum (as opposed to the normal α-form, a body-centered cubic (bcc) structure) tends to be obtained when sput-

TABLE 6.8 Comparison of Forms of Tantalum Films

Property	Bulk Material	Thin-Film α Structure (Sputtered at 4–5 kV)	Thin-Film β Structure (Sputtered at 4–5 kV)	Thin-Film Low Density (Sputtered at 1.5 kV)
Crystal structure	bcc	bcc	Tetragonal	Tetragonal or bcc
Lattice constant	$A = 3.303$	$A = 3.31–3.33$	$A = 5.34$, $c = 9.94$	—
Density (g/cm^3)	16.6	15.6	15.9	12.1
TCR (ppm/°C)	+3800	+500 to +1800	−100 to +100	−100 to +100
Superconductive transition	4.4	3.3	0.5	—
Resistivity (μΩ·cm)	13.48	2.0×10^2	5.0×10^2	2.6×10^4

Source: From [97]; reproduced with permission of The McGraw-Hill Companies.

tering is performed at high pressure and at relatively low substrate temperatures (less than 100°C). If there is any contamination of the argon sputtering ambient, the β form will not be obtained [97]. Interest in tantalum is on its oxide for good, reliable insulating material in a Ta capacitor. Because of the refractory nature of tantalum, the preferred method of deposition is sputtering, accomplished in an inert atmosphere. The chamber is evacuated to 10^{-6} torr and argon is admitted through a control valve to give a pressure of 0.1 torr. The tantalum source is a small-diameter disk of high-purity (99.999%) tantalum. A voltage of few kilovolts is applied for sputter deposition, and the exact value of the voltage depends on the gap between Ta target and the substrate. The technology of tantalum resistors is complicated by the fact that tantalum films can exit in at least three forms [97], and a comparison of the properties of tantalum film is given in Table 6.8. It has been demonstrated that in bcc tantalum, the erosion rate of (111) incident planes is expected to be far greater than that for (001) incident planes [98]. The phase changes in tantalum depend on the way the heat energy is distributed among atoms. If we consider the energy diagrams following Condon–Morse curves for atoms in the two structures (α and β) of tantalum film, the frequency (ν) of the vibrational energy ($h\nu$) of the tantalum lattice is higher in the bcc (α) structure than those of tetragonal (β) tantalum. The higher-frequency phase is more stable at low temperatures, and its lowest energy level must lie below that of the other phase [99].

Low-density Ta films are unstable and heat treatment of 1 to 2 h at 200°C is necessary. In polycrystalline tantalum, the extent to which individual grains convert strain energy into stored cold work depends on the orientation of the crystallites. It has been observed that (111) orientations effectively generate new dislocation line length and will recrystallize readily, whereas (001) crystals do not sufficiently store cold work and will polygomize after exposure to a high annealing temperature [100]. Tantalum films are stabilized by oxidation of the surface in air at a temperature usually between 250 and 300°C. Oxygen has a profound effect on both the resistivity and the TCR values of tantalum films. The presence of a small amount of oxygen in the films can cause the TCR value to increase very rapidly to a negative value on the order of 400 ppm/°C or more, and the presence of nitrogen during deposition can form tantalum nitride and can change the resistivity and TCR value of the film. Tantalum film thickness ranges from 0.0008 to 0.02 mil (5000 to 200 Å) and resistance ranges from 10 Ω to 1 MΩ can be sputtered very easily. Sputtering Ta in an atmosphere of argon can also prepare high-resistivity films. However, these films show negative temperature coefficients of about −150 ppm/°C.

Tantalum Nitride Tantalum nitride films are finding use as thin-film resistors in hybrid circuits. Sputtering technology is used most commonly and films are deposited in an atmosphere of dry nitrogen. Sputtered Ta_xN films seem to be the most popular thin-film resistor material [101,102]. Argon is then leaked in during sputtering to maintain a pressure at 0.1 torr. The nitridation process increases the resistivity of the film from 50 Ω·cm to 250 Ω·cm. Resistors are prepared on glazed alumina substrates from 5 to 20 kΩ. Meander-type resistors is prepared by using regular photolithography. Ta_2N films deposited by magnetron RF reactive sputtering shows excellent long-term stability [103].

Chromium A great deal of work has been carried out on vacuum-deposited chromium films both as possible resistive films and as suitable bonding material for chromium–gold conductors [94,104]. The TCR values of chromium films vary from large negative values to large positive values as film thickness increases. Work on Cr films deposited by sublimation shows a wide range of TCR values. In an ion-beam system, chromium is vaporized from an electrically heated boat and the vapor is ionized by an incident electron beam. These resistors are claimed to be more stable than vacuum-deposited films. Studies of chromium films indicate that they consist of relatively pure islands of metal in a matrix of insulating chromium oxide [105].

Silicon–Chromium Another good resistor material is sichrome (silicon–chromium) film. Sputtered deposited silicon chromium thin-film resistors can attain high resitor values with a near-zero TCR by adjusting the target composition, reactive gas pressure, deposition rate, and annealing temperature [106, 107]. This film is also being used in Bi-CMOS technology. Sputter conditions and the resistor values of sputtered Si–Cr films are presented in Table 6.9.

Cermet Cermet films are composed of a metal and an oxide film such as silicon monoxide. The metal is generally chromium. Several other mixtures of metals and dielectrics have been tried [108], but the chromium–silicon monoxide mix has been found most satisfactory. Flash evaporation is the best method for depositing cermet films, which keeps the composition reasonably uniform. The TCR is negative when the percentage of chromium is low. With a higher

TABLE 6.9 Sputter Conditions and Resistor Values of SiCr Films

Series Name	Cathode Power (W)	Deposition Temperature (°C)	Sheet Resistance (Ω/sq)	TCR Range (ppm/°C)
High-power cold deposition	600	Ambient	1200–1250	−61 to −66
High-power hot deposition	600	200	1200–1250	−49 to −51
Low-power cold deposition	300	Ambient	1120–1200	−40 to −46
Low-power hot deposition	300	200	1260–1300	−31 to −37

Source: From [107].

percentage of chromium (more than 30%), the TCR value of the cermet film is positive. For more stable film, annealing at 400°C is recommended. The film resistance is seen to increase when heated in air. The maximum temperature to which Cr–SiO films can be subjected for any extended period of time appears to be about 500°C. Above this temperature, sheet resistance becomes unpredictable [109].

Carbon Carbon film or cracked carbon resistors have been in use for 40 years, but because of high processing temperatures and difficulties in sheet resistance control, it is not popular for use in thick-film hybrid circuits. The film is formed by cracking or pyrolysis of hydrocarbon, usually methane, at the hot surface of the substrate. The substrate is a cracked carbon rod, and the carbon film is deposited on it at a temperature of 1000°C [110]. Carbon films can also be deposited by a chemical vapor deposition (CVD) method, including hot filament CVD [111] and plasma (including RF plasma CVD and photo-CVD) [112–114] techniques. In the CVD process, carbon is deposited by passing a hydrocarbon gas through the CVD chamber. The TCR values of these resistors are negative. An extensive review of carbon products has been given by Kohl [115]. The resistivity of carbon films is about 10^{-3} Ω·cm [74]. However, these resistors are not used in hybrid circuits.

6.5 THIN-FILM CAPACITORS

6.5.1 Capacitor Properties

Thin insulating films, usually oxides of metals [116], are used as a sandwich material between two conducting plates, forming capacitors. Films are usually amorphous, since polycrystalline films are generally found to be less insulating and single-crystal insulating films are difficult to prepare and are not cost-effective. The important property of such a device is its ability to store electric charge (q) and consequently, electrical energy. The dielectric properties of these insulating films have been studied for a long time [117,118]. The increasing quantity and complexity of electronic systems, due to the miniaturization of the devices and the increased density of the circuits, has stimulated the development of better-quality materials for the fabrication of capacitors. The two most common dielectric materials to have shown promise as thin-film capacitors are evaporated SiO_2 and anodized Ta_2O_5 [119]. Some organic polymer materials have shown very attractive dielectric properties; but they undergo elastic deformation under stress. Another disadvantage of the polymer dielectrics is the lack of precise control of the film materials. Thin-film capacitors usually have a parallel-plate configuration. A layer of suitable dielectric separates the two conducting layers. The magnitude of the capacitance is a function of overlapping electrode area (area $= LW$, where L is the length and W is the width).

Another important phenomenon of dielectric material is the dielectric strength of a capacitor, which is a measure of the quality of the dielectric material. It is defined as the maximum electric field that the dielectric can withstand without breakdown, under given conditions. It is usually measured in V/mm and can be expressed as [120]

$$\text{dielectric strength} = D_s = \frac{U_{\text{br}}}{t} \tag{6.10}$$

where D_s (V/mm) is the dielectric breakdown strength of the material, U_{br} the breakdown voltage of the capacitor, and t the dielectric thickness (mm) (Figure 6.6).

Knowledge of dielectric strength is very important for the design of thin-film capacitors because it determines the voltage at which destructive breakdown of the dielectric will occur. As a rule of thumb, the working voltage is taken as half of the breakdown voltage. Both electrode and lead resistance can also contribute substantially to loss in the planar thin-film capacitor structure. Since

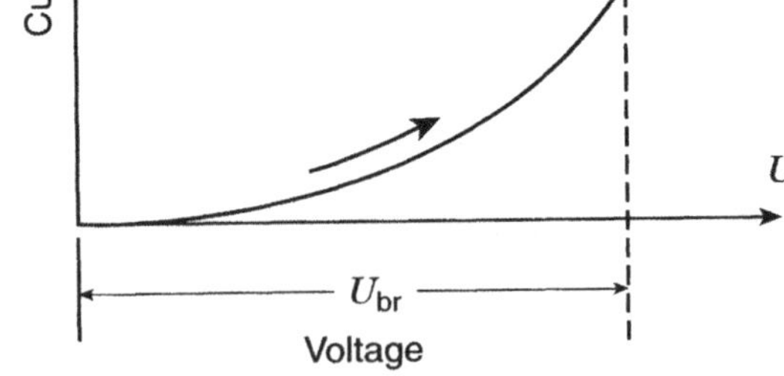

Figure 6.6 Current through the dielectric versus voltage [120].

the electrode resistance and leak resistance is in series with the dielectric, it is convenient to represent the capacitor with a series equivalent circuit [31]. Thus the total dissipation factor of the thin film capacitor is expressed as [119]

$$\tan\delta = \tan\delta' + \omega C R_s \tag{6.11}$$

where $\tan\delta$ is the measured dissipation factor, $\tan\delta'$ the frequency-independent contribution from the dielectric, ω the angular frequency, C the electrode capacitance, and R_s the series resistance. Equation (6.11) can be written as

$$\tan\delta = \tan\delta' + \frac{2\omega C\rho L}{3Wt} \tag{6.12}$$

where $R_s = 2\rho L/3Wt = \frac{2}{3}(L/W)(\rho/t)$; ρ is the resistivity of the electrode, ρ/t is the sheet resistivity of the material (electrodes), and L/W is the aspect ratio.

It is obvious from equation (6.12) that ρ, d, and the aspect ratio strongly affect the high-frequency performance of a thin-film capacitor, and by using thicker and highly conductive materials for capacitor electrodes, the high-frequency performance of the device can be improved. Aluminum has shown much promise for use in conducting electrodes for the capacitor.

6.5.2 Capacitor Materials

Silicon Monoxide Evaporated silicon monoxide is the most widely used dielectric material for thin-film capacitors [117]. The most attractive quality of SiO is its high vapor pressure. It evaporates at a much lower temperature than do either silicon or silicon dioxide. The stability of silicon monoxide as a dielectric, which is closely related to the mechanical stresses, has a profound influence on the film [121]. Silicon monoxide has a tendency to form silicon dioxide when exposed to an ambient of oxygen. The resistivity of the film varies from 10^{11} to 10^{16} Ω·cm. SiO films have high leakage currents and high dissipation factors (between 0.05 and 0.1). The temperature coefficient of capacitance (TCC) of SiO film is about 1000 ppm/°C. To achieve lower values of dissipation factor, leakage currents, and TCC, SiO films should be heated in air at temperatures between 250 and 450°C for 5 to 6 h. The dielectric constant (K) value of SiO is approximately 6.0, and the dielectric strength of the material is 2 MV/cm.

Silicon Dioxide Silicon dioxide film is gaining in importance as insulating material for capacitors. Besides its extensive use as a dielectric material in capacitors, it is also used in IC fabrication. SiO_2 films are deposited mostly by sputtering in a pure oxygen atmosphere [122]. The dc resistivity varies from 10^{14} to 10^{16} Ω·cm. The dielectric constant (K) and dielectric strength are 3.8 to

4.1 (depending on the procedure of film formation) and 10^7 V/cm, respectively. The dissipation factor at 1 kHz is about 0.0003.

Silicon Nitride Silicon nitride films are amorphous and are finding interest as dielectric materials in thin-film capacitors. The dielectric constant of bulk Si_3N_4 is around 9.4. Silicon nitride is generally deposited by low-pressure CVD and the deposition temperature is maintained between 700 and 800°C [123]. To avoid thermal stresses, low-temperature (300 to 400°C) plasma-enhanced CVD is preferred. Plasma-enhanced CVD silicon nitride, however, tends to be non-stoichiometric and contains substantial quantities of atomic hydrogen (10 to 30%). The dielectric constant for Si_3N_4 varies from 6 to 9 with a dielectric strength of 6×10^6 to 1×10^7 V/cm. Thermal stability for high-temperature-grown Si_3N_4 is excellent [124]. Si_3N_4 has a dc resistivity value of about 10^{14} Ω·cm [118,125].

Tantalum Oxide Oxide coatings over metal can be grown by forming metal ions and oxygen ions, which combine either at the metal surface, inside the oxide, or at the oxide–oxygen interface, depending on which ion migrates through the oxide. Among the metal oxide films, tantalum oxide (Ta_2O_5) is the most thoroughly studied dielectric used as a thin-film capacitive dielectric material [126]. A research group from the University of Texas has claimed that a novel H_2–O_2 pyrogenic postanneal of Ta_2O_5 films improve film quality [127]. The capacitor area is defined either by using a proper mask or by selectively etching using standard photolithography technique. The dissipation factor of tantalum film capacitors is between 0.002 and 0.01 in the frequency range 0.1 to 10 kHz. The temperature TCC between 0 and 100°C is about 200 ppm/°C. Capacitors based on tantalum oxide formed by plasma oxidation or by reactive sputtering have not been evaluated as extensively as units using anodic oxide. However, films produced by aqueous anodization are superior to those formed by reactive evaporation and reactive sputtering. The dielectric constant for reactively sputtered film is about 16. The resistivity of the film is 10^4 Ω·cm [128].

Barium Titanate $BaTiO_3$ is a ferroelectric material and a nonlinear dielectric. The essential features of this class of dielecrics are (1) hysterisis under the action of alternating current, (2) very high values of permittivity, (3) pronounced dependence of dielectric parameters on temperatures, and (4) the presence of spontaneous polarization without an external field [129]. The dielectric material for ceramic capacitors sometimes contains barium titanate as the dielectric material [130]. The electrodes are usually made from precious metals such as platinum, gold, or palladium–gold. The dielectric thickness varies from 1 to 5 mils (25 to 125 μm), relatively thicker than that of tantalum capacitor film dielectric. The $BaTiO_3$ powder particle size is generally between 2 and 5 μm. A dielectric sheet of 1 to 5 mils (25 to 125 μm) is prepared from a mixture of ceramic powder with an organic binder. The dielectric constant of this material is 420 at room temperature and the dielectric strength is 0.35 MV/cm. The resistivity of the material is about 10^{10} Ω·cm [131].

Titanium Dioxide Titanium dioxide occurs in three crystalline polymorphs: rutile (tetragonal), anatase (tetragonal), and brookite (orthorhombic). Rutile is the most stable [132]. The electronic transport properties of transition metal oxides are a topic of long interest [133,134]. Titanium dioxide has recently gotten some attention as a dielectric material [135]. A research group from Yale University reported work with TiO_2 for high-K dielectric applications [127] using a jet-vapor deposition technique which combines titanium vapor generated by 1 torr dc sputtering in argon and atomic oxygen generated from a microwave discharge sustained by a fast flow of oxygen through a quartz nozzle. A tetragonal anatase phase [136] and dc resistivity for reactive triode sputtered TiO_2 has been reported by Tang and co-workers to be about 5×10^2 Ω·cm at room temperature and 6×10^6 Ω·cm at 100°C [137]. The dielectric constant of titanium dioxide can vary from 14 to 110, and the dielectric strength of the material is 0.04 to 0.084 MV/cm [118].

Zirconium Oxide Zirconium oxide (ZrO_2) has become a popular material in the electronic industry because of its high refractive index, high melting temperature, hardness, and corrosion-resistant properties. The difficulty with ZrO_2 is that neither of its high-temperature phases (tetragonal and cubic) is quenchable. As a result, thin-film coatings of ZrO_2 cracks upon cooling. This effect is minimized by alloying ZrO_2 with either CeO_2, Y_2O_3, or Al_2O_3 [138,139]. This material is being used as a dielectric in thin-film capacitors [140].

The high-K dielectric behavior of ZrO_2 films doped with 10% aluminum has gained some attention. The material is prepared by co-sputtering. Sharp Laboratories of America has reported preparing this material from a target of aluminum and zirconium with a 1:5 power ratio in oxygen and argon at room temperature. They have reported that the leakage current is <0.64 A/in^2 (0.1 A/cm^2) for 0.00012-mil (3-nm)-thick Zr–Al_2O_3 film [127]. The resistivity of ZrO_2 films sputtered in RF-excited, rare gas discharges (Kr, Ar, and NeO) containing 4% O_2 exceeds 10^{10} μΩ·cm [141]. The dielectric constant value for the reactively sputtered film is about 20, and the dielectric strength is 4 MV/cm.

6.6 THIN-FILM INDUCTORS

Fabrication of a microminiature inductor has presented a formidable problem [142]. The difficulties associated with presently available chip inductors are their sizes. Hybrid and microwave design engineers tend whenever possible to eliminate this device from their circuits. Thus inductors remain the stepchildren in hybrid microcircuits. However, thin-film technology has made it possible to

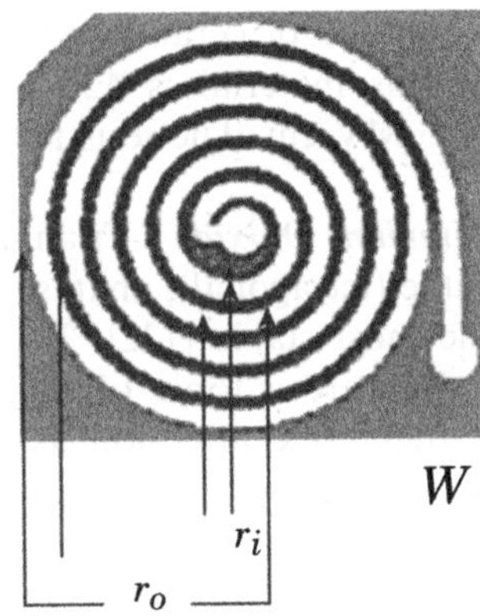

Figure 6.7 Deposited thin-film spiral inductor. The test pattern is formed by photolithography.

develop low-value inductors at low processing cost. Inductors presently in use are generally planar. Inductors with larger values occupy a larger surface area. The important factors relating to inductors are that it should be small and should be able to show all the electrical properties of a low-cost chip inductor [143–145]. Figure 6.7 shows a thin-film spiral inductor where the film is deposited by vacuum evaporation and fine-line structures are fabricated following photolithography.

The inductance L of a circular spiral is given by [146]

$$L = \frac{0.8a^2N^2}{6a + 10C} \quad \text{nH} \tag{6.13}$$

where $C = r_o/4 - r_i/4 =$ breadth of the coil and $a = r_o/8 + r_i/8$, and the quality factor Q is given by

$$Q = \frac{WL_s\{[1 - (f/f_r)^2]\}}{R_s} \tag{6.14}$$

where L_s is the low-frequency inductance, f is the frequency of interest, and $f_r = 1/(2\pi\sqrt{L_sC_s})$, where C_s is the distributed capacitance. Inductors are generally fabricated by depositing a good conductor in a closely spaced spiral configuration. A multiple coil with fine lines >1 mil (25 μm) and gap widths down to 0.8 mil (20 μm) ± 1 is achieved by taking advantage of advanced photolithography, which can produce a submicrometer-level structure [146]. Film thickness is maintained at about 0.8 mil (20 μm) to avoid the skin effect, although a film thickness of less than 0.4 mil (10 μm) can be deposited very easily. The spiral thin-film inductor has been used successfully in a 225 to 440-MHz tuned amplifier with 200 W output power and 6 dB gain [147–149].

6.7 TECHNOLOGIES OF THE TWENTY-FIRST CENTURY

As feature size is decreasing (from 0.18 to 0.10 μm), circuit density is increasing. As a result, capacitive, inductive, and resistive couplings between neighboring circuits are growing [150,151]. At the same time, the wire thickness of interconnecting terminals is not reduced as much as the line width, which accounts for the increased coupling due to capacitive reactance. The *IR* drops across the power distribution lines increase significantly as the current density increases and the feature size decreases [152]. A deep submicrometer-level feature size brings the power lines and associated routing closer, and crosstalk is expected as a result of mutual capacitance and inductance between neighboring routing. The longer and closer the lines, the higher the crosstalk will be.

As the feature size decreases, the circuit element density and I/O count increase. The self-inductance of the power supply distributed lines causes a large amount of chip current change within a very short period of time, and this (dI/dt) introduces noise within the chip. This noise can affect the circuit in three ways [153]: (1) it may increase chip-to-chip delay, (2) it may affect the overall operation of the receiving chip, and (3) it may affect the gate of the sending chip. Fluctuations in the power line due to inductive coupling can be modeled after Faraday as

$$\text{self-inductace emf} = -L\frac{dI}{dt}$$

where L is the self-inductance and I is the current.

A number of methods can be used to minimize dI/dt noise, such as a decoupling capacitor, multiple power and ground pins, and to tailor driver turn-on characteristics [154]. In addition to transient noise, resonance effects at the chip and board levels must be taken into account. Module pin inductance and chip capacitance and resistance may be used for a resonant circuit, and if the resonance frequency is near the resonant frequency or one of its harmonics, a large voltage fluctuation will build up. Thus the circuit impedance should be kept to a minimum. To deal with the resistance offered by a particular metal used for global and local interconnections, copper is by all odds the best choice (considering resistivity; thermal conductivity; electromigraton; adhesion to alumina, silicon, and silicon dioxide; and cost). To further improve the resistance × capacitance (RC) delay in interconnects, a copper low-K system is thought to be the answer for present deep-submicrometer-level devices.

The difficulty with copper is that dry etching is almost impossible, and it diffuses very fast through silicon and its oxides. The rate of diffusion is a function of temperature and ionization energy, and it is believed that donor atoms are fast diffusers. The interstitial impurity ions of copper are believed to diffuse through tunnels in the open structure of elemental semiconductors [155]. Copper can diffuse about 600 μm in 1 minute at 900°C in silicon [156]. Electrical activity due to copper diffusion can degrade device operations by causing larger leakage currents and lower breakdown voltage. Leakage currents are the source of noise in solid-state imagers and dynamic random access memories (DRAMs) [157]. In GaAs, copper is a deep triple acceptor. It has a high solid solubility (6×10^{18} cm^{-1}) and is very effective at reducing the diffusion length of *n*-type GaAs. Most transitional elements (Co, Cu, Au, Ni, Pt, and Ag) diffuse by an interstitial–substitutional mechanism and end up in both types of sites [158].

For barrier layer material, besides those properties, the structural orientation of the material is also an important factor (because diffusion through the grain boundaries or lattice defect centers is minimum when the barrier film is amorphous). Single-crystal barrier layers are impractical, because of easy diffusion paths through grain boundaries, lattice, and defect centers, so amorphous barrier layers will be ideal. Amorphous layers are metastable in nature, and during annealing at high temperatures, these amorphous layers transform into crystalline or polycrystalline layers. As a result, grain boundaries are formed and some defect centers are likely to form. Thus the crystalline structure of the film is more vulnerable than the amorphous counterpart.

The main principle of a barrier layer should be that the layer keep its amorphous state until a high temperature is reached (which is not detrimental to the device). Instead of nitrides, silicides of a metal (transition) might be a better choice from the standpoint of interaction with silicon. When the barrier material is saturated with silicon, it may be possible to minimize the outdiffusion of silicon atoms through the grain boundaries and formation of copper silicide (higher resistivity than that of copper alone).

When copper is deposited over silicon or its oxide, silicon may diffuse inside and can form copper silicides (CuSi). As a result, the resistivity of the copper film will increase, which will increase the *RC* (time constant) value, which is detrimental for speed and electrical and heat conduction. To prevent this diffusion, a barrier layer is deposited in between the copper and the silicon or its oxide. This diffusion barrier should keep the chemical, mechanical, metallurgical, and morphological integrity at high temperatures. The chemical integrity means that it should not degrade chemically and should maintain its resistivity and thickness during processing. Mechanical and metallurgical integrity means that the material should have controllable polish properties (during chemical mechanical polishing time), should adhere to the contacting films, and should have enough mechanical strength to ensure film integrity in subsequent processes all the way through packaging of the final product. Morphological integrity means that it should stay in an amorphous state at high temperatures.

Tantalum and its nitrides (Ta_2N and TaN) are attracting more attention because the materials are very stable and do not form Cu–Ta or Cu–N alloy [159–161]. XRD analysis of as-deposited sputtered tantalum (Ta) film shows the growth of β-tantalum along the ⟨002⟩ direction with an additional shoulder along ⟨202⟩. The barrier layer remains unchanged (no outdiffusion of silicon) unto 600°C. After annealing at 650°C and upward, forms $TaSi_2$ and growth is seen along (111). However, at 800°C there is secondary growth along (200) and (112) as well as (111). Some observers have also reported the formation of Ta_5Si_3, an amorphous layer at the Ta–Si interface. Annealing above 700°C shows $TaSi_2$ crystallites (annealing and its effect on the films are presented in Chapter 5). Sputtered tantalum nitride film 300 nm thick shows a cubic structure and the growth along (200) is strong, with secondary growth along (111). The structure remains unaltered below 800°C, which shows better integrity over Ta film on silicon [162]. As-deposited sputtered Ta_2N film is amorphous in nature (XRD studies show a broad peak with a full width at half maximum of 5°). The TEM diffuse diffraction pattern indicates that the as-deposited Ta_2N film is dominated by an amorphous phase. The plain-view TEM dark-field mosaic spread within the matrix indicates that roughly 3-mm scattered Ta_2N crystallites are embedded in amorphous film. Annealing for 5 minutes at 600°C, XRD pattern confirmed that the mosaic structure was transformed into a crystalline structure with sharp rings which resembles a hexagonal structure. We know that crystalline-structured films help in diffusion. So we can expect silicon diffusion above this temperature. However, it has been found that a sputter-deposited second layer of TaN, which is initially an amorphous film, can prevent diffusion below 700°C. Nitrides of other transition materials are being tried and are competing with each other.

To address the dry etching problem, a new, dual damascene method has been developed. The single greatest advantage of this damascene technology is that neither a wet nor dry etching process is needed for selective etching of the metal layers, which are notorious for process problems: corrosion, resist burn, time critical with resist, formation of foreign materials, and galvanic effect with copper [163]. This damascene process involves copper as the main interconnecting metal for global and local metallization. Planarization is done with chemical mechanical polishing, and all patterning is done with dielectric etching (dielectric materials may be SiO_2, nitrides, oxides of metals, or low-K porous materials).

Dielectric films are an integral part of semiconductor devices and are integrated into circuits for several reasons. Their properties are critical in determining the electrical characteristics of the components and selection of the appropriate techniques for forming, for deposition (spin-on/CVD/PVD, etc.) and patterning them are key parameters that processing engineers deal with from time to time. As the circuits and devices have evolved and become more sophisticated with increased circuit density and submicrometer-level device size, processes and techniques for forming dielectric films have become more refined and controlled. The dielectric films can be extended from inorganic (oxides, nitrides silicides, and borides of silicon or transition metals) to organic chemical compounds (Kapton, Teflon, MSQ, polyimides, etc.). These dielectric materials are used for insulation, fabrication of capacitors and barrier layers, or may be an integral part of a device (e.g., a metal oxide semiconductor). Thus material properties are to be controlled so that the device yield does not deteriorate because of their presence. In addition to electrical properties, these dielectric materials should have good adherence properties should be chemically inert, and should be free of pinholes after processing.

At low frequencies (<1013 Hz) polarizibility arises from the ionic and polar components of the material, while at high frequencies, electronic polarizibility is important. Thus dielectric materials having low polarizibities are better in designing low-*K* materials. It has been observed that materials having single C–C and C–F bonds have the lowest electronic polarizibility, making fluorinated and nonfluorinated aliphatic macromolecules potential candidates for low-*K* applications [164,165]. Conversely, materials having double and triple bonds show higher electronic polarizibility because of the increased mobility of π-electrons. In general, materials with low polarizibility having a low dielectric constant show poor adhesion properties. In contrast, aromatic π-bonding configurations have higher bonding strength. Thus, SiLK (a low-*K* dielectric from Dow Chemical) is a cross-linked polyaromatic, a matrix that largely eliminates the nuclear polarizibility component and features high bond strength and thus excellent thermal resistance [166,167].

Materials containing hydroxyl and carbonyl groups are polar functional groups which can attract water via hydrogen-bonding interactions and thus drastically elevate the dielectric constant ($K = 78.5$, the dielectric constant of water). Thus in low-*K* material, polar functionality–containing elements such as oxygen or nitrogen should be avoided. Hydrogen silesquioxanes (HSQ, $HSiO_{1.5}$) condense at elevated temperatures and form a ladder network. During annealing care must be taken to protect hydrogen atoms from moisture as well as oxygen, which for the silonol groups could raise the *K* value. Methyl silsesquioxanes (MSQ, $RSiO_{1.5}$, where R stands for CH_3) and similar materials can minimize this problem by the substitution of lable H atoms by hydrophobic organic R groups. However, when portions of the organic groups are integrated into a mostly inorganic resin matrix, they can segregate at the surface to lower their free energy, possibly leading to adhesion challenges. Thus low-*K* materials have not yet been fully developed.

There is an another technology, carbon nanotubes (CNTs), which is expected to be an alternative to conventional silicon-based microelectronics when the feature size will reaches about 10 nm [168–173]. A CNT is a rolled cylindrical structure with a honeycomb lattice. The length of a CNT can be more than 1 μm and the diameter is about 1 nm (10^{-9} m). According to its diameter and chirality, the electronic structure of a CNT can be semiconducting or metallic. The energy gap of a CNT is dependent on the diameter and can vary from 1 to 0 eV. Laser ablation, arc discharge, and CVD are the most common methods used to synthesize the material. A single-wall CNT can be regarded as a quantum wire in which electrons can move without being scattered by scattering centers. Conductance fluctuations due to the magnetic field are observed in CNTs [174].

REFERENCES

1. Z. M. Meiksin, Thin- and thick-film distributed networks, in *Thin and Thick Films for Hybrid Microelectronics*, Lexington Books/D.C. Heath, Lexington, MA, 1976, Chap. 11.
2. C. E. Jowett, *The Engineering of Microelectronic Thin and Thick Films*, Macmillan, London, 1976.
3. G. M. Messner, Cost density analysis of interconnections, *IEEE Trans. Comput. Hybrid Manuf. Technol.*, Vol. 10, No. 2, 1987, pp. 143–151.
4. C. P. Wong, Materials for electronic packaging, in *Electronic Packaging and Interconnection Handbook*, 3rd ed., C. A. Harper, ed., McGraw-Hill, New York, 2000.
5. C. A. Neugebauer and R. O. Carlson, Comparison of wafer-level integration with VLSI approaches, *IEEE Trans. Comput. Hybrid Manuf. Technol.*, Vol. 10, No. 2, 1987, pp. 184–189.
6. R. Tummala, E. Ramaszewski, and A. Koperstein, *Handbook on Electronic Packaging*, Chapman & Hall, London, 1997.
7. F. W. Kear, Thick- and thin-film circuits, in *Hybrid Assemblies and MCM*, Marcel Dekker, New York, 1992, Chap. 4.
8. J. J. Licari, Thin and thick films, in *Electronic Materials and Processes Handbook*, C. A. Harper and R. N. Sampson, eds., McGraw-Hill, New York, 1994, Chap. 8.
9. D. M. Pozar, Microwave ICs, in *Microwave Engineering*, Addison-Wesley, Reading MA, 1990, Sec. 11.6, p. 646.
10. P. C. L. Yip, *High-Frequency Circuit Design and Measurements*, Chapman & Hall, London, 1990.
11. High-frequency characteristics of thin films, in *Handbook of Thin-Film Technology*, 2nd ed., L. I. Maissel and R. Glang, eds., McGraw-Hill, New York, 1983, pp. 13–23.
12. S. Wolf and R. N. Tauber, *Silicon Processing for the VLSI Era*, Vol. I, Lattice Press, Sunset Beach, CA, 1986, Chaps. 5, 6, 10, and 16.
13. Leon I. Maissel, Electrical properties of metallic thin film (influence of heat treatment), in *Handbook of Thin-Film Technology*, 2nd ed., L. I. Maissel and R. Glang, eds., McGraw-Hill, New York, 1983, pp. 13–26.
14. B. Heinz, Sputter targets and thin-film defects, *Vacuum Thin Film*, Oct. 1999.
15. D. S. Campbell, Mechanical properties of thin films, in *Handbook of Thin-Film Technology*, 2nd ed., L. I. Maissel and R. Glang, eds., McGraw-Hill, New York, 1983, Chap. 12.
16. T. K. Gupta, Comparative studies of the contact angle as a measure of adherence for photoresist, *Thin Solid Films*, Vol. 75, 1981, pp. 319–329.
17. R. Glang and I. V. Gregor, Generation of patterns in thin films, in *Handbook of Thin-Film Technology*, 2nd ed., L. I. Maissel and R. Glang, eds., McGraw-Hill, New York, 1983, Chap. 7.
18. S. P. Murarka and M. C. Peckerar, Pattern transfer and metallization, in *Electronic Materials Science and Technology*, Academic Press, San Diego, CA, 1989, Chaps. 6 and 10.

19. L. L. Hench and J. K. West, Alumina, in *Principles of Electronic Ceramics*, Wiley, New York, 1990, Chap. 6.
20. R. C. Buchanan, *Ceramic Materials for Electronic Processing: Properties and Applications*, 2nd ed., Marcel Dekker, New York, 1991.
21. P. Horowitz and W. Hill, *The Art of Electronics*, 2nd ed., Cambridge University Press, Cambridge, 1998, Chaps. 7 and 13.
22. *Catalog on Laser Processing*, Accu-Tech, San Marcos, CA.
23. B. V. Hiremath, Ceramic thin and thick films, in *Ceramic Trans.*, Vol. 11, American Ceramic Society, Washington, DC, 1980.
24. R. Brown, Thin-film substrate, in *Handbook of Thin-Film Technology*, 2nd ed., L. I. Maissel and R. Glang, eds., McGraw-Hill, New York, 1983, Chap. 6.
25. H. Stenson et al., U.S. patent 3,698,923.
26. J. E. Hove and W. C. Riley, eds. *Ceramics for Advanced Technologies*, Wiley, New York, 1965.
27. J. K. Hagge, Ultra-reliable packaging for silicon on silicon WSI, *Proc. 38th Electron. Components Conf.*, Los Angeles, May 1988, pp. 282–292.
28. A. A. Chernov, ed., *Growth of Crystals* (English translation), Vol. 12, Consultants Bureau, New York, 1984.
29. J. H. Butler et al., Thin-film growth mechanism, *Philos. Trans. R. Soc. London*, Vol. A342, 1993, pp. 209–222.
30. J. J. Licari, Diamond, in *Electronic Materials and Processes Handbook*, C. A. Harper and R. N. Sampson, eds., McGraw-Hill, New York, 1994, Sec. 8.2.6, pp. 8–17.
31. R. C. Sundahl and E. J. Sadora, Effects of alumina substrate surface defects on thin-film interconnect patterns, *Proc. IEEE*, Vol. 59, No. 10, Oct. 1971, pp. 1462–1467; J. J. Licari, Thin and thick films, in *Electronic Materials and Processes Handbook*, C. A. Harper and R. N. Sampson, eds., McGraw-Hill, New York, 1994, Chap. 8.
32. R. D. Jones, *Hybrid Circuit Design and Manufacture*, Marcel Dekker, New York, 1981.
33. A. R. Von Hippel, *Dielectric Materials and Applications*, MIT Press, Cambridge, MA, 1991.
34. J. C. Anderson, *Dielectrics*, Chapman & Hall, London, 1964.
35. D. S. Campbell, Mechanical properties of thin films, in *Handbook of Thin-Film Technology*, 2nd ed., L. I. Maissel and R. Glang, eds., McGraw-Hill, New York, 1983, Chap. 12; D. A. Chance and D. L. Wilcox, Metal ceramic constraints for multilayer electronic packages, *Proc. IEEE*, Vol. 59, No. 10, Oct. 1971, pp. 1455–1462.
36. G. Di Giacomo, *Reliability of Electronic Packaging and Semiconductor Devices*, McGraw-Hill, New York, 1997.
37. K. Azar, ed., *Thermal Management in Electronic Cooling*, CRC Press, Boca Raton, FL, 1997.
38. J. C. Anderson, K. D. Leaver, R. D. Rawlings, and J. M. Alexander, eds., *Materials Science*, 4th ed., Chapman & Hall, London, 1990.
39. I. H. Khan, The growth and structure of single-crystal films, in *Handbook of Thin-*

Film Technology, 2nd ed., L. I. Maissel and R. Glang, eds., McGraw-Hill, New York, 1983, Chap. 10.

40. N. F. Mott, *Proc. Cambridge Philos. Soc.*, Vol. 32, 1938, p. 281, and Vol. 35.
41. N. F. Mott and E. A. Davis, *Electronic Processes in Non-crystalline Materials*, 2nd ed., Clarendon Press, London, 1979.
42. R. S. Muller and T. I. Kamins, eds., *Device Electronics for Integrated Circuits*, 2nd ed., Wiley, New York, 1986.
43. H. B. Huntington and A. R. Grone, Current-induced marker motion in gold wires, *J. Phys. Chem. Solids*, Vol. 20, 1961, p. 76.
44. F. M. Heurle and P. S. Ho, Electromigration, in *Thin Films: Interdiffusion and Reactions*, J. M. Poate, K. N. Tu, and J. W. Mayer, eds., Wiley, New York, 1978.
45. I. A. Blech and E. S. Meieran, *Appl. Phys. Lett.*, Vol. 11, 1967, p. 263.
46. S. P. Murarka and M. C. Peckerar, Metallization, in *Electronic Materials Science and Technology*, Academic Press, San Diego, CA, Chap. 6, 1989.
47. J. E. McDoland and J. G. Eberhart, Adhesion in an aluminum oxide metal system, *Trans. Met. Soc. AIME*, Vol. 233, Mar. 1965, pp. 512–576; D. M. Mattox, Surface and film characterization, *Vac. Technol. Coat.*, Jan. 2001.
48. G. W. A. Dummer, *Materials for Conductive and Resistive Functions*, Hayden Book Co., New York, 1970, pp. 170 and 243.
49. J. A. King, J. Freer, and R. Woodard, *Materials Handbook for Hybrid Microelectronics*, Artech House, Norwood, MA, 1988.
50. A. E. Riad and F. D. Barlow III, Thin-film materials processing, in *Hybrid Microelectronics Handbook*, J. E. Sergent and C. A. Harper, eds., McGraw-Hill, 1995.
51. S. H. Narrow and H. Stipple, A theoretical and experimental study of chemical etching methods of TiW–Au system for microwave hybrid applications, *Proc. Int. Symp. Hybrid Microelectron.*, 1985, p. 268; D. M. Mattox, Influence of oxygen on the adherence of gold films to oxide substrates, *J. Appl. Phys.*, Vol. 37, No. 9, Aug. 1966, pp. 3613–3615.
52. H. Yoenhara, Y. Fukuda, and T. Shirakawa, Development of superminiature and ultrahigh-density packed resistor module, *Proc. Int. Symp. Hybrid Microelectron.*, 1987, p. 147; P. Benjamin and C. Weaver, The adhesion of evaporated metal films on glass, *Proc. R. Soc. London*, Vol. A261, 1961, pp. 512–516.
53. M. P. Lepselter, *Bell Syst. Tech. J.*, Vol. 45, 1966, p. 233.
54. X. W. Lin and D. Pramanik, Future interconnect technologies and copper metallization, *Solid State Technol.*, Vol. 41, No. 10, 1998, pp. 63–79.
55. J. M. Steigerwald, S. P. Murarka, and R. J. Gutmann, *Chemical Mechanical Planarization of Microelectronic Materials*, Wiley, New York, 1997, pp. 68–72.
56. T. Turner, Practical issues in the probing of copper pads, *Solid State Technol.*, Mar. 2001.
57. S. P. Murarka, I. V. Verner, and R. J. Gitmann, *Copper: Fundamental Mechanisms for Microelectronic Applications*, Wiley, New York, 2000.
58. T. K. Gupta, Aluminum alloy as an interconnecting material for integrated circuit fabrication, *Microelectron. Reliab.*, Vol. 19, 1979, p. 337.

59. S. K. Ghandhi, *VLSI Fabrication Principles: Silicon and GaAs*, Wiley, New York, 1983, p. 447.
60. S. P. Murarka and M. C. Peckerar, eds., *Electronic Materials Science and Technology*, Academic Press, San Diego, CA, 1989, p. 320.
61. G. L. Schnable, R. K. Keen, and L. R. Loewenstein, Tech. Rep. RADC-TR-67-311, Rome Air Development Center, Rome, NY, Sept. 1967.
62. *2000 ITRS Road Map*, Solid State Technol., Mar. 2001.
63. S. P. Murarka, I. V. Verner, and R. J. Gitmann, Future direction of copper technology, in *Copper: Fundamental Mechanisms for Microelectronic Applications*, Wiley, New York, 2000, Chap. 10, p. 313.
64. D. J. Thomas, Y. P. Song, and K. Powel, Meeting *ITRS Road Map* challenges with low-*K* dielectric etching, *Solid State Technol.*, Mar. 2001.
65. G. W. A. Dummer, ed., *Materials for Conductive and Resistive Functions*, Hayden Book Co., New York, 1970, pp. 159 and 297–302.
66. F. Huber, Thin films of titanium and titanium oxide for microminiaturization, *Proc. Electron. Comput. Conf.*, Washington, DC, 1964, p. 38.
67. A. Christou, *Electromigration and Electronic Device Degradation*, Wiley, New York, 1993.
68. R. V. Penny, Current-induced mass transport in aluminum, *J. Phys. Chem. Solids*, Vol. 25, 1964, pp. 335–345; S. P. Murarka and M. C. Peckerar, Metallization, in *Electronic Materials Science and Technology*, Academic Press, San Diego, CA, 1989.
69. H. Casey, W. Mulligan, and J. Wood, Resistive metal glaze and its product applications, *Proc. Electronic Comp. Conf.*, Washington, DC, 1963, p. 20.
70. D. E. Harrison and J. D. Harrison, in *Electronic Materials and Processes Handbook*, C. A. Harper and R. N. Sampson, eds., McGraw-Hill, New York, 1994, p. 5.26.
71. L. F. Miller, Ternary alloy, in *Thick-Film Technology and Chip Joining*, Gordon & Breach, New York, 1972, Chap. 4; K. E. Haq, K. H. Behrndt, and I. Kobin, Adhesion mechanism of gold underlayer film combination to oxide substrates, *J. Vac. Sci. Technol.*, Vol. 6, 1969, p. 148.
72. K. L. Chopra, *Thin-Film Phenomena*, McGraw-Hill, New York, 1969.
73. R. W. Wilson and L. E. Terry, IC metallization system: a comparison in electronic packaging and production, Oct. 1970, Sec. 2, pmc 29; N. F. Mott, *Proc. Cambridge Philos. Soc.*, Vol. 32, 1938, p. 281.
74. L. I. Maissel and R. G. Glang, eds., *Handbook of Thin-Film Technology*, 2nd ed., McGraw-Hill, New York, 1983.
75. C. Neugebauer, J. B. Newkirk, and D. A. Vermilyea, eds., *Structure and Properties of Thin Films*, Wiley, New York, 1959; N. F. Mott and E. A. Davis, *Electronic Processes in Non-crystalline Materials*, Clarendon Press, Oxford, 1979.
76. A. F. Mayadas and M. Schatkes, *Phys. Rev.*, Vol. B1, 1970, p. 1382.
77. J. W. Mayer and S. S. Lau, Energy bands and wave behavior, in *Electronic Materials Science for Integrated Circuits in Si and GaAs*, Macmillan, New York, 1990, Chap. 12, p. 365.
78. H. M. Naguib, *Proc. Int. Symp.*, 1977, p. 53; J. S. Fisher and P. M. Hall, Termination materials for thin-film resistors, *Proc. IEEE*, Vol. 59, No. 10, Oct. 1971, pp. 1418–1424.

79. R. T. Galla and W. C. Vergara, *Microelectron. Reliab.*, Vol. 7, Aug. 1968, p. 185.

80. D. A. Bell, *Noise and the Solid State*, Wiley, New York, 1985, p. 47.

81. R. H. Fowler and L. W. Nordheim, *Proc. R. Soc. London*, Vol. A119, 1928, p. 173; J. G. Simmons, Electronic conduction through thin insulating films, in *Handbook of Thin-Film Technology*, 2nd ed., L. I. Maissel and R. Glang, eds., McGraw-Hill, New York, 1983.

82. A. M. Goldstein and N. B. Grover, *Semiconductor Surfaces*, North-Holland, Amsterdam, and Wiley, New York, 1965.

83. M. Ohring, Yield, reliability and failure time, in *Reliability and Failure of Electronic Materials*, Academic Press, San Diego, CA, 1998, Chap. 1.

84. L. Maissel, Thin-film resistors, in *Handbook of Thin-Film Technology*, 2nd ed., L. I. Maissel and R. Glang, eds., McGraw-Hill, New York, 1983, p. 12-5; V. E. Kane, Process capability indices, *J. Qual. Technol.*, Vol. 18, Jan. 1986, pp. 44–51.

85. M. Pecht, *Product Reliability, Maintainability, and Supportability Handbook*, CRC Press, Boca Raton, FL, 1995; M. E. Goldberg and J. Vaccard, eds., *Physics of Failure Electronics*, Spartan Books, New York, 1965.

86. J. M. Juran and F. M. Gryna, eds., *Juran's Quality Control Handbook*, 4th ed., McGraw-Hill, New York 1988.

87. Office of the Assistant Secretary of Defense, Reliability of military electronic equipment by the advisory group of reliability of electronic equipment, R&D, June 1957.

88. J. M. Juran and F. M. Gryna, *Quality Planning and Analysis*, 2nd ed., McGraw-Hill, New York, 1980, p. 179.

89. E. J. Dudewicz, Basic statistical methods, in *Juran's Quality Control Handbook*, 4th ed., J. M. Juran and F. M. Gryna, eds., McGraw-Hill, New York, 1988; A. K. Sinha, *Proc. IEEE Reliab. Phys. Symp.*, New York, 1980, p. 165.

90. H. E. Boyer and T. L. Gall, eds., *Metals Handbook*, American Society for Metals, Metals Park, OH, 1985.

91. R. H. Alderson and F. Ashworth, *Br. J. Appl. Phys.*, Vol. 8, 1957, p. 205.

92. L. Holland, *Vacuum Deposition of Thin Films*, Chapman & Hall, New York, 1961.

93. T. K. Lakhmann, Properties of the nickel chromium films, *Trans. 8th Natl. Vac. Symp.*, Vol. 2, Pergamon Press, New York, 1961, p. 868.

94. L. I. Maissel, Thin-film resistors, in *Handbook of Thin-Film Technology*, 2nd ed., L. I. Maissel and R. Glang, eds., McGraw-Hill, New York, 1983, Chap. 18, p. 18-11; J. W. Swanson and D. S. Campbell, Structural and electrical properties of 80:20 NiCr thin films, *Thin Solid Films*, Vol. 1, 1967, pp. 183–202.

95. J. F. Krasucki, *Int. J. Hybrid Microelectron.*, Vol. 5, No. 2, 1982, p. 183; W. A. Pliskin, *The Effect of Moisture on RF Sputtered and Fused Glass Films*, American Ceramic Society, Washington, DC, May 1966.

96. J. D. Madden, *Hybrid Circuit Technol.*, June 1985, p. 27; M. H. Monnier, *IEEE Trans. Parts, Maker, Package*, June 1966, p. 44.

97. L. Maissel, Application of sputtering to the deposition of film, in *Handbook of Thin-Film Technology*, M. I. Maissel and R. Glang, eds., McGraw-Hill, New York, 1970.

98. H. E. Roosendaal, Sputtering by particle bombardment, I, in *Physical Sputtering of Single-Element Solids*, R. Behrisch, ed., Springer-Verlag, Berlin, 1981, pp. 219–256.
99. J. C. Anderson, K. D. Leaver, R. D. Rawlings, and J. M. Alexander, eds., *Materials Science*, 4th ed., Chapman & Hall, London, 1990, p. 132.
100. R. A. Vandermeer and W. B. Snyder, *Met. Trans. A*, Vol. 10A, Aug. 1979, p. 1031; E. E. Wood, *MAC Short Course Handbook*, Mineralogical Association of Canada, Toronto, 1982.
101. J. J. Licari, Thin and thick film, in *Electronic Materials and Processes Handbook*, 2nd ed., C. A. Harper and R. M. Sampson, eds., McGraw-Hill, New York, 1974, p. 8.43; G. W. A. Dummer, Resistive films, in *Materials for Conductive and Resistive Functions*, Hayden Book Co., New York, 1970, Sec. 18.2.4, p. 293.
102. J. Baliga, *Semicond. Int.*, Vol. 23, No. 2, 2000, p. 98; D. A. McLean, N. Schwartz, and E. D. Tidd, Developments in tantalum nitride resistors, *IEEE Int. Conv. Record*, Pt. 9, Mar. 1964, p. 128.
103. H. Yoenhara, Y. Fukuda, and T. Shrinkawa, *Development of Superminiaturized and UltraHigh-Density Packed Resistor Module*, ISHM, Reston, VA, 1987, p. 147; R. W. Berry, W. H. Jackson, W. H. Parist, G. I. and A. H. Schafer, A critical evaluation of tantalum nitride thin film resistors, *Proc. Electron. Comput. Conf.*, Washington, DC, 1964, p. 86.
104. D. D. Zimmerman, *Int. Symp. Microelectron.*, 1965, p. 2C-7; G. L. Schnale and R. S. Keen, Metallization and bonds: a review of failure mechanisms, *Proc. IEEE 6th Reliab. Symp.*, Nov. 1967, Vol. 6, pp. 170–192.
105. P. A. Gould, *Br. J. Appl. Phys.*, Vol. 16, 1965, p. 481; R. M. Chapman, The effect of fabrication on variables on chromium thin-film resistors, *Vacuum*, Vol. 13, 1963, p. 213.
106. F. Wu, A. W. McLaurin, K. E. Herson, D. G. Managhan, and S. L. Thomson, *Thin Solid Films*, Vol. 332, 1998, pp. 418–422; R. K. Waits, Silicide resistors for integrated circuits, *Proc. IEEE*, Vol. 59, No. 10, Oct. 1971, pp. 1425–1429.
107. V. V. Felmetsger, Controlled sputtering enables better SiCr films, *Semicond. Int.*, Oct. 2000, p. 181; R. K. Waits, Sputtered silicon chromium-resistive films, *J. Vac. Sci. Technol.*, Vol. 6, Mar./Apr. 1969, pp. 308–315.
108. M. Beckerman and R. E. Thun, *Trans. 8th Natl. Vac. Symp.*, 1961, p. 905; R. Wagner and K. M. Mertz, Stable high-range Cr/SiO film resistors, *Trans. IEEE Comput. Hybrid Manuf. Technol.*, Vol. 11, 1964, p. 97.
109. L. Maissel, Thin-film resistors, in *Handbook of Thin-Film Technology*, 2nd ed. L. Maissel and R. Glang, eds., McGraw-Hill, New York, 1983, p. 18-18.
110. C. L. Wellard, ed., *Resistance and Resistors*, McGraw-Hill, New York, 1960, Chap. 16.
111. C. J. Chu, M. P. D'Evelyn, R. H. Hauge, and J. L. Margrave, Mechanism of diamond film growth by hot filament, *J. Mater. Res.*, Vol. 5, No. 11, 1990, p. 2405.
112. J. Mort and F. Jansen, eds., *Plasma-Deposited Thin Films*, CRC Press, Boca Raton, FL, 1986.
113. A. Bubenzer, B. Dischler, G. Brandt, and P. Koidl, Carbon, rf plasma CVD, *J. Appl. Phys.*, Vol. 54, 1983, p. 4590.

114. M. Hanbasa and T. Kaneoka, Role of ions in carbon film deposition in windowless hydrogen lamp, *J. Vac. Sci. Technol.*, Vol. B5, No. 5, 1987.

115. W. H. Kohl, *Handbook of Materials and Techniques for Vacuum Devices*, Reinhold, New York, 1967.

116. J. W. Diggle, ed., *Oxides and Oxides of Metals*, Vol. I, Marcel Dekker, New York, 1972.

117. J. R. Szedon and W. J. Takei, Dielectric films for capacitor applications in electronic technology, *Proc. IEEE*, Vol. 59, No. 10, Oct. 1971, pp. 1434–1439; J. C. Anderson, ed., *Dielectrics*, Chapman & Hall, London, 1964.

118. P. J. Harrop and D. S. Campbell, Dielectric properties of thin films, in *Handbook of Thin-Film Technology*, 2nd ed., L. I. Maissel and R. Glang, eds., McGraw-Hill, New York, 1983.

119. D. Gerstenberg, Thin film capacitors, in *Handbook of Thin-Film Technology*, 2nd ed., L. I. Maissel and R. Glang, eds., McGraw-Hill, New York, 1983, Chap. 19.

120. B. Tareev, *Physics of Dielectric Material*, Mir Publishing, Moscow, 1975, p. 175.

121. G. A. John and S. W. Chaikin, *IEEE Trans. Master Packag.*, Vol. 2, 1966, p. 29; C. E. Drumheller, *Trans. Vac. Symp.* Pergamon Press, New York, 1960, p. 306.

122. R. S. Clark, *Trans. Met. Soc. AMIE*, Vol. 233, 1965, p. 592.

123. T. Makino, Composite and structure control by source gas ratio in LPCVD SiN_x, *J. Electrochem. Soc.*, Vol. 130, 1983, p. 450.

124. K. B. Schwartz, D. J. Rowelffe, Y. D. Blum, and R. M. Laine, in *Better Ceramics through Chemistry*, Vol. II, C. Binker, D. Clark, and D. Ulrich, eds., Materials Research Society, Pittsburgh, PA, 1986, pp. 407–412.

125. S. M. Sze, *Semiconductor Devices, Physics and Technology*, Wiley, New York, 1995, Chap. 9, p. 362; C. Blaauw, Preparation and Characterization of Plasma deposited silicon nitride, *J. Electrochem. Soc.*, 131, 1114 (May 1984).

126. N. Schwartz and R. W. Berry, in *Physics of Thin Films*, Vol. 2, G. Hass and R. Thun, eds., Academic Press, San Diego, CA, 1964, p. 308.

127. P. Burggraaf, IEDM conference report, *Solid State Technol.*, Vol. 43, No. 2, Feb. 2000, p. 30.

128. G. W. A. Dummer, *Materials for Conductive and Resistive Functions*, Hayden Book Co. New York, 1970, p. 292.

129. B. Tareev, *Physics of Dielectric Materials*, Mir Publishing, Moscow, 1975, p. 204; A. Von Hippel and W. B. Westphal, *High Dielectric Constant Materials as Capacitor Dielectrics*, Tech. Rep. 145, Laboratory for Instrumention & Research, MIT, Cambridge, MA, Dec. 1959.

130. C. Feldman, *Rev. Sci. Instrum.*, Vol. 26, 1955, p. 463; I. H. Pratt, Characteristics of RF-sputtered barium titanate thin films, *Proc. IEEE*, Vol. 59, No. 10, Oct. 1971, pp. 1440–1447.

131. L. Hozar, Properties of $BaTiO_3$, in *Semiconductor Ceramics*, Ellis Horwood, New York, 1994, Chap. 4; D. R. Ulrich and E. J. Smoke, Diversified barium titanate dielectrics, *J. Am. Ceram. Soc.*, Vol. 49, Apr. 1966, pp. 210–215.

132. F. A. Grant, *Rev. Mod. Phys.*, Vol. 31, 1959, p. 646.

133. N. Tasuda, K. Nasu, A. Yamse, and K. Siratori, *Electronic Conduction in Oxides*, Springer-Verlag, Berlin, 1991.

134. D. Adler, in *Solid State Physics*, Vol. 2, F. Sitz, D. Turnbull, and H. Ehrenreich, eds., Academic Press, San Diego, CA, 1968, pp. 1–113.

135. F. Huber, *Microelectron. Reliab.*, Vol. 4, 1965, p. 283.

136. T. K. Gupta, L. J. Cirignano, K. S. Shah, L. P. Moy, D. J. Kelley, M. R. Squillante, G. Entine, and G. P. Smestad, Screen printed dye sensitized large area nanocrystalline solar cell, *Mater. Res. Soc. Proc.*, Vol. 581, 2000, p. 653.

137. T. K. Lakshmanan, C. A. Wysochi, and W. J. Slegeski, Sputtered Titanium dioxide film for Microcircuit Applications, *Proc. Electronic Computer Conf.*, Washington, DC, 1964, p. 14.

138. E. M. Levin and N. F. McMurdie, in *Phase Diagrams for Ceramists*, M. K. Rester, ed., American Ceramic Society, Westerville, OH, 1976.

139. S. B. Qadri, C. M. Gilmore, C. Quinn, E. K. Skelton, and C. R. Gossett, Structural stability of ZrO_2–Al_2O_3 thin films deposited by magnetron sputtering, *J. Vac. Sci. Technol.*, Vol. A7, No. 3, 1989, p. 1220.

140. A. H. Heuer and L. W. Hobbs, Science and technology of zirconia, in *Advances in Ceramics*, Vol. 3, American Ceramic Society Westerville, OH, 1981.

141. C. Kwok and C. R. Aita, The transition from αZr to αZrO_2 growth in sputter-deposited films as a function of gas O_2 content, rare gas type, and cathode voltage, *J. Vac. Sci. Technol.*, Vol. A7, No. 3, 1989, p. 1235.

142. C. W. T. McLyman, *Transformer and Inductors Design Handbook*, Marcel Dekker, New York, 1988, Chap. 8, p. 289.

143. C. A. Harper, Inductive devices, in *Passive Electronic Components Handbook*, 2nd ed., McGraw-Hill, New York, 1997.

144. W. G. Roberts, T. A. Jennings, and S. T. Chen, Vacuum-deposited inductors, *Int. Symp. Hybrid Microelectron.*, 1975, p. 100.

145. F. W. Grover, *Inductance Calculations, Working Formulas, and Tables*, Dover Publications, Mineola, NY, 1973.

146. A. E. Riad and N. N. Schulz, Thin-film circuit design guideslines, in *Modular Series in Hybrid Microelectronics*, R. W. Johnson, ed., ISHM, Reston, VA, 1991.

147. R. E. Shipley, An RF power hybrid using a thin-film matching network with spiral inductors, *Int. Symp. Hybrid Microelectron.*, 1982, p. 205.

148. J. E. Curran and H. Sewell, The technology of thin-film hybrid microwave circuits, *Int. Symp. Hybrid Microelectron.*, 1975, p. 87.

149. V. G. Welsby, *The Theory and Design of Inductance Coils*, Mcdonald, London, 1978, p. 154.

150. D. B. Javis, The effects of interconnections on high-speed logic circuits, *IEEE Trans. Electron. Circuit Comput.*, Oct. 1963, pp. 476–487.

151. I. Catt, Cross talk in digital systems, *IEEE Trans. Electron. Comput.*, Vol. 16, No. 6, Dec. 1967, pp. 743–763.

152. W. S. Song and L. A. Glasser, Power distribution techniques for VLSI circuits, *IEEE J. Solid State Circuits*, Vol. 21, No. 1, Feb. 1986, pp. 150–156.

153. G. A. Katopis, Delta noise specification for a high-performance computing machine, *Proc. IEEE*, Vol. 75, No. 9, Sept. 1985, pp. 1405–1415.

154. H. B. Bakoglu, *Circuits, Interconnections, and Packaging for VLSI*, Addison-Wesley, Reading, MA, 1990, p. 311.

155. C. S. Fuller, J. D. Struthers, J. A. Ditzenberger, and K. B. Wolfstrin, *Phys. Rev.*, Vol. 93, 1954, p. 1182.

156. W. Zuhlehner and D. Huber, *Czochralski Silicon Crystals*, Springer-Verlag, Berlin, 1982.

157. S. P. Murarka and M. C. Peckerar, eds., *Electronic Materials Science and Technology*, Academic Press, San Diego, CA, 1989.

158. S. K. Ghandhi, *VLSI Fabrication Principles: Silicon and Gallium Arsenide*, Wiley, New York, 1983, Chap. 4.

159. G. S. Chen, *Thin Solid Films*, Vol. 353, 1999, p. 264.

160. K. H. Min, K. C. Chun, and K. B. Kin, *J. Vac. Sci. Technol.*, Vol. B14, 1996, p. 3263.

161. J. C. Chung and M. C. Chen, *J. Electrochem. Soc.*, Vol. 145, 1998, p. 3170.

162. M. T. Wang, Y. C. Lin, and M. C. Chen, Barrier properties of very thin Ta and TaN layers against silicon, *J. Electrochem. Soc.*, Vol. 145, No. 7, 1998.

163. *www.semiconductors.net/technical/damascene_copper.htm.*

164. J. H. Golden, C. J. Hawker, and P. S. Ho, Designing porous low-*K* dielectric.

165. M. Morgan et al., Low dielectric constant materials for ULSI interconnects, *Annu. Rev. Mater. Sci.*, Vol. 30, 2000.

166. J. P. Godschalx et al., Polyphenylene oligomers and polymers, U.S. patent 5,965,679, 1999.

167. S. T. Martin et al., Development of low dielectric constant polymers for the fabrication of integrated interconnects, *Adv. Mater.*, Vol. 12, 2000, p. 1769.

168. P. Ball, *Nature*, Vol. 414, 2001, p. 142.

169. P. G. Collins and Ph. Avouris, *Sci. Am.*, Dec. 2000, p. 38.

170. F. Kreupl, A. Graham, and W. Honlein, A status report on technology for carbon nanotube devices, *Solid State Technol.*, Apr. 2002, p. 59.

171. A. Bechtold et al., *Science*, Vol. 294, 2001, p. 1317.

172. V. Derycke et al., *Nano Letters, ACS Publications*, Vol. 1, 2001, p. 453.

173. S. S. Wong et al., *Nature*, Vol. 394, 1998, p. 52.

174. R. Saito, G. Dresselhaus, and M. S. Dresselhaus, *Physical Properties of Carbon Nanotubes*, Imperial College Publishing, London, 1998.

RECOMMENDED READING

Anderson, J., ed., *The Use of Thin Films in Physical Investigations*, Academic Press, San Diego, CA, 1966.

Berry, R. W., P. M. Hall, and M. T. Harris, *Thin-Film Technology*, D. Van Nostrand, Princeton, NJ, 1968.

Chopra, K. L., *Thin-Film Phenomena*, McGraw-Hill, New York, 1969.

Dummer, G. W. A., *Materials for Conductive and Resistive Functions*, Hayden Book Co., New York, 1970.

Riad, A. E., and F. D. Barlow III, *Thin-Film Technology Handbook*, McGraw-Hill, New York, 1998.

Ghandhi, S., *VLSI Fabrication Principles*, Wiley, New York, 1983.

Hass, G., ed., *Physics of Thin Films*, Academic Press, San Diego, CA, 1970.

Holland, L., *Vacuum Deposition of Thin Films*, Wiley, New York, 1956.

Maissel, L. I., and R. Glang, eds., *Handbook of Thin-Film Technology*, 3rd ed., McGraw-Hill, New York, 1983.

Mayer, H., *Structure and Properties of Thin Films*, Wiley, New York, 1959.

Mott, N. F., and H. Jones, *The Theory of the Properties of Metals and Alloys*, Dover Publications, New York, 1958.

Murarka, S. P., and M. C. Peckerar, eds., *Electronic Materials Science and Technology*, Academic Press, San Diego, CA, 1989.

Murarka, S. P., I. V. Verner, and R. J. Gutman, *Copper: Fundamental Mechanisms for Microelectronic Applications*, Wiley, New York, 2000.

Neugebauer, C. A., J. B. Newkirk, and D. A. Vermilyca, eds., *Structure and Properties of Thin Films*, Wiley, New York, 1959.

Neugebauer, C. A., and R. H. Wilson, *Basic Problems in Thin-Film Physics*, Vandenhoeck and Ruprecht, Gottingen, Germany, 1966.

O'Dwer, J. J., ed., *The Theory of Dielectric Breakdown in Solids*, Oxford University Press, New York, 1964.

Tareev, B., *Physics of Dielectric Materials*, Mir Publishing, Moscow, 1975.

Vossen, J. L., and W. Kern, eds., *Thin-Film Process*, Academic Press, San Diego, CA, 1978.

7

THIN-FILM DEPOSITION TECHNIQUES

The rapid development of hybrid microelectronics industry over the last decade has created exceptional demands for thin film [1]. There are many reasons for the widespread use of thin films in microelectronics industry, including hybrid circuits. Perhaps the most important is that only thin films can be processed to yield the small size, low power, and high circuit density required for multilayer hybrid circuits [2]. It also allows for convenient production of high-purity materials with closely controlled composition. There is no rigid definition of thin film. However, a surface layer between 1000 Å (4×10^{-3} mil) and 0.1 μm (0.004 mil) thick may be considered a thin film. There is no universal thin-film process, just as there is none for monolithic circuits. Thin-film processing is more a science than an art because of the greater diversity of materials involved. Cost and reliability weigh heavily in the process used [3].

The variety of processes and materials required to meet adequately the total needs of the thin-film hybrid industry has necessitated the development of several deposition techniques. Vacuum evaporation [4], sputtering [5], chemical vapor deposition (CVD) [6], electroplating, and anodization [7], for example, are all in volume manufacturing, and the technology of each of these techniques has been improved significantly during the past decade [8].

7.1 PHYSICAL VAPOR DEPOSITION

Deposition of films by thermal evaporation is the simplest method available for thin-film deposition [9,10]. In thermal evaporation the material is heated by a

Handbook of Thick- and Thin-Film Hybrid Microelectronics, By Tapan K. Gupta
ISBN 0-471-27229-9

resistive heating element, usually tungsten or molybdenum, and then condensed over the substrate. The substrate can be heated during deposition to achieve structural change in the material deposited [11]. The problem of finding a non-reactive support for materials evaporating above 1000°C can be circumvented in those cases where the 10^{-2} torr vaporization temperature is near the melting point of the evaporant. The elements chromium (Cr), molybdenum (Mo), palladium (Pd), vanadium (V), fluorine (F), and silicon (Si) are examples of these materials. For transition metals, especially the refractory metals (molybdenum, tungsten, tantalum, etc.), evaporation by electron-beam gun is very popular. Metals such as gold, silver, copper, and aluminum, which have low melting points, can be evaporated by any of the methods noted above. Resistive heating is, however, limited in application because of lower throughput associated with the smaller metal charge generally used in the systems. On the other hand, beam evaporations might cause radiation damage and therefore require a high-temperature predeposition heat treatment to anneal out the radiation damage. The main advantage of the physical vapor deposition (PVD) method is its ability to work at considerably lower pressures than the sputtering system, where gas entrapment is a serious problem. Radio-frequency (RF) heating of the evaporating source in an appropriate crucible could provide the best compromise for both large through-put and a clean environment [12].

Physical vapor deposition of thin films is carried out in a vacuum chamber made of glass or stainless steel with a quartz window [13]. A roughing pump system reduces the pressure inside the chamber from atmospheric to a level where a high-vacuum pump can work [14]. A high-vacuum pump [15] that can reach the ultimate pressure ($<10^{-7}$ torr) must work with sufficient speed to accommodate the outgassing loads from the work being carried out in the chamber (Figure 7.1). The most common type of high-vacuum pump is the oil diffusion pump (Figure 7.2).

In an oil diffusion pump, oil is being heated and the vaporized oil molecules collide with the gas molecules trapped inside the walls of the chamber, giving them a downward momentum and directing them out of the chamber [17]. The pumping speed of the pump is defined as the ratio Q/P in units of L/s at 20°C. Q is the quantity of gas that flows through the intake cross section of the pump in unit time and is pressure dependent. P is the partial pressure of gas at a point near the intake point of the pump. The pump downtime t in seconds can be written as

$$t = 276 \frac{V}{S_0 + S} \ln \frac{P}{P_0} \tag{7.1}$$

where V is the volume of the chamber and S_0 and S are pumping speeds in L/s at starting and final pressures P_0 and P, respectively [18]. For faster pumping the oil diffusion pump is replaced with either a turbo molecular pump [16] or a cryogenic pump [15,17].

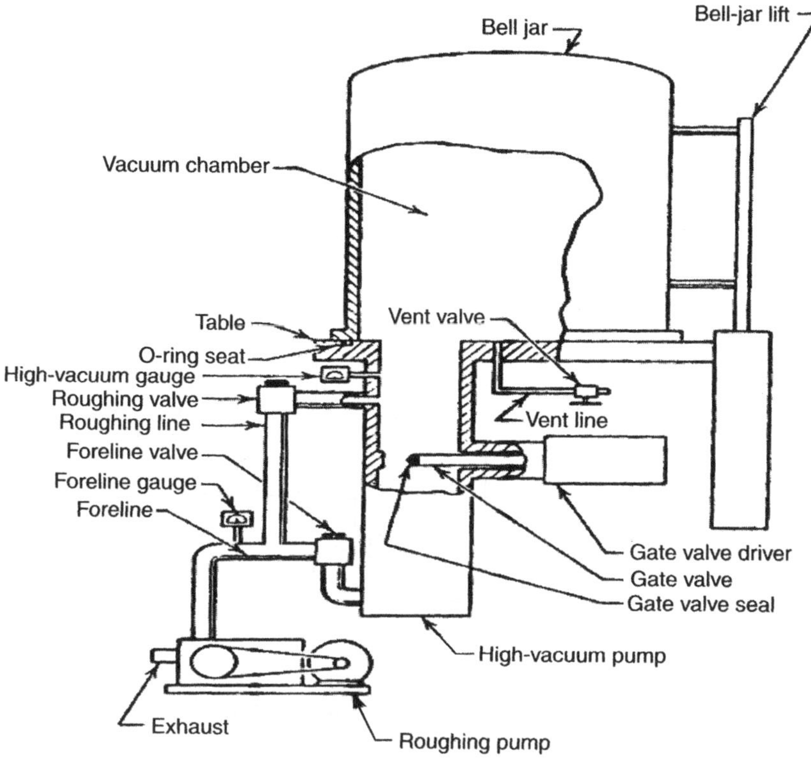

Figure 7.1 Physical vapor deposition system with rotary and high-vacuum oil pumps [16]. (Reproduced with permission of The McGraw-Hill Companies.)

7.1.1 Turbo Pump

A turbo molecular pump is a mechanical pump in which a turbine rotates fast enough to drive molecules. It is a type of high-vacuum pump which functions in a manner analogous to a diffusion pump with a backing pump to keep the fore pressure low enough to sustain molecular flow. Instead of imparting momentum by collisions with streaming high-speed vapor molecules, turbomolecular pumps cause momentum transfer impacts with high-speed (2400 to 3600 rpm) rotating blades. If the fore pressure is allowed to rise to the point of viscous flow (1 Pa), the pump experiences a sudden decrease in rotor speed and thereby, also, pumping speed. A turbo pump will operate at constant speed over normal operating ranges, and their ultimate pressure (10^{-8} to 10^{-9} Pa) is limited by their compression ratio for light gases and the effects of outgassing. Hybrid turbo pumps are used to produce ultrahigh vacuum at the inlet because of their high efficiency [19–21]. Figure 7.3 shows a compound turbo pump.

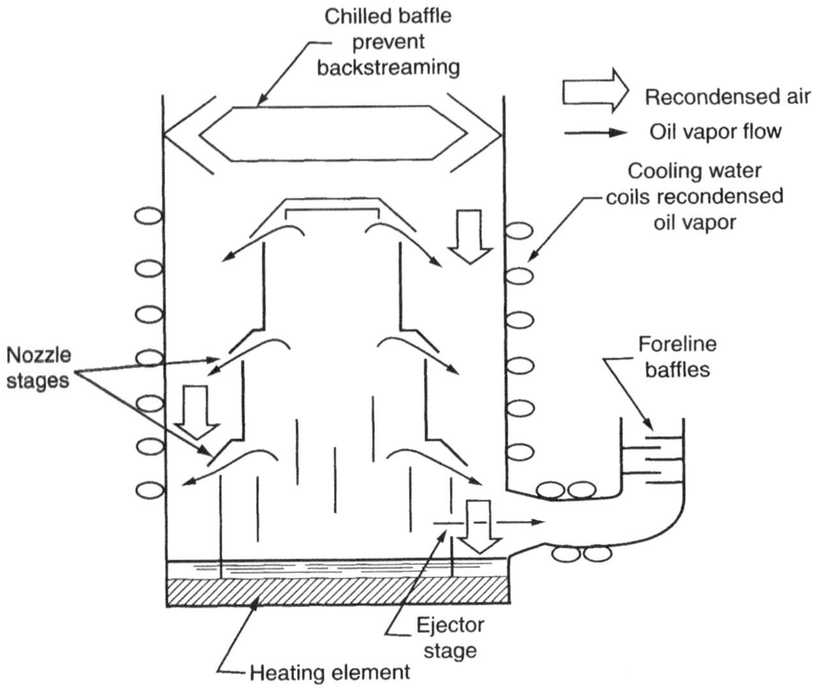

Figure 7.2 Cross section of the high-vacuum oil diffusion pump [16]. (Reproduced with permission of The McGraw-Hill Companies.)

7.1.2 Cryogenic Pump

A cryogenic pump (or cryo pump or sputter ion pump) [23] is popular in microelectronic use because it cannot backstream. The operating parts of cryo pumps are cooled (to liquid helium) metal surfaces installed inside the vacuum chamber. The pump operates at low pressures where the mean free path is on

Figure 7.3 Compound turbo pump. The lower section comprises a Holweck design. (With permission of Kurt J. Lesker Company, Clairton, PA.)

Figure 7.4 Cryogenic pump. (With permission of Kurt J. Lesker Company, Clairton, PA.)

the order of the chamber dimensions. Cryo pumps collect rather than remove gas and must be regenerated periodically by warming them up and pumping away the gas collected. Cryo pumps are free from oil and thus highly appealing for carbon-free film deposition. When cleanliness, bakeability, low power consumption, vibration-free operation, long pumping life, and high pumping speed are major concerns for thin-film deposition, a titanium sublimation pumping (TSP) system is considered to be ideal. Figure 7.4 shows a cryo pump, and Figure 7.5 is the cross section of a two-stage closed-loop helium cryo pump.

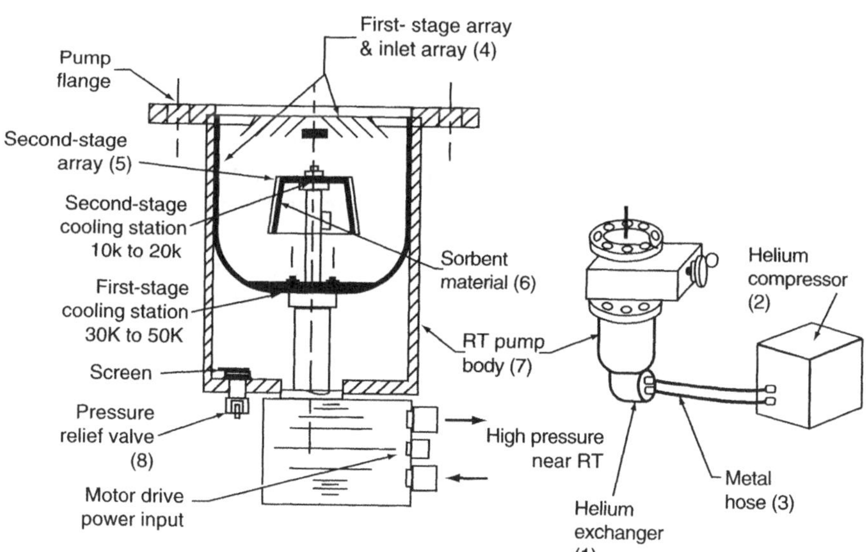

Figure 7.5 Cross section of a typical two-stage closed-loop helium cryo pump [24].

Cryo pumps [25] are in wide use in industrial applications where the presence of hydrocarbons cannot be tolerated. Any mechanism that captures gases on a cold surface is a cryo pump; a common example is frost buildup on an ice cube tray. Although cryo pumps are used in high-vacuum systems, they are considered more sophisticated and expensive than oil pumps. Nonetheless, their cost can be justified on the basis of their performance. According to an observer [26], there are a number of ancillary considerations that affect a decision between oil-free and oil-sealed pumps. Oil-sealed pumps are extremely reliable because they have been built and improved over many decades, and except for periodic oil changes, require little maintenance. On the flip side, used oil is considered in most cases to be toxic waste, which makes it difficult and expensive to dispose of. Oil is messy and potentially hazardous and unfriendly to the atmosphere. Bearing this in mind, the possibility of contamination is the most common decision driver.

A PVD system can be provided with more than one resistive heating source for evaporation of different materials in the same vacuum system. By operating the sources simultaneously, it is possible to deposit multiconstituent films or multilayer films by evaporating one material at a time in a single vacuum evaporation. This is a cost-effective way of making thin-film circuits [27].

7.2 FLASH EVAPORATION

Flash evaporation is another method for deposition of films whose constituents have different vapor pressures. In contrast to two-source evaporation, it does not require provisions to monitor the vapor density, nor is control of the source temperature particularly critical. Powder is fed into the boat by various feeding mechanism [28]. This technique is applicable for the evaporation of alloys, metal–dielectric mixtures, and compounds. Co-evaporation from a single source containing alloy constituents is generally not possible, because fractionation occurs as evaporation proceeds. Fractionation is a phenomenon by which one component evaporates preferentially, leaving the source rich in the other components and thus leading to films richer in the preferentially evaporated component. The choice of filament material is an important factor for flash evaporation. Tungsten, which is a popular choice for the filament material, is not good for perovskite materials, and is replaced by iridium. Another drawback for this flash evaporation method is that evaporation generally takes place under poor vacuum conditions. This is attributable to the high gas content of the evaporant powder during evaporation [29].

7.3 SPUTTERING

When atoms, ions, or molecules bombard a solid surface, many phenomena can occur. The kinetic energy of the impinging particles largely dictates which

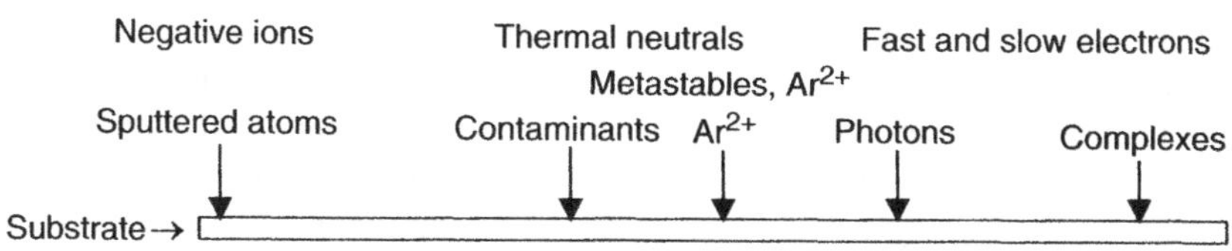

Figure 7.6 Particles bombarding a substrate inside a sputter chamber [30].

events are most likely to take place. When the energies are less than 5 eV, the bombarding particles will reflect from the surface. When energies are between 5 and 10 eV (which is more than the binding energies of the target), surface migration or surface damage effects can take place. When the energies are higher than 10 keV, impinging particles may embed in the target. When the energies exceed 10 keV, two effects may arise: (1) either a fraction of the energy of the impinging ions is transferred to the solid in the form of heat and damages the lattice, or (2) another fraction of the energy causes atoms from the surface to be dislodged and ejected as gas (sputtering). In sputter deposition, the target material (purity 99.998%) is bombarded by energetic ions [5,30,31]. This method is sometimes called *impact evaporation*, since a surface is bombarded with ionized particles which transmit energy to the atoms or molecules of the surface so that the most energetic atoms are emitted. The particles emitted then diffuse to an adjacent surface, where they build up a thin-film deposit. Figure 7.6 shows the mechanism of a sputtering system, and Figure 7.7 is a schematic of a typical sputtering system.

Unlike evaporation, sputtering processes are very well controlled and are generally applicable to all materials: metals, alloys, semiconductors, and insulators. RF, dc, and dc magnetron sputtering can be used for metal deposi-

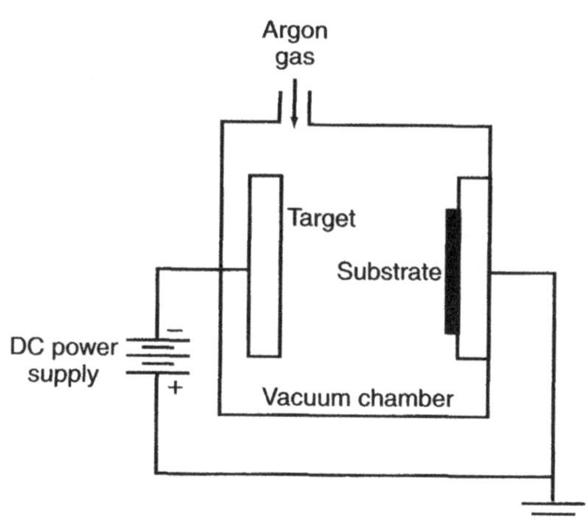

Figure 7.7 Schematic of a dc sputtering system [30].

tion. For insulating material, deposition pulsed dc reactive sputtering is now often used to avoid the arcing that results when insulating deposits on target surface become charged [32]. Sputtering from a composite target would appear to be ideal for compositional and desired stoichiometry. Alloys can also be deposited by sputtering metal in a reactive environment. In reactive sputter deposition an elemental material is sputtered and the sputter-deposited film reacts with the gaseous environment or a codeposited chemical species to form a film of compound material [33].

Three basic modes of sputter deposition have been developed. In the *diode mode*, the target material is attached to the cathode, and the substrate upon which a sputtered film is to be deposited is placed on the anode. The *triode mode* adds a hot filament to the diode configuration and generates high-density plasma independent of target excitation. The *magnetron mode* employs a magnetic field to increase the distance traveled by electrons, thus increasing the probability of their colliding with gas molecules to produce the ions needed for sputtering. The deposition rate is high at low pressure in the magnetron mode and it is the most widely used method for the commercial fabrication of thin films. Both dc and RF excitation are used for sputtering. With dc excitation, target materials are conductors and ac or RF excitation sputtering can be used for both conductors and dielectrics or insulators. The applied power of a RF system has a frequency of approximately 13.56 MHz. because it is one of the frequencies designated by international communication authorities. Figure 7.8 shows the schematic of an RF sputtering system. In magnetron sputtering, a magnetic field is applied at right angles to the electric field. This results in a force to the electrons perpendicular to the direction of motion, causing them to spiral through the gas and increasing their efficiencies in ionization. Thus a magnetron sputtering system consumes lower power because the participating ions are concentrated, due to the magnetic field applied.

Magnetron sputtering is a technique commonly used to deposit thin films for a variety of applications, including metallization [34]. Collimated magnetron sputtering has been demonstrated to be very useful in IC fabrication for high-aspect-ratio coating [35,36]. In sputter deposition, care is needed regarding the angle of deposition because it can cause defects in film nucleated by small substrate particles [34].

Conformal coverage (step coverage) is very important, especially when vias and trenches with high aspect ratios are to covered. One way to perform conformal coverage of high-aspect-ratio trenches and vias is dc bias sputtering. The basic principle is used to cover sidewalls to resputter already deposited material and redeposit this on the sidewalls [37]. *Step coverage*, defined as the percentage ratio of the minimum thickness to the nominal thickness, can be written as

$$\text{step coverage (\%)} = \frac{t_s}{t_n} \times 100 \tag{7.2}$$

where t_s is the minimum thickness of a film as it crosses a step and t_n is the

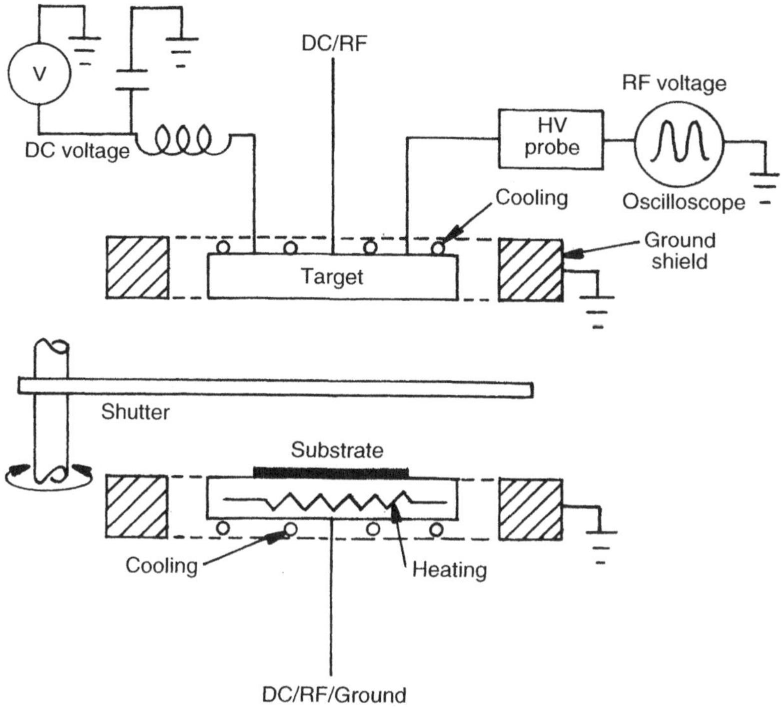

Figure 7.8 Schematic of an RF sputtering system [30].

nominal thickness of a film on flat regions. Step coverage depends on several factors, such as the aspect ratio [length of the step to the width of the step (L/W)], the step shape, and the step slope. The sputtering process can be adjusted to attain better step coverage by (1) sputter etching of substrate to reduce step slopes before deposition, (2) optimizing the shape of the target and controlling the target regions where maximum sputtering occurs, and (3) heating the substrate.

Bias sputtering is considered to be a better sputtering process when conformal coverage is important [38,39]. In bias sputtering, RF bias is applied to a substrate to enhance acceleration of the impinging ions, thereby transferring energy to the surface atoms. This causes them to be transported to the step sidewalls. Bias sputtering can cause more surface atom migration and resputtering. Under cosine distribution, resputtered atoms can be deposited toward the wall, resulting in more efficient step coverage.

Another type of sputtering used to obtain chemical compounds from an element is *reactive sputtering*, in which in addition to argon as an inert gas, nitrogen or oxygen is introduced to form the oxides and nitrides of the respective element. Ionized metal plasma deposition techniques has been verified experimentally to be a better technology for step coverage than conventional sputtering, due to enhanced directionality of the sputtered atom [40]. As-deposited Ta film over SiO_2 is amorphous. However, annealing at 500°C crystallizes the film, showing a hexagonal close-packed structure. In the presence of oxygen, however, the integrity of Cu/Ta_xN interconnect begins to degrade at 200°C, due to the formation of copper oxide with high-density voids. Sputter-deposited tantalum carbide film on silicon dioxide is amorphous in nature. After annealing at 800°C, TaC in the presence of oxygen forms crystalline Ta_2O_5 material. Due to the crystalline nature of Ta_2O_5, copper diffuses through the barrier layer and forms copper silicide.

The preferred atom ejection pattern for face-centered cubic metals is in the ⟨110⟩ direction [41,42]. However ejection along ⟨100⟩, and under favorable conditions, ejection along ⟨111⟩, are also not rare occasions. For body-centered cubic metals, the closest packed ⟨111⟩ is the preferred ejection direction [43]. However, ejection along ⟨100⟩ is also possible if the bombarding energy is high. For semiconductors the preferred ejection direction is along ⟨111⟩, with secondary ejection along ⟨100⟩.

7.4 CHEMICAL VAPOR DEPOSITION

Chemical vapor deposition is defined as the formation of a nonvolatile solid film on a substrate by the reaction of vapor-phase chemicals (reactants) that contain the required constituents [44,45]. According to the conditions of the processes involved in the deposition of films, CVD processes can be categorized as (Table 7.1) atmospheric pressure CVD, low-pressure CVD, and plasma-enhanced CVD [17]. CVD processes are often selected over competing deposition techniques because they offer (1) high-purity deposits, (2) a great variety of chemical compositions, (3) fairly good film adherence, and (4) good economy and process control. CVD process generally follows these sequential steps: (1)

TABLE 7.1 Characteristics and Applications of CVD Reactors

CVD Process	Advantages	Disadvantages	Applications
Atmospheric pressure	Simple, fast deposition rate, low-temperature operation	Poor step coverage, particle contamination	Low-temperature oxides
Low-pressure	Excellent purity and uniformity, conformal step coverage, large wafer capacity	High-temperature operation, low deposition rate	High-temperature oxides, Si_3N_4, W, WSi_2, poly-Si
Plasma-enhanced	Low-temperature operation, fast deposition rate, good step coverage	Chemical (H_2) and particulate contamination	Low-temperature insulators, passivation layer

The inert gases are introduced into the reaction chamber with a given composition and flow rate; (2) the gas species are driven to the surface of the substrate; (3) reactant gases are adsorbed on the substrate; (4) the adatoms form films by migration; and (5) dissolved gaseous by-products are removed from the reaction chamber.

In an atmospheric CVD system, the pressure in the reaction chamber is not connected to the vacuum system and an RF generator is not required. The temperature within the reaction chamber is controlled. Gas flow is measured in units of volume/unit time, where the volume measurement assumes standard temperature and pressure. The mass flow m is given by

$$m = CW\,\Delta T \tag{7.3}$$

where C is a constant that depends on heat transfer coefficients, the specific heat, the density, and thermal conductivity of the gas; W is the heater power; and ΔT is the temperature difference between the points where the sensors are located. The reactors used in an atmospheric pressure CVD system may be a horizontal tube or gas injector type or a plenum-type continuous processing reactor.

In low-pressure CVD the reaction chamber is connected to the vacuum chamber and the pressure inside the chamber is pumped down to the required value. High-vacuum cryopumps are generally used because these pumps do not use backing pumps, and since the operation of cryo pumps is based on an oil-free pumping mechanism, there is no oil backstreaming from such pumps. In low-pressure CVD reactors, the system operates in a reaction-rate-limited mode. In low-pressure CVD systems, pressure is generally maintained between 0.25 and 2.0 torr and temperature between 550 and 600°C. In low-pressure operation, gas-phase reactions are lowered, and the film is subjected to less contamination. Two types of reactors are used in low-pressure CVD systems: hot wall horizontal tube and isothermal cold wall vertical flow. The first category has wide applications for depositing polysilicon, silicon nitride, phosphor silicate glass, borophosphate silicate glass, tungsten, and undoped and doped silicon dioxide.

The third and last of the major CVD deposition methods, plasma-enhanced CVD, is categorized not only by its pressure regime but also by its method of energy input. A lower substrate temperature is the major advantage of plasma-enhanced CVD. The plasma is generated by the application of an RF field to the low-pressure gas, thereby creating free electrons within the discharge region. Three types of plasma-enhanced CVD reactors are generally available: parallel plate, horizontal tube type, and single wafer type. A plasma-enhanced CVD system operates at a low substrate temperature and can enhance the deposition rate over thermal reactions alone. Film properties such as adhesion to substrate, low pinhole density, better step coverage, good electrical characteristics, and compatibility with fine-line pattern transfer are the main features of a plasma-enhanced CVD system [17].

Chemical vapor deposition is a process that provides conformal coating of metals. The low-temperature CVD process has proved to be an especially good method for highly conformal deposition [46]. The interconnections between levels of wiring, designated as *vias*, have been filled successfully with copper without shadowing [47]. The greatest advantage of low-temperature thermal CVD over PVD is selectivity, the ability of the precursor molecule to chemically recognize the desired growth surface. Perhaps the greatest advantage of CVD is in the production of metastable materials. Most CVD processes operate far from equilibrium, so that kinetically rather than thermodynamically favored products are produced.

The availability of a suitable gaseous source for CVD is one of the most important factors. Tungsten and other refractory metals can be deposited to fill submicrometer-level via and trenches that have high aspect ratios [48]. However, in different types of CVD systems, the processing temperature and pressure inside the chamber are different. As a result, the mean free paths of the reactant atoms are different. Suppose that the arrival flux forms an angle θ and $\theta + d\theta$; then the arrival of the total flux at any point can be written as [17]

$$\int_0^{\pi} P(\theta)\, d\theta \tag{7.4}$$

For atmospheric pressure CVD the mean free path of the reactant gas molecules is shorter and the frequent collisions of gas molecules completely randomizes their velocity vectors. Under such conditions, $P(\theta)$ is constant and independent of the θ value. However, $P(\theta)$ is zero for values of θ that are blocked by the substrate. If the arrival angle at (a) is 180°, at (b) is 270°, and at (c) is 90°, the resulting thickness for atmospheric pressure CVD film at (a) will be $180/180 = 1$; at (b), $270/180 = 1.5$ times greater than at (a); and at (c), $90/180 = \frac{1}{2}$ of the thickness of the film that is obtained at (a). So we can see that the bottom part is growing at 50% of the rate at (a) and 1.5 times that at (b) compared to (a). In low-pressure and plasma-enhanced CVD, the reactants have more energy than in atmospheric pressure CVD, resulting in increased adatom migration. Thus improved step coverage can be expected.

An ordered film of copper on tantalum or on its nitride, grown over silicon dioxide on silicon (100), grows randomly and the sheet resistance of the film shows higher values than those of CVD-grown Cu film on a previously grown PVD or sputtered Cu film. When Cu film is grown by CVD over a previously grown thin Cu film either by PVD or by sputtering, the CVD-grown film grows preferentially along the $\langle 111 \rangle$ direction and shows a face-centered cubic structure [49]. High-resolution transmission electron microscopy shows high-density micro twins that traverse completely through the grains. It has been observed that copper deposited with hexafluoroacetylacetonate with 3,3-dimethyl-1 butane between a substrate temperature of 100 and 275°C, in an ambient of hydrogen, shows $\langle 111 \rangle$-oriented strong growth compared to $\langle 200 \rangle$ growth in the ambient of nitrogen or argon. Films grown in a hydrogen atmosphere show a lower resistivity than that of films grown under an inert atmosphere [50].

7.5 ION-BEAM DEPOSITION

Although sputtered deposition provides very good resistive films for hybrid circuits, the kinetic energy provided for the deposition is inadequate to control the morphology and nucleation growth of the film [51,52]. This has led to the development of a technology of ion-beam deposition (IBD). Ion-beam technology has been employed successfully for a variety of critical applications in thin-film processing. Ion-beam sources have several advantages in thin-film applications [53]. Compared with thermal energy processes such as evaporative deposition, beam ions are very energetic. Compared with plasma processes, ion sources permit independent control of beam energy, current density, direction, and background pressure. In general, ion sources can be operated at low background pressures and are therefore suited for processes with low contamination from background impurities. Endowed with great versatility, an ion-beam system can be used for in situ cleaning, etching of low-volatile materials, reactive etching, chemically assisted etching, sputter deposition, reactive sputter deposition, and ion-beam-assisted deposition of evaporated films. Diamond coatings and modification of material properties or surface morphologies of thin films can also be achieved using IBD. Several configurations of the ion-beam apparatus can be utilized to deposit high-quality thin films. Some applications are given below.

7.5.1 Ion-Beam Sputter Deposition

In ion-beam sputter deposition (IBSD), two independently operated ion-beam sources are mounted in vacuum. Typically, this is referred to as *dual-ion-beam deposition*. The first ion beam is aimed at a target to generate the sputtered material, which will continue deposition of the thin film [54]. The second ion beam is focused on the substrate to provide ion assistance, which greatly aids in providing outstanding film properties. An IBSD system is a low-pressure process usually conducted at 10^{-4} torr, where the sputtered atom retain their kinetic energy due to minimal scattering in the low-pressure environment [55]. The result of dual-ion-beam deposition is smooth, high-quality, pinhole-free films with enhanced adhesion and microstructural control. The technique also yields excellent coverage on thin high-aspect-ratio features. Typically, film properties from dual-ion-beam deposition exceed those of films deposited by evaporation or magnetron sputtering. It is the independent control of ion-beam parameters that allows users to engineer for the film properties desired.

Evaporative metal oxide films and magnetron or plasma sputtering processes provide reasonably good temperature coefficients and tolerances of 0.01% or better. A major disadvantage of these resistors is their poor long-term stability. Nickel–chromium film deposited by ion-beam deposition provides an approximate one decade improvement in absolute TCR (less than 5 ppm/°C), with a variation in TCR value below 20 ppm per 1000 h [56]. Controlling for the effect of grain boundary scattering, low TCR and high stability values have been achieved with these resistors. Sputter deposition methods provide kinetic energy to the atoms between 1 and 20 eV, which is not sufficient to have optimum control of film morphology and nucleation growth, whereas ion-beam deposition, which provides a kinetic energy between 10 and nearly 1000 eV, is able to control both morphology and nucleation growth.

7.5.2 Ion-Beam-Assisted Deposition

Ion-beam-assisted deposition (IBAD) is generally used in conjunction with other (source) types of thin-film deposition to enhance the properties of thin film without the need for further heating of the substrate. IBAD has been used successfully in evaporative, pulsed laser, and molecular beam epitaxy as well as in ion-beam sputtering applications. IBAD is used to enhance film properties, particularly to increase film density, resulting in a reduction in film stress. Additional enhancements in film quality have been manifested by controlling film microstructure, orientation, and temporal environmental stability.

7.6 PULSED-LASER DEPOSITION (LASER ABLATION)

Pulsed laser deposition (PLD) has the advantage of producing high-quality thin films of materials having high melting temperatures, which is difficult or impossible using conventional physical vapor deposition [57–62]. As a matter of fact, the PLD technique has been applied successfully to a wide variety of chemically and structurally complex materials [63]. Essentially, PLD uses a high-power laser, usually a krypton fluoride excimer, operating at a wavelength of 248 nm with high energy, typically from 0.1 to 1.0 J and pulses of 15 to 45 ns duration with repetition rates of 1 to 100 Hz. The laser beam is scanned across the target while being rotated so as to have uniform ablation from the surface. As all the segments of the target material are vaporized in less than 1 ns, the stoichiometry of the materials deposited remains very much the same as the target material. Large-area (75-mm-diameter) films with good uniformity (within $\pm 1.5\%$) and without contamination has been grown successfully by Greer and co-workers [64]. In the late 1980s laser-assisted deposition of electronic materials were investigated extensively [65,66]. This process is now being used for depositing superconductors [67,68] and for graphite and electro-optical materials [69].

7.7 HIGH-DENSITY PLASMA-ASSISTED DEPOSITION

High-density plasma-assisted deposition (HDPAD) of thin films is used widely in low-temperature processing of ultralarge-scale integrated circuits, solar energy cells, flat panel displays, etching systems for photolithography, antireflective coating, and optical coatings [70]. HDPAD has been proven to be an excellent coating process, especially for high-aspect-ratio trenches and vias. A high degree of local planarization (chemical mechanical planarization) has been achieved simply by adjusting the deposition/sputter etch ratio. By adjusting the wafer temperature to 400°C or less, intermetallic dielectric can be deposited without putting any thermal stress on the metal. The HDPAD technique usually includes simultaneous deposition and etching. The system is generally provided with an inductively coupled plasma source, an electron cyclotron resonance system, low reactor pressure (10 millitorr), and electron density inside the chamber as high as 5.0×10^{11} cm^{-3}.

7.8 ELECTROPLATING

7.8.1 Electrode Electroplating

Electroplating is based on electrolysis, a process of decomposition of compounds either in a state of fusion or solution when electric current is passed through. Electrolysis can also be called *ionic* or *electrolytic dissociation* as a result of which metals and hydrogen are liberated at the cathode while the nonmetallic portion of the electrolyte is liberated at the anode. Ions charged with positive electricity and liberated at the negative electrode (i.e., the cathode) are called positive ions or *cations*, and ions charged with negative electricity and liberated at the positive terminal of the electrode are called negative ions or *anions*. Thus, when a direct electric current (having low voltage) is passed through two electrodes dipped in an electrolyte dissolved in a solvent (either in water or any ionizing solvent), a molecule of the electrolyte, which is electrically neutral, breaks up into two or more ions of different kinds (positive and negative ions). The positive ions will move toward the negative electrode and will deposit over the electrode (if all the conditions are satisfied), and vice versa. If a metal or other electrical conductor is made the cathode in an electrolysis cell, it can be plated with another metal to protect it against corrosion, to decorate it, or to purify the metal deposited.

The transition from Al to Cu interconnects is under way or under development at most major semiconductor manufacturers [71]. As evidenced by numerous press announcements and technical papers [72,73], interest in copper interconnects is now widespread. Use of a dual-damascene process flow [74] allows construction of a multilevel Cu stack with potentially fewer steps and reduced cost than are available using conventional Al-based interconnects. A new process technology for damascene interconnect fabrication is the electrodeposition of Cu for feature fill. This process is unique among copper deposition methods because it allows filling of high-aspect-ratio features from the bottom up, avoiding the introduction of a center seam or void. Existing process chemistries developed for printed circuit board applications are providing sufficient capability to fill sub-0.25-μm-generation IC features. Many manufacturers are, however, utilizing copper at smaller device dimensions, where such process chemistries do not provide void-free filling.

Copper ion is deposited on the cathode primarily when a dc current is passed through an acidic solution of copper sulfate. Electrodes can be platinum or can be copper, one being an impure copper bar and the other a pure copper bar constituting the two electrodes. The pure copper bar will be the cathode. The current breaks down the acidic copper sulfate solution as

$$CuSO_4 \leftrightarrow Cu^{2+} + SO_4^{2-} \tag{7.5}$$

Copper ions (Cu^{2+}) will be deposited on the cathode, and sulfate ions (SO_4^{2-}) will be deposited on the anode. The pure copper has less entropy than the mixture of copper and impurities; hence energy must be supplied to overcome the entropy decrease. To initiate the nucleation, the cathode of the electroplating system must be a conductor [75]. The amount of copper deposited on the seed plate (metallic cathode) is proportional to the magnitude of the current. The thickness d of the copper deposited is given by

$$D = t\frac{JE\alpha}{\rho} \tag{7.6}$$

where t is the time of deposition, J the current density, E the electrochemical equivalent, α the current efficiency ratio of actual/theoretical weight deposited, and ρ the density of the material deposited (copper). The morphology and composition of the material deposited will depend on the presence of impurities, current density, bath temperature, concentration of the electrolyte, and the physical and chemical nature of the surface on which the ions will be deposited. The deposition of metal on a metal surface electrons are transferred with an energy corresponding to the Fermi energy; however, when the metal is being electroplated on semiconductor surfaces, electron transfer occurs either by means of conduction or valence bands or surface states. As a matter of fact, a key feature of electrochemical deposition of metals on semiconductor surfaces is that to achieve external control over deposition processes, deposition should occur through electron transfer to the metal ion from the majority carrier band (i.e., the conduction band for an n-type and the valence band for p-type semiconductor) [76,77]. The interaction energy between many semiconductor and metals is relatively week, so that metal deposition on semiconductor surfaces follows a three-dimensional island growth (Volmer–Weber) mechanism. Consequently, deposition of continuous thin films requires detailed understanding of the mechanism of nucleation growth and the role of parameters such as surface morphology, solution chemistry, and applied potential. The electrodeposition of silver, gold, nickel, and chromium can be carried out in the same manner as copper.

7.8.2 Electroless Electroplating

Another deposition technology for deposition of thin film from a chemical solution is electroless plating, which does not required electrical contact or a dc electric field. The history of electroless plating began when manufacturers plated printed circuit boards. In electroless plating (autocatalytic process) electrons are supplied by a reducing agent such as alkali borohydrides, alkali diboranes, hypophosphoric acid, or formaldehyde [78]. The deposition rate, adhesion, and surface morphology of the films are strongly dependent on the pH value and bath temperature of the solution. This process has been used successfully to deposit optical thin films for solar cells [79,80], layered superconductors [81], and the copper seed layer before copper plating in the dual-damascene process.

7.9 SOL-GEL COATING

Sol-gel coating is the deposition of a suitable material in the form of a mixture of suspended precursor particles. Destabilization of the sol changes the viscosity of the material as it is being dried out, typically in air at 100 to 150°C. To promote adhesion, films are annealed at high temperatures (250 to 300°C). Annealing changes the film morphology, too. The most common types of sol gel material is based on metal alkoxide $M(OR)_x$, where R represents an alkoxy group. The alkoxide is dissolved in a solvent with water and is destabilized by hydrolysis. The advantage of this process include high purity and a very high degree of homogeneity over large surface areas [82,83].

7.10 ATOMIC LAYER DEPOSITION

Atomic layer deposition (ALD) technology was first developed about 25 years ago in Finland. This process is very similar to the chemical vapor deposition process, where chemical reactants are introduced in pulses, separated from each other in the flow stream of a purge gas. Each reactant pulse reacts chemically with the wafer surface and is capable of achieving precise monolayers growth [84,85]. ALD has been used to deposit a variety of materials, including II–VI and III–V compound semiconductors, elemental silicon, silicon dioxide, and metals and their oxides. The reaction temperatures are very low (between 200 and 400°C), and the deposition rate is on the order of 1 Å per cycle, where each cycle lasts a few seconds. ALD's first and foremost advantage is conformality and uniformity wafer to wafer. It is hoped that matured ALD will be able to solve some of the problems of dual-damascene copper processing (e.g., high-aspect-ratio trench and via fillings). However, at this stage the deposition rate of the ALD process has raised several questions about its throughput [86,87].

7.11 SUMMARY

The physical vapor deposition system, one of the oldest systems, is used very frequently for its simplicity and cost. Old-fashioned PVD systems coupled with mechanical and diffusion pumps (both of them are having oil) are being replaced by dry pumps coupled with either cryo or turbo molecular pumps, to eliminate potential oil contamination. Several dry pumps are available: one, the screw pump, which can avoid big temperature and gas velocity changes that can quickly cause particles to become trapped in the complex pumping mechanism and cause the pump to seize, due to particle entrapment. Moreover, screw pumps have gas paths that can be five times shorter than other types of pumps [88]. By controlling substrate temperature, rotation, and spinning, and making an oblique or glancing angle of incidence (GLAD), films of different morphologies can be achieved [89]. As a matter of fact, GLAD PVD can control material properties and microstructures and thus can be used to deposit thin films which can be suitable for helicoidal bianisotropic media deposition for the fabrication of laser mirrors and notch filters [90,91], humidity sensors and thermal barriers [92–94], field emission devices [95–97], and rectangular arrays of pillars as perpendicular magnetic storage media [98,99].

PVD copper seed layer deposition before copper plating is a routine step in the copper damascene process [100]. Sputtering is also a member of the PVD family. Bias sputtering and magnetron sputtering (cylindrical magnetron system) are used widely for uniform coating [101,102]. High-density plasma CVD is another CVD system that is very promising for depositing thin films inside a very small and reentrant gap [103]. This process might be a very useful tool for high-aspect-ratio gap filling in the copper damascene process.

REFERENCES

1. R. G. Glang, Application of thin films, part IV, in *Handbook of Thin-Film Technology*, L. I. Maissel and R. Glang, eds., McGraw-Hill, New York, 1970.
2. R. W. Johnson, R. K. F. Teng, and J. W. Blade, *Multichip Modules System Advantages, Major Construction, and Materials Technologies*, IEEE Press, Piscataway, NJ, 1991.
3. Thin-film deposition, in *Electronic Materials Science and Technology*, S. P. Murarka and M. C. Peckerar, eds., Academic Press, San Diego, CA, 1989, Sec. 6.4, p. 296.
4. L. Holland, *Vacuum Deposition of Thin Films*, Wiley, New York, 1961.
5. L. Maissel, Application of sputtering to the deposition of film, in *Handbook of Thin-Film Technology*, M. I. Maissel and R. Glang, eds., McGraw-Hill, New York, 1970.
6. W. Kern and V. Ban, Chemical vapor deposition of inorganic thin films, in *Thin-Film Processes*, J. L. Vossen and W. Kern, eds., Academic Press, San Diego, CA, 1978.
7. D. S. Campbell, The deposition of thin films by chemical method, in *Handbook of Thin-Film Technology*, L. I. Maissel and R. Glang, eds., McGraw-Hill, New York, 1970.
8. J. F. O'Hanglong, *A Users' Guide to Vacuum Technology*, Wiley, New York, 1980.

9. R. R. Lapalle, *Practical Vacuum System*, McGraw-Hill, New York, 1972.
10. R. W. Berry, P. M. Hall, and M. T. Harris, *Thin-Film Technology*, Krieger Publishing, Melbourne, FL, 1979.
11. J. P. Herth and G. M. Pond, *Condensation and Evaporation, Nucleation, and Growth Kinetics*, Macmillan, New York, 1963.
12. W. H. Kohl, *Handbook of Materials and Techniques for Vacuum Devices*, Reinhold, New York, 1967.
13. C. F. Powel, J. H. Oxley, and J. M. Blocher, eds., *Vapor Deposition*, Wiley, New York, 1966.
14. S. Dushman, *Scientific Foundations of Vacuum Technology*, Wiley, New York, 1962.
15. R. Glang, R. A. Holmwood, and J. A. Kurtz, High-vacuum technology, in *Handbook of Thin-Film Technology*, L. I. Maissel and R. Glang, eds., McGraw-Hill, New York, 1970, p. 2-112.
16. W. S. Ruska, Vacuum science and technology, in *Microelectronic Processing*, McGraw-Hill, New York, 1987.
17. S. Wolf and R. N. Tauber, Vacuum technology for VLSI applications, in *Silicon Processing in the VLSI Era*, Vol. I, Lattice Press, Sunset Beach, CA, 1986, Chap. 3.
18. J. F. Hanlon, *A User's Guide to Vacuum Technology*, Wiley, New York, 1980.
19. A. Roth, *Vacuum Technology*, North-Holland, Amsterdam, 1982; S. Wolf and R. N. Tauber, High-vacuum pumps, III: Turbomolecular pumps, in *Silicon Processing in the VLSI Era*, Vol. I, Lattice Press, Sunset Beach, CA, 1986, p. 95.
20. M. H. Hablanian, *Vac. Technol. Coat.*, Vol. 1, No. 7, Sept. 2000.
21. S. Kruger and A. H. Shapiro, *Proc. 7th Nat. Symp. Vac. Technol.*, C. R. Meissner, ed., Pergamon Press, New York, 1991.
22. Courtesy of Lebold vacuum USA, with permission of Kurt J. Lesker Clairton, PA.
23. R. Scholl, Cryopumping in semiconductor applications, *Solid State Technol.*, Dec. 1983, p. 83; S. Wolf and R. N. Tauber, High-vacuum pumps, II: Cryopumps, in *Silicon Processing in the VLSI Era*, Vol. I, Lattice Press, Sunset Beach, CA, 1986, p. 92.
24. K. M. Welch, Closed-loop gaseous helium cryopump, *Vac. Technol. Coat.*, Sept. 2000.
25. J. Balliga II, Cryopump applications and operation, *Microelectron. Manuf. Test.*, Oct. 1984, p. 24.
26. P. Danielson, Back streaming from oil-sealed pumps, *Vacuum Thin Film*, July 1999, p. 8.
27. D. G. Powell, A multiple resistivity thin-film process, *Proc. Int. Symp. Microelectron.*, 1975, p. 114; *Product and Vacuum Technology Reference Book*, Leybold-Hereaus Vacuum Products, Export, PA.
28. R. Glang, Vacuum evaporation, in *Handbook of Thin-Film Technology*, L. I. Maissel and R. Glang, eds., McGraw-Hill, New York, 1970, Chap. 1, pp. 92–97.
29. E. K. Muller, B. J. Nicholson, and G. L. Turner, *J. Electrochem. Technol.*, Vol. 1, 1963, p. 158.
30. B. Chapman, Sputtering, in *Glow Discharge Processes*, Wiley, New York, 1980, Chap. 6.

31. K. Wasa and S. Hayakawa, *Handbook of Sputter Deposition*, Noyes Publications, Park Ridge, NJ, 1991.
32. D. A. Glocker, An estimate of the potentials developed on coated anodes during pulsed DC reactive sputtering, *Vac. Technol. Coat.*, Aug. 2000, p. 52.
33. D. M. Mattox, Sputter deposition, *Vac. Technol. Coat.*, Nov./Dec. 2000, p. 46.
34. P. K. Mirkarimi, S. L. Baker, M. A. Wall, and P. A. Kearney, Sputter deposition process, *Solid State Technol.*, Nov. 2000, p. 96.
35. S. Rossanagel, D. Mikalsen, H. Kinoshita, and J. Cuomo, Collimated magnetron sputtering deposition, *J. Vac. Sci. Technol.*, Vol. A9, 1991, pp. 261–265.
36. S. Rossnagel, Sputter deposition for semiconductor manufacturing, *IBM J. Res. Dev.*, Vol. 43, 1999, pp. 163–179.
37. J. L. Vossen, *J. Vac. Sci. Technol.*, Vol. 8, 1971, p. S12.
38. H. J. Geippel et al., *IEEE Trans. Electron. Devices*, Vol. 27, 1980, p. 1417.
39. S. P. Murarka, *Silicides for VLSI Applications*, Academic Press, San Diego, CA, 1983.
40. K. P. Yap, H. Gong, J. Y. Dai, T. Osipowicz, L. H. Chan, and S. K. Lahiri, Integrity of copper–tantalum nitride metallization under different ambient conditions, *J. Electrochem. Soc.*, Vol. 146, No. 4, 1998, p. 1256.
41. R. S. Nelson and M. W. Thomson, *Proc. R. Soc. London*, Vol. A259, 1961, p. 458.
42. G. S. Anderson and G. K. Wehner, *J. Appl. Phys.*, Vol. 31, 1960, p. 2305.
43. R. Nelson, *Philos. Mag.*, Vol. 8, 1963, p. 693.
44. W. S. Ruska, Thin films, mainly non-metals (CVD), in *Microelectronic Processing*, McGraw-Hill, 1987, Chap. 8; D. E. Rosner, *Transport Processes in Chemically Reacting Systems*, Butterworth Publishers, Boston, 1986.
45. V. Bean, Chemical vapor deposition of inorganic thin films, in *Thin-Film Processes*, J. L. Vossen and W. Kern, eds., Academic Press, San Diego, CA, 1978, pp. 251–331.
46. M. A. Green and R. A. Levy, Chemical vapor deposition of metals for IC applications, *J. Met.*, Vol. 37, 1985, p. 67.
47. D. B. Beach, Design of low-temperature thermal CVD process, *IBM J. Res. Dev.*, Vol. 34, No. 6, Nov. 1990, p. 795.
48. E. K. Broadbent, *Tungsten and Other Refractory Metals for VLSI Applications*, Vols. 1 and 2, Materials Research Society, Pittsburgh, PA, 1987.
49. R. Kroger, M. Eizenberg, D. Cong, N. Yoshida, L. Chen, R. Ramaswami, and D. Carl, Properties of copper film prepared by chemical vapor deposition for advanced metallization of microelectronic devices, *J. Electrochem. Soc.*, Vol. 146, No. 9, 1999, pp. 3248–3252.
50. K. K. Choi and S. W. Rhee, Effect of carrier gas on CVD of copper with hfac and DMB, *J. Electrochem. Soc.*, Vol. 146, No. 7, 2001, pp. 473–478.
51. S. P. Murarka and M. C. Peckerar, Chemical vapor deposition, nucleation, and growth of thin films, in *Electronic Materials Science and Technology*, Academic Press, San Diego, CA, 1989.
52. W. Kern and G. L. Schnable, Low-pressure chemical vapor deposition for VLSI processing: a review, *IEEE Trans. Electron. Devices*, Vol. 26, 1979, p. 647.

53. W. R. Sobie, Ion-beam technology for thin film applications, *Vac. Thin Film*, Apr., 1999, p. 36.
54. G. Stengel, R. Kaitna, H. Loschner, P. Wolf, and R. Sacher, Ion projection system for IC productions, *J. Vac. Sci. Technol.*, Vol. 16, 1979, p. 1883.
55. G. Carter and J. S. Collington, *Ion Bombardment of Solids*, American Elsevier, New York, 1968.
56. J. S. Logan, Control of RF-sputtered film properties through substrate tuning, *IBM J. Res. Dev.*, Vol. 14, 1970, p. 172.
57. P. D. Reader and H. R. Kaufman, Optimization of an electron-bombardment-ion source for ion machining applications, *J. Vac. Sci. Technol.*, Vol. 12, 1975, p. 1344.
58. P. R. Simon and J. Krebill, State-of-the-art third-generation thin-film resistive films, *Proc. Int. Symp. Micrelectronics*, Atlanta, GA, 1986, p. 525.
59. J. E. Harry, *Industrial Lasers and Their Applications*, McGraw-Hill, New York, 1974; R. A. McMahon and H. Ahmed, in *Laser and Electron Beam Processing of Electron Materials*, C. L. Anderson, G. K. Celler, and G. A. Rozgonyi, eds., Electrochemical Society, Pennington, NJ, 1980, p. 123.
60. J. Wilson and J. F. D. Hawker, *Lasers: Principles and Applications*, Prentice Hall, Upper Saddle River, NJ, 1987.
61. M. Ross, ed., *Lasers Applications*, Academic Press, San Diego, CA, 1972; T. O. Yep, R. T. Fulks, and R. A. Powel, in *Laser and Electron-Beam Solid Interactions and Materials Processing*, J. F. Gibbons, L. D. Hess, and T. W. Sigmons, eds., North-Holland, New York, 1981.
62. A. E. Siegman, *Lasers*, University Science Books, Mill Valley, CA, 1986, p. 1004; R. M. Osgood and H. H. Gilgen, *Annu. Rev. Mater. Sci.*, Vol. 15, 1985, p. 459.
63. D. B. Chrisey and G. K. Hubler, eds., *Pulsed Laser Deposition*, Wiley, New York, 1994.
64. S. Boughaba, M. U. Islam, J. A. Greer, and M. Tabat, Large-area pulsed laser deposition, *Vac. Thin Film*, Vol. 2, No. 10, 1999, p. 14.
65. S. P. Murarka and M. C. Peckerar, Future beam processes and in-situ processing, in *Electronic Materials Science and Technology*, Academic Press, San Diego, CA, 1989, Chap. 11, p. 536.
66. T. Shibata, A. Wakita, T. W. Sigmon, and J. F. Gibbons, Processing of silicon and other semiconductors, in *Semiconductors and Semimetals*, Vol. 17, J. F. Gibbons, ed., Academic Press, San Diego, CA, 1984.
67. A. J. Basovich et al., *Thin Solid Films*, Vol. 228, 1993, pp. 193–195; F. Sanchez et al., *Appl. Sci.*, Vol. 69, 1993, pp. 221–224.
68. R. Sobolewski and W. L. Kula, Book of abstracts, conference on the science and technology of thin film superconductors, Nov. 14–18, 1988, Colorado. Springs, Co.
69. H. Takata, O. Murata, T. Kawai, S. Kawai, and M. Okuyama, *Appl. Phys. Lett.*, Vol. 64, No. 4, 1994, p. 428.
70. D. R. Cote, S. V. Nguyen, A. K. Stamper, D. M. Armburst, D. Tobben, R. A. Conti, and G. Y. Lee, Plasma-assisted CVD for ULSI semiconductor circuits, *www.research.ibm.com/journal/rd/431/cote.html*, 2001.

71. J. Reid and S. Mayer, Factors influencing fill of IC features using electroplated copper, Novellus Systems, Portland Technology Center, Wilsonville, OR, private communication.
72. T. Taylor, T. Ritzdrof, F. Lindberg, B. Carpenter, and M. LeFebre, *Solid State Technol.*, Nov. 1997, p. 47.
73. S. S. Kruglikov, N. T. Kdudryavtsev, and V. I. Sinyakov, *Trans. Inst. Met. Finish.*, Vol. 44, 1966, p. 152.
74. *www.semiconductors.net/technical/damascene_copper.htm.*
75. E. K. Broadbent et al., Experimental and analytical study of seed layer resistance for copper damascene electroplating, *J. Vac. Sci. Technol.*, Vol. B17, Nov./Dec. 1999, p. 2584.
76. C. Ji, G. Oskam, and P. C. Searson, Electrodeposition of copper on silicon from sulfate solution, *J. Electrochem. Soc.*, Vol. 148, No. 11, 2001, pp. 746–752.
77. G. Oskam, J. G. Long, A. Natarajan, and P. C. Searson, *J. Phys.*, Vol. D31, 1998, p. 1927.
78. F. Barlow, A. E. Riad, and R. Brown, Film deposition techniques and processes, in *Thin-Film Technology Handbook*, A. E. Riad and F. D. Barlow III, eds., McGraw-Hill, New York, 1999, Chap. 1.
79. T. K. Gupta, Optical properties of polycrystalline CdSe films, *J. Mater. Res.*, Vol. 7, No. 5, May 1992, p. 1243.
80. T. K. Gupta, Anomalous transport properties in CdSe films, in *Solar Energy Mater. Solar Cells*, Vol. 27, 1992, p. 327.
81. T. K. Gupta, Preparation and characterization of layered superconductors, *Phys. Rev.*, Vol. B43, No. 7, 1991.
82. C. J. Brinker and G. W. Scherner, *Sol-Gel Science*, Academic Press, San Diego, CA, 1990; D. Segal, *Chemical Synthesis of Advanced Ceramic Materials*, Cambridge University Press, Cambridge, 1989.
83. L. C. Klein, ed., *Sol-Gel Technology for Thin Films, Fibers, Preforms, Electronics, and Specialty Shapes*, Noyes Publications, Park Ridge, NJ, 1988.
84. T. Suntola and M. Antson, U.S. patent 4,058,430, 1977.
85. T. Suntola and M. Simpson, eds., *Atomic Layer Epitaxy*, Blackie & Sons, London, 1990.
86. S. M. Bedar, Atom layer epitaxy and deposition processes, *J. Vac. Sci. Technol.*, Jan./Feb. 1994; S. M. George, A. W. Ott and J. W. Klaus, Surface chemistry of atomic layer growth, *J. Physical Chemistry*, Vol. 100, 1996, p. 13121.
87. A. E. Braun, ALD breaks materials conformality barrier, *Semicond. Int.*, Oct. 2001, p. 52.
88. S. Munley, I. McNaught, D. Mrotek, and C. Y. Lin, Vacuum pump, *Semicond. Int.*, Oct. 2001, p. 77.
89. M. Malac, R. Egerton, and M. Brett, Thin films deposited at glancing incidence and their applications, *Vac. Technol. Coat.*, July 2001, p. 49.
90. K. Robbie, M. J. Brett, and A. Lakhtakia, *J. Vac. Sci. Technol.*, Vol. A13, No. 6, 1995, pp. 2991–2993.
91. H. A. Macleod, *Thin-Film Optical Filters*, 3rd ed., American Institute of Physics, New York, 2001.

92. A. T. Wu, M. Seto, and M. Brett, Capacitive SiO humidity sensors with novel microstructures, *Sensors Mater.*, Vol. 11, 1999, pp. 493–505.

93. K. D. Harris, D. Vick, E. J. Gonzales, T. Smy, and K. Robbie, Porous thin films for thermal barrier coatings, Surface Coatings Technol., Vol. 138, 2001, pp. 185–191.

94. M. J. Brett, *Surf. Coat. Technol.*, Vol. 138, 2001, pp. 185–191.

95. M. J. Colgan and M. Brett, Photonic bandgap materials, *Thin Solid Films*, Vol. 391, 2001.

96. M. Tokushima, H. Kosaka, A. Tomita, and H. Yamada, *Appl. Phys. Lett.*, Vol. 76, No. 8, 2000, pp. 952–954.

97. A. Chutinan and S. Noda, *Phys. Rev.*, Vol. B57, 1998, p. 2006.

98. B. Dick, M. J. Brett, T. J. Smy, M. Freeman, M. Malac, and R. Egerton, *J. Vac. Sci. Technol.*, Vol. A18, No. 4, 2000, p. 1838.

99. K. Fluri, G. Fitzpatrick, N. Chiem, and D. J. Harrison, Electronic microchip for biological applications, *Anal Chem.*, Vol. 68, 1996, p. 4285.

100. J. Reid et al., *Solid State Technol.*, July 2000.

101. R. V. Stuart, *Vacuum Technology, Thin Films, and Sputtering*, Academic Press, New York, 1983.

102. P. J. Kelly and R. D. Arnell, Magnetron sputtering: a review of recent developments and applications, *Vacuum*, Vol. 56, 2000, p. 159.

103. G. Y. Lee et al., A low deposition rate high-density plasma CVD process for high aspect ratio, 175-nm technology and beyond, *Proc. IEEE Tech. Conf.*, 1999, p. 152.

RECOMMENDED READING

Brown, S. C., *Introduction to Electrical Discharges in Gases*, Wiley, New York, 1966.

Chapman, B. N., *Glow Discharge Process*, Wiley, New York, 1980.

Chapman, B. N., and J. C. Anderson, eds., *Science and Technology of Surface Coating*, Academic Press, London, 1974.

Dushman, S., *Scientific Foundations of Vacuum Techniques*, 2nd ed., Wiley, New York, 1962.

Holland, L., *Vacuum Deposition of Thin Films*, Wiley, New York, 1956.

Kohl, W. H., *Handbook of Materials and Techniques for Vacuum Devices*, Reinhold, New York, 1967.

Lapalle, R. R., *Practical Vacuum System*, McGraw-Hill, New York, 1972.

Maissel, L. I., and R. Glang, eds., *Handbook of Thin Solid Films*, 2nd ed., McGraw-Hill, New York, 1983.

O'Hanlon, J. F., *A User's Guide to Vacuum Technology*, Wiley, New York, 1980.

Townsend, P., and J. C. Kelly, *Sputtering and Ion Implantation*, Academic Press, London, 1976.

8

COMPONENT ASSEMBLY AND INTERCONNECTIONS

8.1 COMPONENT ASSEMBLY

Surface-mounting technology (SMT) has been emerging as one of the most significant technologies in the electronics industry. As the name implies, SMT is the process of mounting discrete active and passive components on the surface of a hybrid circuit board. The technology is basically the application of science and engineering. SMT eliminates the trouble of making holes on the substrate to make interconnections. Although SMT is a straightforward method, its impact on the production floor is enormous not only on components and design but also on materials and equipments [1]. Starting in the late 1970s through to the 2000s, SMT had established itself as an engineering tool to fabricate the lowest-cost packaging units, with a fine pitch [ca. 16 mils (406 μm)] [2] and highest function card assembly format. The arduous path of developing a widespread and universally accepted assembly technology infrastructure from tooling assembly processes, including package formats, has been forged. The results of these efforts are formidable; SMT continuously and routinely produces high assembly yields and reliable subassembly products [3]. Now at the beginning of the twenty-first century, SMT has reached a critical crossroads in defining a technology road map that will capture packages with a high number of interconnections (>200 I/Os) [4].

The placement of components such as capacitors and resistors has been the key topic of discussion in the assembly industry. The demand for smaller size, light weight, and low-powered hybrid microcircuits by many board assemblers

Handbook of Thick- and Thin-Film Hybrid Microelectronics, By Tapan K. Gupta
ISBN 0-471-27229-9

has led to even smaller components and advanced technology in these products. For machine manufacturers, the challenge is to develop assembly equipment that is more resistant to obsolescence in an age of dynamic technological change [5,6]. According to an observer, an automated selective assembly system capable of automatic assembly of surface-mounting components (SMCs) with high quality and high-volume production is the main concern for SMT manufacturers [7]. As a matter of fact, component assembly requires an automated selective assembly system that can dispense solder paste selectively, place the missing components, and reflow solder without affecting previously assembled components [7]. The impact and benefits of this type of component reassembly to the industry in terms of both cost and performance are enormous [8]. With the increased use of SMT; active and passive elements in chip form have experienced a dramatic increase in use. Nevertheless, techniques for putting these chips over thick- and thin-film hybrid substrates that have a high density of circuitry pose a real challenge in the hybrid circuit industry. The mechanical assembly of a thick- and thin-film hybrid circuit consists primarily of affixing active or passive components in chip form to conductor regions of an already established conductor or resistor pattern [9]. To achieve the most satisfactory, repeatable, and reliable results in bonding alloy placement by the use of a metal/chemical system, selection of a particular type of bonding system is critical. A typical procedure for mounting active and passive devices that has the required temperature range must have the metallurgical capability of absorbing stress generated by the thermal mismatch between substrate and components. An additional consideration is that the bonding alloy selected must have a lower flow point than that of alloy used to affix the component [10].

8.1.1 Silicon–Gold Eutectic Bonding

A eutectic diagram results when the addition of a component to a melt lowers its overall freezing point, as shown in Figure 8.1. Here the freezing point of the

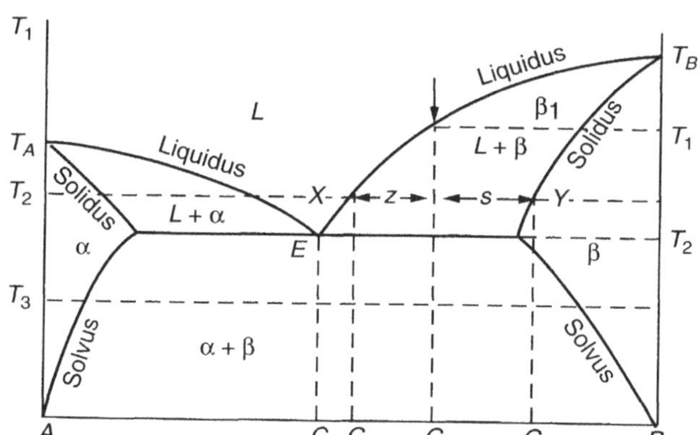

Figure 8.1 Binary-phase diagram [10].

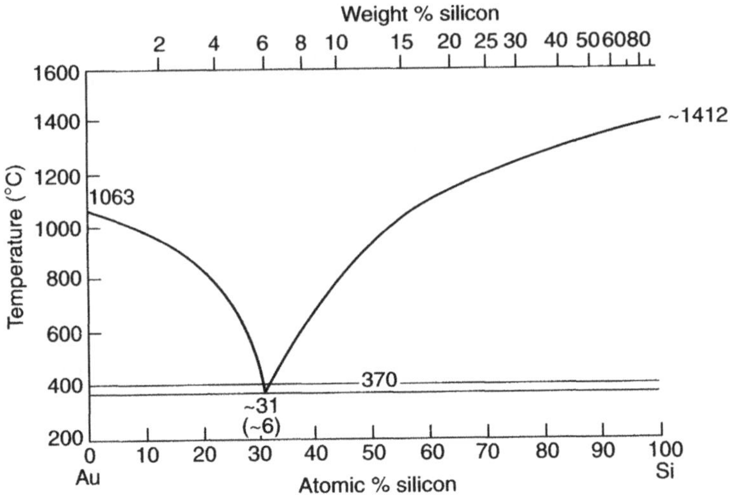

Figure 8.2 Gold–silicon system [11].

molten mixture has a minimum value T_E, below T_A and T_B. This minimum value is known as the *eutectic point* and the corresponding mixture C_E is called the *eutectic composition*. Eutectic systems usually occur when two components are completely miscible in the liquid phase but are only partly soluble in the solid state.

One of the most common techniques for silicon (Si) chip attachment is the use of gold to form an Au–Si eutectic alloy [11,12]. The melting point of gold is about 1063°C, and that of silicon is about 1414°C, yet the melting point of 81.4 Au together with 18.6 at % Si alloy is 363°C [12]. A gold–silicon eutectic system is shown in Figure 8.2. In eutectic bonding a small amount of alloy acts as a mortar between the silicon chip and the gold bond pad on the substrate. Thus around 363°C, both surfaces are joined together, ensuring better thermal characteristics and stability [13].

During bonding, the unit is usually agitated or scrubbed ultrasonically to ensure uniform contact between preform and chip. To prevent oxidation, the surrounding area is covered with an inert atmosphere. Another consideration is the stress developed due to thermal expansion mismatch between the die, the gold eutectic interface, and the lead frame material. This results in microcracks on the die. Although more expensive, the eutectic die attach process (primarily, a gold–silicon eutectic) is a reliable method of die attachment that is still used widely in military applications and most hermetic packages, where the demand for high performance and reliability overrides the cost.

8.1.2 Adhesive and Epoxy Bonding

Polymers The current worldwide market for die-attach adhesives, especially plastic dual-inline packages (DIPs), is based primarily on epoxy thermoset technology [14]. Presently, the semiconductor market requires solventless, high-purity (<10 ppm each of Na^+, K^+, NH_4^+, and Cl^-), fast-cured (10 to 30 min at 110°C), long-shelf life/pot life, low-stress, and highly reliable single-component epoxy materials, that show good adhesive properties at wire bonding temperatures. Epoxies are now used for assembly of high-reliability hybrid microcircuits. Many commercial applications are now common [15], but recent advances in epoxy technology permit the use of epoxy for device bonding as long as there is adequate documentation of process control. The compatibility of an epoxy material is determined not only by its mechanical stability but also by its electrical behavior [16]. Since the bonding of the device with epoxy is a surface-dependent phenomenon, the cleanliness and mechanical integrity of the bonding is extremely important.

Electrical conduction of conductive epoxy as a function of both cure schedules and shelf age may be significant in applications involving saturation pulse current. DuPont has developed a thermoplastic ribbon die-attach adhesive and claim that it will meet some of the requirements of the electronic industry [18]. These QL series adhesives are aromatic polyester, polyimide, or acrylic. The coefficient of thermal expansion of these epoxies varies between 5.3 and 21.3×10^{-5} m/m·°C, the thermal conductivity varies between 0.23 and 0.63 W/m·K, and the thermal stability varies between 0.005 and 0.003% weight loss per second at 470°C. The adhesive flip chip is one of the best developed and most promising technologies [18–20]. The adhesive assembly processes can be stenciled or dipped. In the stenciled adhesive process conductive adhesive is applied through metal stencils and the typical thickness ranges from 1 to 4 mils (25 to 100 μm). Depending on pad size and pitch, the stencil openings can vary from 1.5 to 4 mils (37 to 100 μm). On the other hand, in the dipped adhesive assembly system, a thin layer of conductive adhesive is spread with a doctor blade and the test chip is dipped into the adhesive layer. The dipped adhesive connection is better suited to closely spaced contacts, but it has a drawback: The contacting surface gets less conductive adhesive and requires more careful control of uniform spreading of the adhesive to achieve a satisfactory electrical and mechanical joint [21]. A polymer adhesive (EP38) developed by Masterbond [22] is described as a resin-based coating composition which can be used as an epoxy coating system and can be cured at 20 to 25°C. According to NASA space application specifications, a cured epoxy must not exhibit condensable volatile exceeding 0.1% of its original cured weight during exposures to 275°F (125°C) for 24 h at a pressure of 1×10^6 Torr. Figure 8.3 shows an automated die-attach instrument.

Silver–Glass Epoxy Conductive epoxies are formulated to meet practically any processing and application requirement [23]. Gold and silver conductive epoxies are noted for their excellent electrical and mechanical properties [24,25]. An excellent feature of conductive epoxies is the fact that they are solvent free and are able to withstand temperature cycling from −55 to +125°C without physical or electrical degradation. Cured epoxies show low shrinkage, low creep, and are virtually insensitive to moisture and most solvents. Either

Figure 8.3 Automated die-attach instrument [17]. (Courtesy of Ablestick, Gardena, CA.)

gold or silver (70%) is used as a filler and blended approximately with 30% resin. Conductivity is dependent on the filler concentration. Some epoxies do not contain any solvents. Therefore, after cure no outgassing occurs to corrode the circuit.

Organic adhesive bonds are cured at low temperatures, are easy to rework, and have outstanding potential usefulness for attaching semiconductor chips to hybrid circuits [24,25]. By eliminating the need for a solderable plate, potentially large cost savings can be realized. In addition, the low curing temperatures of epoxies make them attractive for attaching chips which are sensitive to normal die-bonding temperatures. Epoxy die attachment combined with ultrasonic wire bonding comprises a low-temperature assembly and packaging process. Silver-filled epoxies have great potential for use in low-cost die bonding to an inexpensive metal surface. Their mechanical, electrical, and thermal properties are consistently reliable after very long exposures at 150°C and even show good performance after nearly 1000 h at 200°C. However, temperature-humidity-bias testing shows corrosion due to residual mobile impurities such as chloride, which remains after processing. Strengths have been found to decrease significantly after storage at 200 and 300°C. Silver-filled epoxies with a low ion concentration are available in the market and are said to have a 24-h pot life. Different low-ion silver-filled epoxies are available in the market. These conductive epoxies develop strong, durable, void-free, thermally conducting bonds and can be screen-printed easily [26].

Recent development of epoxy application techniques has helped the microwave engineers to fabricate hybrid circuits up to 60 GHz. Traditional eutectic bonding methods limit the substrates available to the microwave engineer to those that can tolerate the high temperatures required in eutectic die bonding, limiting the choice of dielectric constants and substrates. Epoxies suitable for use at microwave frequencies provide a wide choice of dielectric constants. The flexibility, brittleness, and thermal sensitivity of various substrates present some challenging fabrication problems, but epoxies have proven their effectiveness in fabricating microwave hybrid devices.

Whether it is the precise (within ± 0.003 in. or 0.0075 cm) placement of a PIN diode in an assembly of a 15-GHz attenuator inside a rigid wave cavity, or the attachment of a gallium arsenide (GaAs) beam-leaded Schottky barrier diode, or thermocompression or any type of wire bonding, an epoxy bonding system is a much simpler and more cost-effective way to bond a component in a complex hybrid microcircuit. However, much of its success will depend on precisely how one can dispense the epoxy. In fabricating a 60-GHz mixer using Kapton, a gallium arsenide (GaAs) beam-leaded Schottky barrier diode is to be affixed to the metallized kapton. Similarly, in fabricating a 5- to 6-GHz amplifier, a 160-Ω impedance line between the drain of the first and second stages and the drain bias supply is realized by welding a 0.0007-in.-diameter wire to the bias line. In all these cases, epoxy bonding is the simpler solution to thermocompression or any wire bonding.

With the advent of GaAs as a practical semiconductor material, new and even stricter demands are placed on the conductive epoxy materials. Ionic purity of electrically and thermally conductive die-attach adhesives used in hybrid manufacturing still continues to be a key issue [27]. Another problem commonly encountered with conductive die-attach adhesives is the *raining* of volatilize.

Rheological properties play a critical role in the formulation of adhesives [28]. A new adhesive formulated by DuPont is a silver–glass die-attach adhesive (DITAC 4621) which consists of 69 wt % silver flake, 17 wt % glass frit, 13% organic solvent, and 1% acrylic binder. When the composition is fired for bonding, the organic volatilizes and the frit sinters to form the cohesive silver–glass matrix [29].

8.1.3 Solder Joint Bonding

In an assembly of hybrid microcircuits, passive and active discrete chip components are attached by a wide variety of methods, ranging from soldering and eutectic die bonding to epoxies and other cements. Soldering is one of the oldest metallurgical joining methods and has been adopted in industrial practice for a long time [30,31]. Solder joints in thick-film circuits are expected to show reliability at least equal to thick-film component reliability. Frequently, the reliability of a solder joint is determined superficially by visual inspection. This may or may not give the true functionality of the solder joint. True solder joint integrity can only be demonstrated from qualitative studies of its mechanical, physical, and chemical properties.

In the solidified state, 60% Sn/40% Pb is comprised of grains that contain alternating lamellae of nearly pure tin (Sn) and lead (Pb) phase [32]. The lead phase constitutes 30% of the solder volume, but the number of lamellae varies with solidification. In lead-free soldering, solder material consists of 3.6 vol % silver mixed in a nearly pure tin matrix [33]. The lead–tin binary system is

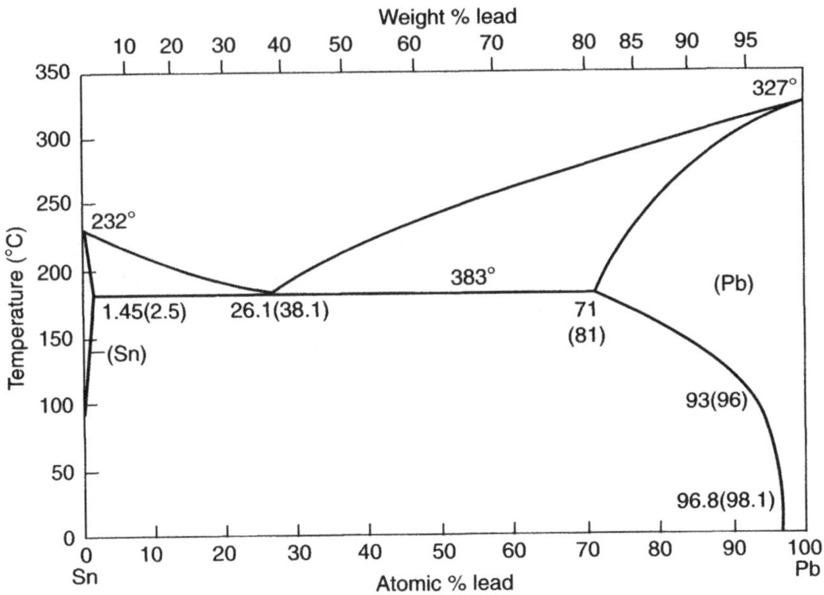

Figure 8.4 Lead–tin system [12].

shown in Figure 8.4. The eutectic point has 38 wt % lead and 62 wt % tin and the eutectic temperature is 183°C. The freezing points of pure tin and pure lead are 232 and 327°C. Tin–lead alloys exhibit superplastic behavior in which strains of several hundred percent can be achieved prior to fracture [34]. When the strain rates supersede the superplastic behavior of the material, a tensile sample develops localized neck fracture. Interconnected dimples, sharp tear ridges, and delamination at grain boundaries characterize the fracture surface [35] and broken solder connections of the hybrid circuit assembly [36]. The solder joint is not the only limiting factor that controls circuit reliability. It is important to test the reliability of thick-film components to obtain reliable solder attachment [37]. Many different soldering methods are in use today for soldering the attachment of surface-mounting devices to a hybrid microcircuit [31].

Manual soldering underneath a microscope is always possible but is very tedious and labor intensive. Six soldering methods are available for SMDs: hand, hot belt, infrared, wave, vapor phase, and laser. Current reflow solder attachment processes for surface-mounting and through-hole components (THC) may provide economic and product reliability advantages. However, the *pin in paste* process has gained popularity as surface-mount technology matures and the quantity of THCs diminishes. As a matter of fact, some THCs, such as connectors, transformers, and inductors, as well as other soldering processes, have been successful in reducing the number of soldering operations and labor-intensive component insertion operations through the pin in paste process. Consequently, an increase in productivity is experienced through reduction in manufacturing cost and increased throughput [38,39].

A life prediction approach has been developed for solder joints under thermal fatigue. Two failure criteria are used to define the final failure of the solder joint. The first is defined by a critical crack length which is adjusted to the arbitrary fraction of the joint width, and the second is the conventional approach that failure occurs when $J_{integral} \geq Jc$, where Jc is the fracture toughness of the solder joint and is a property of the material [40,41].

Hand Soldering Surface-mounting hand soldering is sometimes more challenging than through-hole soldering because of the smaller pitches and higher lead counts. Contact and heated hand soldering are the most common hand-soldering methods adopted in surface-mount technology. It is the most cost-effective way of attaching a discrete device [42]. It requires excellent operator skill and good tools and is very time consuming. However, automated robotic soldering systems with technology that monitors the performance of the soldering tips are now available. Figure 8.5 shows an automated robotic soldering machine.

In contact soldering the tip or collar is attached to a soldering tool which transfers the heat to a solder alloy and the joint simultaneously. Soldering tips need to be cleaned from time to time, depending on the temperature and number of uses. The cost of contact soldering equipment can vary according to the additional tools with which it is equipped. For surface-mount application the operating temperature for a constant-temperature system is in the range 335 to 365°C, and for a restricted temperature system it is 285 to 315°C [44]. In heated gas soldering, heated air or an inert gas is directed to the solder and joints [45]. The heated gas temperature range varies from 300 to 400°C. Nozzle design is very important in heated gas soldering. Surface-mount soldering generally requires a solder wire diameter in the range 20 to 30 mils (500 to 750 μm), whereas through-hole solder wire diameter ranges from 50 to 62 mils (1.3 to 1.6 mm) [46]. A heated gas soldering system avoids the localized heat stress that

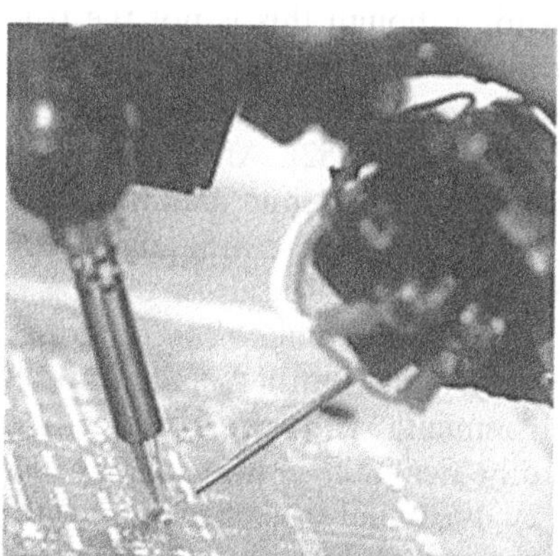

Figure 8.5 Automated robotic soldering machine [43]. (Courtesy of Automated Unlimited, Woburn, Massachusetts.)

may occur with a contact solder system. The heat transfer mechanism in hand soldering follows Fourier's law of heat conduction [47]. Typically, input energy is high when the heat source is in intimate contact with the pad, and as a result, the reflow rate of the solder metal is high. The high flow rate cools the solder metal faster, and under a high cooling rate, lead-rich dendrites disappear because there are many more nucleation sites for lead. Under a fast cooling system, growth kinetics is generally not established and the resulting microstructure consists of a fine eutectic matrix with supersaturated lead.

Hot Belt Soldering Hot belt soldering was developed by the hybrid microelectronics industry for soldering SMDs to ceramic substrates. Although this method is still popular, it has limitations of maximum substrate size, throughput, and adequate heat distribution to devices with widely differing heat requirements, heat tolerance, and belt vibration [48]. Predominant heat transfer mechanism for a hot belt reflow system is convection. Since the convection fluid is a gas (generally N_2), convective heat transfer coefficients are typically an order of magnitude lower than those encountered in vapor-phase soldering [47]. This means that the total system temperature must be maintained much higher than in a condensation-type system.

Infrared Soldering Focused and unfocused infrared (IR) soldering have not enjoyed wide use because of difficulty with obtaining adequate heat distribution to devices without causing heat damage. In IR soldering, heat transfer is accomplished primarily by radiation and the system is energy rich. The heat transfer depends on the fourth power of the differences in absolute power between the radiation source and the workpiece [49,50].

Wave Soldering The concept of wave soldering was developed in the early 1980s and the technology is still used widely by engineers and those engaged in surface-mount assembly [51]. Theoretical kinetics are still basic and simple. The attachment method is the same as the basic principle of soldering with flux, heat, and solder to create a metallurgical connection. The preferred alloy for wave soldering is eutectic 63% tin and 37% lead, although this is not the true theoretical eutectic composition. Wave soldering has a shortened contact time, and the inclination of the conveyor belt allows molten solder to peel off into the pot, reducing bridging between neighboring solder joints [52]. After the solder pots in wave soldering machines are run for long periods of time, oxidation and intermetallic formation change the pot composition [53]. Approximately 60,000 metric tons of solder are used annually worldwide [54]. Recently, there has been an increased worldwide focus on the use of lead-free solder alloy to reduce lead exposure [55]. The U.S. Occupational Safety and Health Administration (OSHA) strongly recommends a lead-free environment. Lead-free solder should meet the following criteria: (1) it should be readily available and the cost should be comparable to that of lead–tin solder; (2) it should be compatible with existing lead–tin solder as to good joint strength wettability; and (3) it should

Figure 8.6 Commercial computer-controlled motorized conveyor wave soldering machine. (Courtesy of Novastar Technologies, Huntington Valley, Pennsylvania.)

have thermal and electrical conductivity comparable to that of existing lead–tin solder. However, wave soldering has limited use with SMDs because of their need for adhesive or other mechanical attachment and thermal shock to large ceramic packages. Figure 8.6 shows a commercial automated wave soldering machine.

Vapor-Phase Soldering Since its introduction in 1975, vapor-phase soldering has become popular. Basically, this method involves exposure of a wire assembly (WA) to hot vapor over a fluid whose boiling point is slightly higher than that required to melt the solder. Absorption of heat by the WA causes the solder to fuse and solidify. In vapor-phase soldering solder joint forms a grainy structure, which may be shiny or dull, depending on several conditions [56]. The vapor condenses on the WA, and the condensate drains back into the boiling liquid for reuse. The most commonly used fluid is perfluorotriamylamine, a common tradename being Fluorinate FC-70 (3M Company). This material boils at 215°C, which exceeds the melting point of tin–lead eutectic alloy (62% tin and 36% lead with 2% silver with a melting point of 179°C, or 60% tin and 40% lead with a melting point between 183 and 188°C) [57]. Two types of vapor-phase soldering systems are commercially available: batch type and inline. In the batch type [30], the PWA is lowered vertically into the hot vapor zone and pauses until solder fusion has occurred. Then the PWA is slowly raised out of the hot vapor to allow the molten solder to solidify. Inline carriers carry the PWA horizontally in one direction on a continuous open weave. It is then coated on a metal belt through the hot vapor zone. During its passage through the cool-down zone, the molten solder solidifies. Vapor-phase SMDs have a number of advantages and disadvantages. The advantages are: (1) soldering can be accomplished at a constant process temperature; (2) simultaneous solder melting and solder joint is possible; and (3) due to simultaneous solder fusion, the surface tension of the molten solder allows self-aligning orientation of 64-pad devices. The disadvantages of this method are: (1) it cannot be used with heat-sensitive SMDs; (2) due to formation of intermetallic compound and long heat cycles, solder joints are dull and brittle; and (3) excessive heat sometimes put stress on the circuit.

Laser Soldering Two types of lasers are very much in use for this purpose: yttrium–aluminum–garnet (YAG) and carbon dioxide (CO_2). Laser soldering has a number of advantages: (1) it provides highly localized heating and minimizes the chance of damage to heat-sensitive devices; (2) it forms ductile and shiny solder joints; and (3) it develops less stress on the circuits. The energy from a YAG laser comes from the light emitted by a 1.06-μm wavelength, which is better absorbed by metals rather than by glass and plastics. On the other hand, a CO_2 laser emits light of 0.6 μm, which is absorbed by metals and plastics. Solder joints are made by a reflected laser beam which is scanned on an unfused tin–lead plate. The CO_2 laser does not appear to melt solder once the initial rosin flux had been vaporized. Rosin flux also absorbs 1.06-μm radiation and effectively couples this energy into solder [53].

One group of researchers felt that the fatigue properties of a solder joint are related to the chemistry of the solder materials [58,59], while others felt that microstructure plays a predominant role in fatigue consideration, with failure occurring along grain boundaries of larger grains. Regardless of chemical composition, microstructure and reflow method are the two major contributors to premature failure due to fatigue [60]. On the other hand, stress in leadless SMDs is less than that in leaded devices [61].

Solder Joints [39] The most common joint forms used in the hybrid industry today are lap and butt joints. In a *lap joint*, a flat or round foot is placed at the edge, extending across a conductor solder pad. The foot is usually the maximum size that will fit within the confines of the solder pad. In a *butt joint*, leads are cut short without forming a foot over the pad and contacts are made perpendicular to the plane of the solder pad. It is observed that for dual-inline packages and TO-18 transistors, lap solder joints work better, whereas in TO-3 diodes, butt solder joints are preferable. However, further improvements in surface mount technology in the areas of interconnection materials and processing techniques will lead to significant increases in the service life of SMDs. Paramount success of this new technology in high reliability environments or in future high-density hybrid technology, including very large scale and ultralarge scale integrated applications, is its ability to fabricate reliable soldered interconnections [37,57]. Presently, the most significant potential reliability problems associated with surface-mount solder joints include the following: (1) thermal cycling of circuit assemblies when there is a large mismatch in the temperature coefficient of expansion between components, packages, interconnections, and substrate materials [62]; (2) cyclic stress occurring in soldered interconnections while packages are power cycling; and (3) operational temperatures are high enough to significantly enhance the effect of creep, corrosion, and additional intermetallic compound formation [63].

Control variability of the heating rate is critical in the reflow of surface-mount assemblies. Surface-mount bonds are often populated with electronic devices that are sensitive to high-temperature heating rates (thermal shock). Thermal shock is more critical with SMDs, as there is less thermal resistance between solder joint and chip than in leaded through-hole components. Short- to medium-wavelength infrared (IR) radiation used to dry and reflow solder paste for SMDs offers a high degree of product heating rate variability. The heating chamber is divided into zones to preheat, dry, and reflow solder while maintaining high process speeds. Independent top and bottom emitter control allows energy to be input from either side of the board to avoid components that may be thermally sensitive. The three most common reflow methods used presently are vapor phase, shortwave IR, and medium-wave IR. Vapor phase utilizes a convective mode heat transfer with a change of phase fluid. The speeds for vapor-phase and short-wavelength IR are typically between 30 and 90 s, whereas speeds for medium-wavelength IR are between 3 and 5 s. Both short- and medium-wavelength IR methods have low operating costs (power only). Vapor phase requires additional cost for fluid. Vapor phase and medium-wave IR heat components to the solder joint temperature, whereas short-wave IR can maintain lower component temperatures when necessary. The vapor-phase heating rate is high, whereas IR methods can control the heating rate and prevent thermal shock.

The basic failure processes in solder bonding are related primarily to metals and alloys used in soldering and the failures may be due to one or more of the following: creep, mechanical fatigue, thermal fatigue, corrosion-enhanced fatigue, intermetallic compound formation, detrimental microstructure development, and joint void. Mismatched thermal expansion between the solder materials and the materials to be soldered, inadequate wetting, and leaching also contribute to solder joint failure [37].

Solder joints are often desirable and sometimes required in hybrid microelectronic circuits. Microcircuit metallization often contains several micrometers of gold, which is easily wet by any low-melting-point solder. The usefulness of lead–tin eutectic and off-eutectic solders is limited by their propensity to dissolve large quantities of gold rapidly [64]. This dissolution, commonly called *scavenging*, can lead to complete and irreparable destruction of Au conductor patterns. After it has been proven that the solder chosen does not scavenge substrate metallization, one must determine those soldering conditions that promote good wetting and spreadability. Wetting parameters such as flux type, substrate temperature, and solder size must ultimately be given careful consideration. The third factor, that must be considered is the nature and rate of solid-state reactions which occur between solder and substrate metallization. Single and multiphase reaction layers contain various intermetallic compounds which might appear during aging. Due to excessive electrical resistance or low ductility, such compounds may limit or terminate component usefulness. Pb–In (indium) solder does not dissolve Au metallization. It is recommended that Pb–In solder not be used on Au film thicker than 10 μm.

8.1.4 Solder Alloys [64]

High-Temperature Solders A number of tin–lead (Sn–Pb) or tin–lead–silver (Sn–Pb–Ag) solders used in soldering hybrid microcircuits have high melting point temperatures (Table 8.1). These are used as step soldering or for some

TABLE 8.1 Commonly Used High-Temperature Solder Alloys

Alloy	Liquidus (°C)	Solidus (°C)	Shear Strength	
			At 20°C	At 100°C
5% Sn–93.5% Pb–1.5% Ag	301	296	23.8	15.7
10% Sn–88% Pb–2% Ag	290	268	—	—
10% Sn–90% Pb	234	183	34.3	13.7

Source: From [3, p. 171].

component assembly. However, these solder alloys suffer from several drawbacks, such as poor wetting, dull surface finish, and low mechanical strength [12,59].

Low-Temperature Solders Low-melting-point solders generally melt below 183°C. The alloys typically used in these solder materials are tin and lead; and bismuth (Bi), Sn, and Pb; and indium (In), Sn, and Pb. Sn–Bi solders are used for higher strength and are comparable to Pb–Sn solders. Sn–In solders have lower mechanical strength, higher cost, but excellent wetting properties [65] and higher ductibility [66]. The Sn–Pb–Bi ternary solder alloy results from a small Bi dopant (ca. 2 to 5%) and has shown higher strength and fatigue life with reduced melting temperature (175 to 181°C) [67]. With increased adoption of surface-mounting techniques, the properties of the joining material used to attach SMDs and/or SMCs in thick- and thin-film hybrid circuits are crucial to the quality and reliability of assembly. The intricate design and high density of circuitry impose different and/or high demands on the properties and performance of solder alloys.

In the low-temperature operation, solder in the form of paste is becoming more popular for easy operation and handling. However, such paste suffers from pad bridging, inadequate solder joint standoff height, vehicle/flux residue cleaning difficulty, and solder balling [53]. In terms of functionality, a solder paste is composed of three major components: solder alloy powder, vehicle system, and flux system. 62% Sn–36% Pb–2% Ag solder alloy powder is generally used with organic vehicle systems containing 85 wt % alloy powder. Performance can be grouped into four categories: applicability, solderability, residue characteristics, and joint integrity. *Applicability* includes the compatibility of a paste with a paste-applying technique such as screen/stencil printing or dot dispensing at a specific mesh size/thickness or gauge size. *Solderability* includes the capability of a vehicle–flux system to clean the solder powder under a given set of conditions; it also includes the flow property of the paste during heating. *Residue characteristics* cover the physical state and chemical activity of residue and its compatibility with a specific cleaning process. *Joint integrity* refers to the mechanical, thermal, and electrical properties of the resulting joint under a certain service environment.

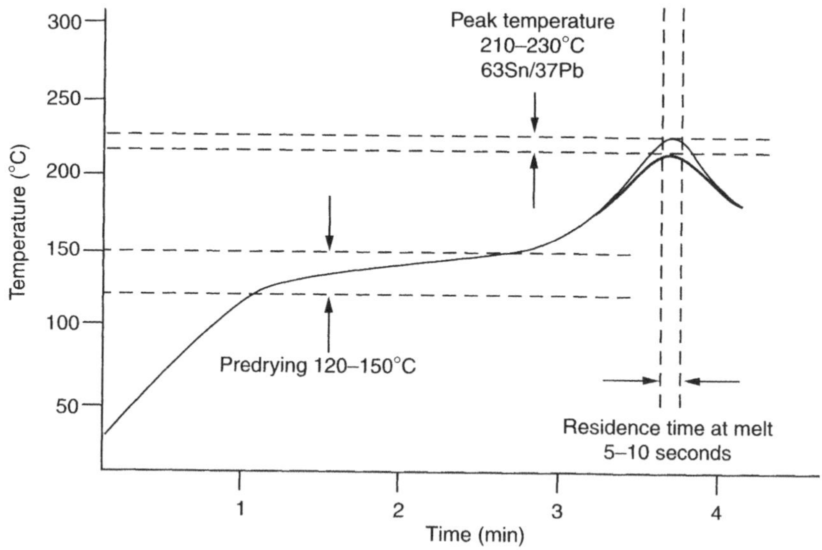

Figure 8.7 Reflow profile comprising three stages of heating [1]. (Reproduced with permission of The McGraw-Hill Companies.)

8.1.5 Solder Reflow System

The integrity of a solder joint technology depends mostly on the reflow of the solder on the surfaces of the contacts. This reflow of the solder material on the surface of the joint depends on the wetting ability of the solder material, which is directly related to the dwell time at the specific temperature in the proper range and the specific temperature being set. However, the failure mechanism of solder joint is a complex phenomenon and related to different conditions of the soldering, including the materials and the cleanliness of the surface. A simulated reflow profile is shown in Figure 8.7. From the curve we can easily understand that a reflow profile has three different temperature stages at three different timings: natural warm-up time, preheating or soaking time, and spike and reflow time of heating. The recommended time for the second stage is between 45 and 150 s in the temperature range 120 to 150°C. The third stage is to spike quickly to the peak reflow temperature at a rate of 1.0 to 4.0 C/s. The rule of thumb in setting the peak temperature is 20 to 50°C above the liquidus or melting temperature [1].

Figure 8.8 shows Novastar's reflow solder machine. The reflow system includes exchangeable heater technology, with viewing windows; cool-down station, real-time graphic temperature profiler, and forced-air heating system.

8.1.6 Lead-Free Interconnects

The replacement of traditional lead-bearing solders with lead-free alternatives has brought enormous change in the market of process control and assembly of

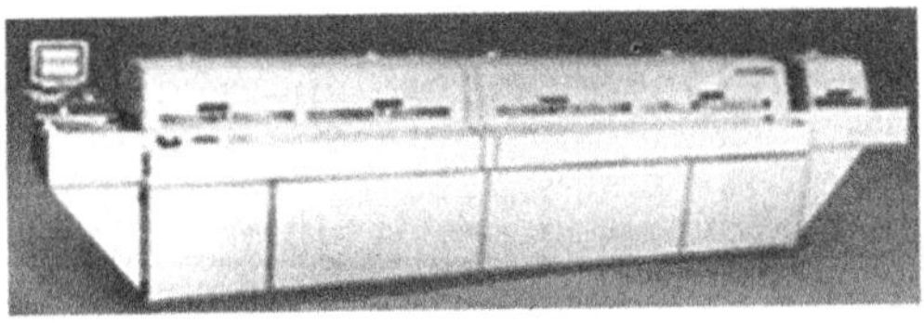

Figure 8.8 Reflow solder machine. (Courtesy of Novastar Technologies, Huntington Valley, Pennsylvania.)

hybrid circuit components [68,69]. Surface solderability (solder, fluxes, and process parameters), thermal damage (components, connectors, organic laminates etc.), postassembly cleaning practices, and solder joint reliability (consumer electronics, automotive controls, and military hardware) are just a few of the areas that must be addressed by design, manufacturing, and cost engineers in the course of this transition. Many of the technical hurdles are unique to the subassembly or product line in a particular company [70–72]. Ongoing research in lead-free interconnections shows that this emerging technology can be used for ball grid arrays, flip-chip and chip-scale packaging, and for ultrahigh pitch connections.

Reflow conditions of 63% tin–37% lead solder pastes have a melting temperature of about 183°C, whereas lead-free pastes with tin–silver compositions have a melting temperature, between 216 and 221°C. This result is that for proper wetting, large components are to be heated to 230°C. A combination of IR heating and a forced-air convection system, heating each element individually and precisely, provides the means required to process lead-free assemblies reliably. Experimental data show that metal compositions of solder pastes generally feature poor spreadability [70]. Moreover, lead-free solder paste with oxidizable metal components such as copper and zinc are to be heated in an inert ambient. It is hoped that lead-free solder paste will offer the potential for zero-defect production when an assembly line can be combined with automated process monitoring.

Another strong lead-free candidate for solder attachment is a Sn–Ag–Bi–In quaternary system. All the elements in this system affect the melting temperature of the resulting solder alloy [72]. To minimize the reflow temperature, an adequate Ag dosage is essential, and bismuth should be maintained at a low level to prevent fatigue [69].

8.2 INTERCONNECTIONS

The increased circuit density and circuit performance success that modern hybrid microcircuit processing technology can provide depends not only on the surface-mount technology but also on the ways that interconnections between these devices and the outer world are made. The impedance of the interconnecting wires plays a major role in the circuit performance [73,74]. To minimize the circuit impedance and *RC* delays, the lengths of the circuit wiring are to be shortened. Unfortunately, the lengths of chip-to-chip interconnections on various packaging levels have not shrunk at the same rate as the size of SMDs. The use of bonding wire is still the predominant way to connect semiconductor dies electrically to package termination [75]. The requirement for certain properties, including uniformity and consistency of the parameters required for bonding wire, is extremely critical [76]. A significant portion of failure experienced has been in electrical connections. To avoid problems such as sagging, ion bonding strength, and wire fracture, certain mechanical and metallurgical properties must be maintained during the wire fabrication process.

Interconnections will gain importance as we continue reducing feature sizes of the SMDs and enlarging substrate dimensions. They may eventually be the limiting factors. Consequently, understanding the on-chip and chip-to-chip wiring issues and subtleties of the electronic packaging will be the key to harnessing the full potential of future hybrid microelectronic circuits. The wiring capacitance and resistance provide additional delays of interconnections and introduce crosstalk and simultaneous switching noise [77]. To minimize the negative effects of the long wires, the floor planning and placement of the functional subblocks must be designed such that long wires are avoided in the circuit path [78]. To reduce the resistance of the interconnection line, a low-resistive material compatible with hybrid microcircuit substrate has to be addressed [79]. Most of the metal suffers from either one or more than one of the following: poor adherence, metal migration, incompatibility with high-temperature processing, and high cost. Copper is the most successful successor, because it has significantly lower resistivity than gold, fewer complications than silver, and is cost-effective [80]. Another important area is the selection of dielectric low-*K* materials, which will reduce interconnection capacitance. However, low-*K* materials can be porous, hygroscopic, and deteriorate at lower temperatures than do moderate- or high-*K* materials. So there will always be a trade-off between the materials and the *RC* effect. Multilevel interconnections with wider and thicker lines at the upper levels are seen to reduce interconnections and *RC* effect [81]. In a thin-film multichip module the lengths of the interconnections between chips are considerably shortened [82]. Thus by reducing the percentage of the circuit area by a factor of 7, one can reduce the power consumption by 30% and the operating frequency by a factor of 3 [74,82].

The wires used in interconnections from chip to chip or from chip to outer world are generally made of gold, aluminum, and an alloy of Al with 1% silicon [83]. The wire diameter varies from 0.7 to 1 mil (18 to 25 μm). Gold wire is often used in thermocompression bonding, while aluminum is generally used with ultrasonic bonding. For ball bonding, gold wire is used. To minimize the wiring capacitance and resistance, one must reduce the average interconnection length [84].

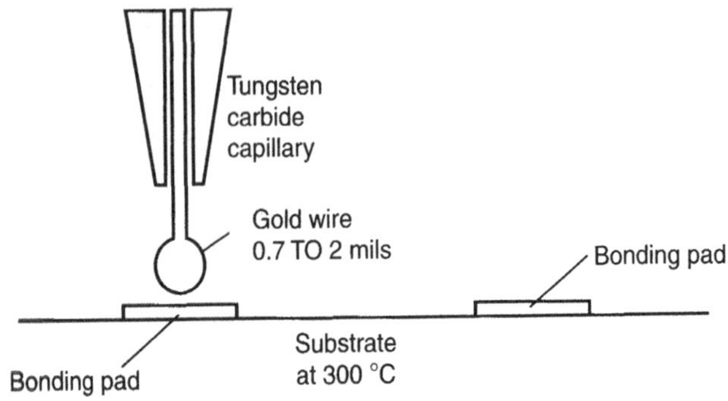

Figure 8.9 Initial setup for thermocompression wire bonding [90]. (Courtesy of Marcel Dekker, Inc.)

8.2.1 Thermocompression Wire Bonding

Thermos is a Greek word meaning *heat*, and *compression* is the state of being compressed. In other words, during thermocompression wire bonding, *heat* and *pressure* are applied simultaneously [85,86]. Pressure is applied normal to the point of contact to form a metallurgical bond at a temperature considerably below the lower melting point of the metals. Figures 8.9 and 8.10 show the initial and final setup diagrams of thermocompression wire bonding system. In thermocompression bonding, a gold wire is joined to a gold or aluminum substrate contact pad by a process in which heat and pressure are applied to the wire and to the contact to make a diffusion-bonded cohesive connection. Thermocompression wire bonding is a flexible technique for making interconnections between chips of hybrid microcircuit and can be accomplished by three basic methods: ball bonding, stitch bonding, and wedge bonding.

In the ball bonding technique, a gold wire of small diameter [0.5 to 3 mil (13 to 76 μm)] is fed through a tungsten carbide capillary which is heated to form a ball on the end of the wire [87]. In most cases, the substrate is heated to 300°C to obtain proper bonding temperature. During bonding time about 100 g of pressure is applied normal to the wire. Upon completing one ball bond, the capillary tube containing the gold wire is landed to another metallized pad and a bond is made without a ball. The second bond without the ball is referred to as a *stitch bond* [88]. This procedure can be repeated several times, and finally,

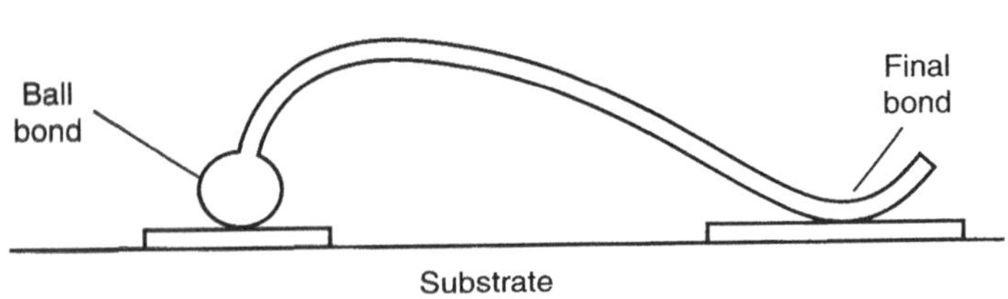

Figure 8.10 Final stage after thermocompression wire bonding [90]. (Courtesy of Marcel Dekker, Inc.)

the wire can be cut with a hydrogen flame. In thermocompression wedge bonding a wedge-shaped tungsten carbide tool is used to put normal pressure on the wire. This force causes plastic deformation of the wire as well as cohesion between wire and the land. This force sometimes deforms and weakens the wire, resulting in a shorter bonding life [89].

Thermocompression wire bonding requires a high level of cleanliness at the interface. Plasma cleaning prior to wire bonding resulted in assemblies with substantially increased adhesion of ball bonds [91], whereas a noncleaned surface results in voids and premature bond failure during accelerated heating. High temperature (ca. 300°C) sometimes produces a brittle gold–aluminum bond with poor electrical conductivity, which degrades with time. This causes eventual failure of bonding and is called a *purple plague defect* [89,92]. Careful bonding over a large area can avoid this type of defect. For complex hybrids, prolonged heat can result in drift in the parameters of both active and passive components. However, a thermocompression bond is reliable and very strong. Wire bonders use an electronic flame-off (EFO) subsystem to form gold balls at the end of the gold wire. The end of the wire melts in the glow discharge and a ball forms due to surface tension in molten metal. Flame-off subsystems generally use a positive-charged electrode, where gas ions formed in the discharge and accelerated toward the wire, and the negative terminal return to ground. Electric spark melts the gold wire and forms a ball. Figure 8.11 shows various stages of operation of ball and stitch bonding systems. Today, most advanced wire bonders use negative force, which eliminates sputtering of the wire, reduces capillary clogging, and extends capillary life. Negative EFO also controls precisely the ball size.

One of the biggest problems with fine pitch bonding relates to the bonding hardware. Conventional 30° capillaries simply cannot do a bond on 4.5-mil (114-μm) centers without destroying the neighboring bond. Special thinner *bottleneck capillaries* might be the answer. These fine capillaries leave enough room to bond, but there is a trade-off between the thinner capillaries and the lifetime of the capillaries.

Examinations of materials with wide range of physical and mechanical properties indicate that excessively hard film results in weak, nonsticking bonds. However, the wire bondability of such films can be improved by removing surface oxides and roughness. From the standpoint of good wire bondability, very high adhesion of the film to the substrate is not required. Surprisingly, doped rather than pure gold metallization exhibited the highest ball shear and wire pull strength [91]. Figure 8.12 shows a typical aluminum and gold wedge bonder system capable of making 70-μm pad pitch bonding (Figure 8.13).

8.2.2 Thermosonic Wire Bonding

Solid-phase bonding of metals involves the adhesion of contacting surfaces at temperature below their melting points. Thermosonic or hot-wire ultrasonic

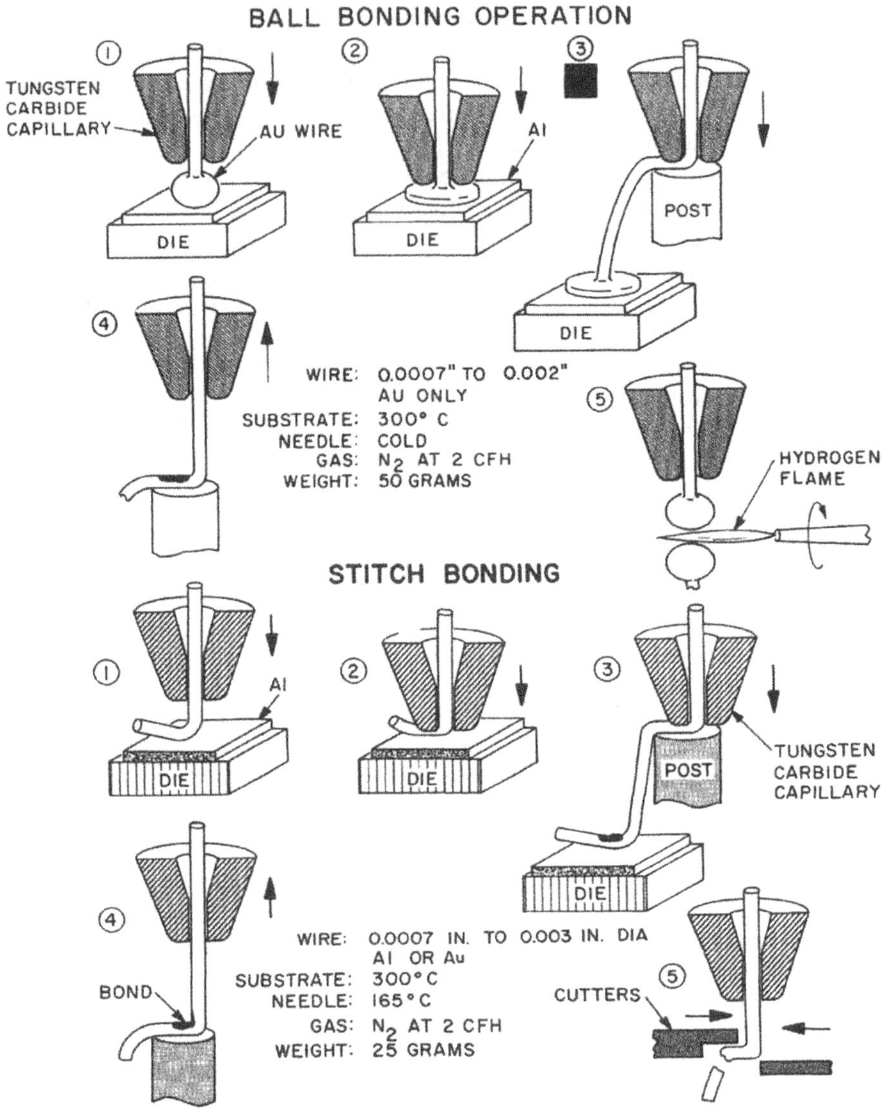

Figure 8.11 Steps in the ball and stitch bonding process [93]. (Courtesy of Integrated Circuit Engineering Corporation, Phoenix, Arizona.)

bonding employs the variable of thermocompression bonding (i.e., force, temperature, and time), and in addition, introduces vibrating motion of one surface against the other [95]. The thermosonic wire bonder machine utilizes a variable heating system (from 25 to 600°C). An unheated capillary bonding tool is attached directly to an ultrasonic horn to which vibratory energy is furnished by a dual ultrasonic generator. The ultrasonic generator includes a 60-kHz

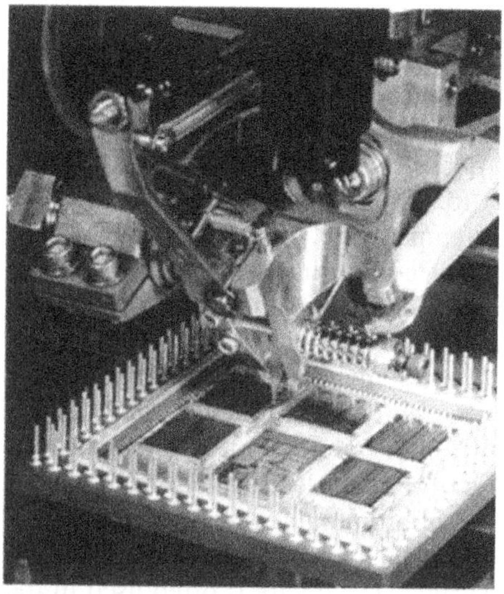

Figure 8.12 Aluminum and gold wedge bonder [94]. (Courtesy of Kulicke & Soffa Industries, Willow Grove, Pennsylvania.)

oscillator, a transducer, and a velocity coupling system (horn). The piezoelectric type of transducer converts the electrical energy into vibratory motion of the bonding tip. The bonding system, with dual clamping, helps to provide alignment of the bonding tool directly over the part or to the pad to be bonded. The substrate temperature is measured by a miniature Pt/Pt thermocouple, connected to a digital readout temperature indicator. A Holdex force gauge is provided to measure the clamping force (varied over the range 15 to 135 g). A magnetic pickup and oscilloscope are used to calibrate time width, ultrasonic power amplitude, and frequency [96].

The bonding tools include a titanium carbide capillary, quartz capillary, and ceramic capillary. Ball size is maintained at approximately $2\frac{1}{2}$ times the wire diameter, which are approximately 2.4 mils (0.06 mm). The advantages of thermosonic bonding over thermocompression bonding can be summarized as follows: (1) The bond forces required for thermosonic bonding are relatively low compared to the forces required for thermocompression bonding; (2) thermosonic wire bonding can be performed at reduced substrate temperatures; (3) thermosonic bonding essentially eliminates the need for special prebond cleaning; and (4) thermosonic bonding shows little or no change in tensile strength as a result of thermal environmental aging tests.

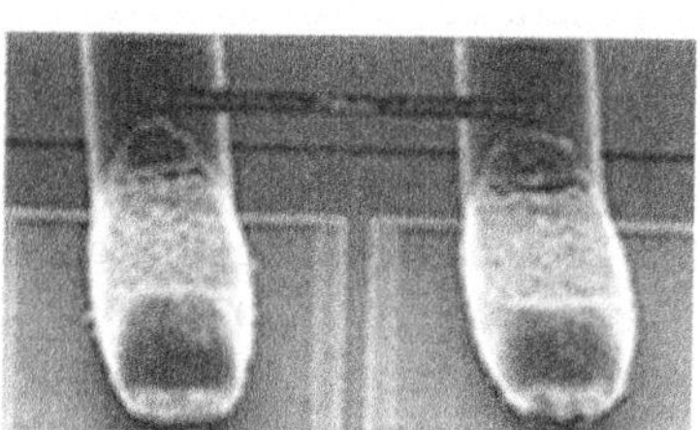

Figure 8.13 A 70-μm pad pitch wedge bond [94]. (Courtesy of Kulicke & Soffa Industries, Willow Grove, Pennsylvania.)

Thermosonic bonding accelerates the diffusion process for gold and aluminum and, in times as short as a few milliseconds, creates a strong, welded intermetallic structure. The duration of the ultrasonic energy applied and the level of energy absorption have a significant influence on reactivity and material properties. The area of intermetallic phase formation between Au and Al is observed to increase from about 20% to 100% when 60 kHz standard is approximately doubled with a bonding temperature of 175°C. The use of bonding wire is still the predominant way of connecting semiconducting dies electrically to package terminations in hybrid circuits. Bonding wire diameter range from 0.7 mil (18 μm) to approximately 3 mils (76 μm), 1-mil (25-μm) diameter being the most popular. The requirement for certain properties, including uniformity and consistence of the required parameters for bonding wire, is extremely critical. A significant portion of failures experienced has been in electrical connections [97]. The wires used for such electrical connections are generally made of gold, aluminum, or alloy of aluminum silicon. Thermocompression and ultrasonic bonding are the two principal processes used in hybrid microelectronics to bond wire to a metal surface. Gold wire is often used in the thermocompression bonding, whereas aluminum is generally used with ultrasonic bonding. It has been shown that silver bearing conductors have bonding yields comparable to gold conductors and the bond strengths of wire bonds made to silver bearing conductors do not degrade during high temperature aging [90]. Gold and silver are both *fcc* structure with lattice parameters 4.09 Å and 4.08 Å, respectively. This indicates that when a clean silver surface and a clean gold surface are brought into atomic contact, the driving force to reduce surface energy should result in a gold silver bond with no lattice distortion.

Wire bonding, which is basically a cold-welded process, has not changed significantly throughout the many years during which microelectronic devices have been manufactured. The techniques and equipment have evolved to a high state of sophistication. Modern wire bonders can make as many as 5 to 10 bonds per second with great precision and reliability. Accommodating a great variety of lead frames and packages, they can be programmed quickly to satisfy the wire bonding requirements of almost any microelectronic device, including those with hundreds of leads. In the near future, we can expect to see wire bonders with artificial intelligence.

8.2.3 Ultrasonic Wire Bonding

In ultrasonic wire bonding, the ultrasonic energy generated (20 to 60 kHz) by the oscillator is applied to the wedge [90,98]. The bonding wire is fed through the capillary and passes under the wedge. As the wedge vibrates under the influence of the ultrasonic energy, the wire underneath is rubbed against the metal surface, causing local heating and ultimately resulting in a metallurgical weld. Figure 8.14 shows the ultrasonic bonding process. Aluminum wire of 1 mil (25.4 μm) diameter is generally used for bonding, although 0.7-mil (18-μm) gold wire can be used. Any oxide on the tip of the aluminum wire is broken

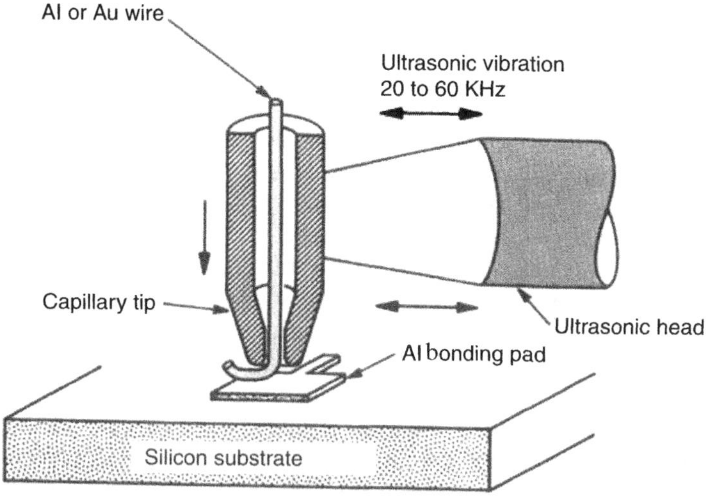

Figure 8.14 Ultrasonic bonding process mechanism [93]. (Courtesy of Integrated Circuit Engineering Corporation, Phoenix, Arizona.)

through and the oxide tends to help the friction heating process, producing exceptionally reliable bonds. It has been demonstrated that with adequate understanding and control, this process can give a very reliable bond [90]. The compatibility of ultrasonic wire bonding depends on several factors (e.g., power of the ultrasonic generator, force exerted by the capillary, the duration of ultrasonic vibration, and the frequency of the ultrasonic generator). By forcing a wire against a metallized land and then vibrating the wire at an ultrasonic frequency, usually around 60 kHz, ultrasonic bonding makes the contact for an aluminum-to-aluminum bond (Figure 8.15). In ultrasonic bonding, high-frequency scrubbing reduces the surface roughness and eliminates any surface oxides, thus facilitating molecular contact. The higher amplitude of ultrasonic vibration is detrimental for two reasons: (1) the hardened asperity bridges will not be able to relieve the tensile stress created by the elastic change of the bond, which will result in a low-adhesion bond; and (2) the metal immediately outside the bond region might result in cracking. The greatest advantage of ultrasonic bonding is its ability to denude the metal surface from any organic and inorganic contaminants and clean the metal surface covered with oxide. The breaking of the metal oxide is assisted by tangential stress at the contact.

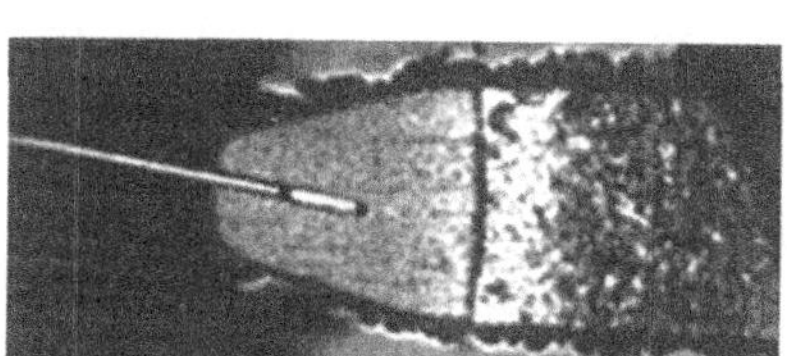

Figure 8.15 Ultrasonic bond (flatpack) aluminum wire [93]. (Courtesy of Integrated Circuit Engineering Corporation, Phoenix, Arizona.)

Ultrasonic vibration also improves instantaneous plasticity and ductility of metal, and it is especially recommended when bonding multiple numbers of terminals simultaneously at low temperature. A major disadvantage of ultrasonic bonding is that the substrate has to be rotated after each bond to set up the direction of the next bond. To minimize the possibility of mechanical damage to the aluminum wire during feeding, or the occurrence of microcracking, and in a further attempt to increase uniformity in wire loop characteristics, a wire depooling method is developed. A magnetic chuck in the form of a pot can prevent the entire packaging system from vibration, and such a system has been reported [99]. Figure 8.15 shows typical ultrasonic aluminum wire bonding.

After the first bond the substrate is moved in the direction of the wire being fed until the wire is over the second termination pad. On completion of the second bond the wire is clamped and tagged away from the bond, leaving a short tail. Ultrasonic bonding requires no local heating, and aluminum wire can be used for wire bonding, which is more economical than gold wire bonding. The major disadvantage of ultrasonic bonding is that the substrate has to be rotated after each bond. In ultrasonic bonding two useful metals, gold and aluminum in the form of wire (1 to 0.7 mil), are used.

8.2.4 Automated Single-Point Tape Automated Bonding

Tape automated bonding (TAB) is another bonding system [100,101] used frequently in hybrid microcircuit fabrication. Compared to wire bonding, TAB bonders are capable of bonding very closely spaced leads. Despite its higher cost for materials and assembly, TAB has become the most cost-effective choice for many high-density applications. In addition to low-temperature ultrasonic bonding, it is possible to use materials other than gold for tape automated bonding. TAB is becoming a practical solution for high-lead-count devices because wire bonding reaches its limit with pads spread on 6-mil (152-μm) centers. The disadvantages of TAB are (1) it produces round bonds with irregular spacing; (2) the process is slow, especially in the case of single-point TAB; and (3) it is not a low-cost method [95,102].

TAB is considered as a replacement of wire bonding [104]. In some cases TAB is recognized as the only workable technical solution for high-pin-count, high-performance semiconductor devices [48]. Figure 8.16 shows various plastic package configurations, and it has been found that among all these packages, it provides the maximum I/O counts [105,106]. However, many TAB equipment manufacturers believe that automated single-point TAB will be standard equipment in the near future [107].

Special single- or double-layered tape has become available that will be able to meet the challenge of tape automated bonding. Two-layer tapes have the advantage over single-layer tape that of chip testability after bonding. Thin tape is used for a low pin count, and thick tape is used for high-pin-count devices [108]. Many factors must be considered when designing TAB tape, including lead forming, plating, stress relief, and pretestability. As cost is

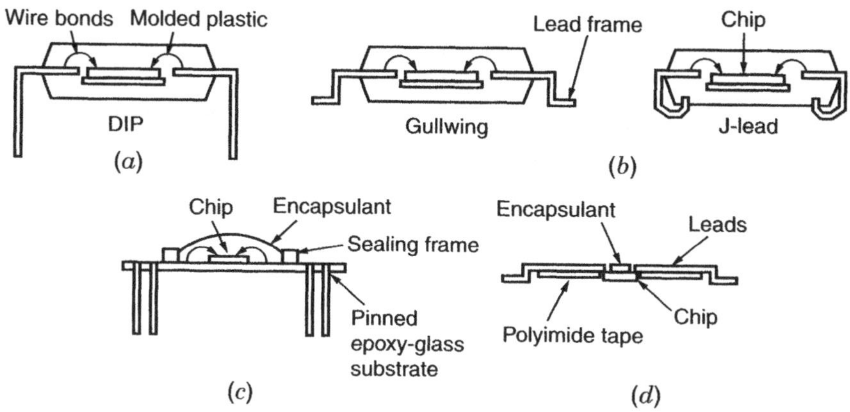

Figure 8.16 Plastic package configurations: (*a*) pin-in-hole DIP; (*b*) surface-mounted molded plastic packages showing gullwing and J-lead styles; (*c*) plastic PGA, compatible with pin-in-hole technology; (*d*) TAB package [103]. (Reprinted from the June 1988 edition of *Solid State Technology*. Copyright 1988 by PennWell Publishing.)

related directly to the circuit area, a designer should design the smallest possible circuit area.

Tape automated bonding has experienced a resurgence of interest because (1) it can be used to pretest and burn devices prior to final assembly, (2) the electrical properties of TAB leads at high frequencies are superior to those of wire-bonded leads, and (3) it has very dense bonding patterns [109]. The traditional method used in TAB is gang bonding, a technique in which force and heat are used to bond all the leads to the die simultaneously. This method is very fast, and the single-point TAB (SPT) technique is useful in many applications [51].

SPT is a high-speed operation in which one TAB lead at a time is bonded using a small wedgelike tool. It can be used equally on both inner and outer lead bonds. It shares the basic advantages of TAB and gang bonding. Its unique features include (1) the ability to measure each lead's height during bonding; (2) the ability to make each bond independently, with independent force; (3) the ability to perform inner and outer lead bonding with the same equipment; (4) the ability to perform multiple and hybrid TAB designs bond in one pass; (5) minimal setup time and device type changeover; and (6) the ability to package with limited access; and (6) the ability to rework with individual leads. Outer lead bonding is typically used for PCBs, and as with inner lead bonding, can be accomplished using single-point or gang bonding. Bond pad geometries for TAB should be optimized, which means designing a passivation opening smaller than that of the bond pad metal. Figure 8.17 shows a typical bond pad layout for TAB, and Figure 8.18 shows the outside connections from a TAB package.

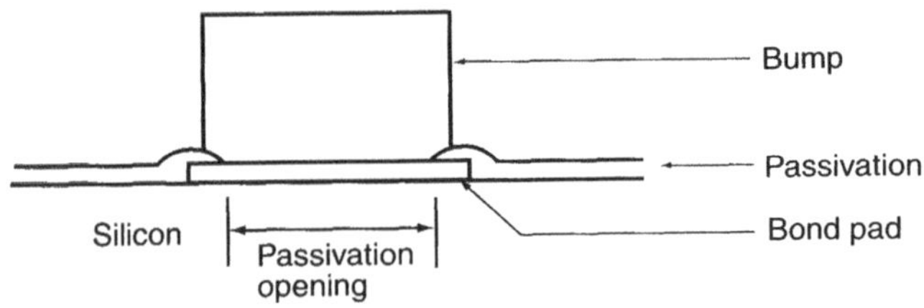

Figure 8.17 Bond pad layout for tape automated bonding [110]. (Reprinted from the June 1988 edition of *Solid State Technology*. Copyright 1988 by PennWell Publishing.)

8.2.5 Laser Wire Bonding

Discrete wire bonding is used for surface interconnections in multilayer ceramic (MLC) packaging [111]. Typically, there are 12,500 wiring pads and 2000 wire channels per MLC module. To handle the complicated task of discrete wire bonding, a fully automated bond using arc plasma is applied locally, which bonds the wire to the pad using ultrasonic energy. For ultrasonic bonding to work, the bonding pads on the MLC module must be covered with a thick layer of gold several micrometers thick. On the other hand, the chip solder attachment process limits the gold thickness on the chip attachment pads to a very thin layer, to avoid intermetallic formation. This incompatibility gives rise to a dual-pad surface metallurgy, which significantly affects the processing involved in MLC manufacturing. The use of laser energy to achieve bonding to a thin gold layer is well established [112].

Laser wire bonding consists of a laser beam (continuous Nd:YAG at a wavelength of 1.06 μm) transmitted through an optical pad. The force applied on the top of the wire ranges from 50 to 100 g. A shutter situated inside the cavity is activated to generate a laser pulse. The duration of the pulse is typically between 20 and 50 ms, and the power level ranges from 20 to 50 W. The heating and wire bonding mechanism is related to temperature. The temperature rise between the surface and the wire depends on the amount of absorption of laser energy. Initially, the wire absorbs very little radiation and is heated mostly by heat conduction from the tip. Once the wire reaches a high temperature, the predominant heating mechanism switches to laser energy absorption by the wire, which then undergoes very rapid localized heating and melting.

Figure 8.18 Outer leads bonded to a TAB board [110]. (Reprinted from June 1988 edition of *Solid State Technology*. Copyright 1988 by PennWell Publishing.)

8.2.6 Flip-Chip Bonding

Portable products such as cellular phones and laptop computers are important segments of the electronic industry. Their small size and light weight provide significant advantages to the consumer. A fundamental limit on the size of many of these products is the tight control of the substrate space and overall reduction of the devices [113]. Assembly engineering of the die and the bonding of the die to the substrate present a challenge in future generations of products as they continue to decrease in size and increase in complexity. Hybrid microcircuit technology offers a means to allow reasonably sized mixed technology ICs and discrete components to be combined into a smaller common package [114].

Four methods of die attach and bonding are in common use in hybrid microcircuits: direct die attachment and wire bonding (the most common), flip chip using solder bumps, tape automated bonding, and various forms of thin-film top-side interconnections. An advantage of direct die attachments and traditional wire bonding techniques is that they are well established and well understood. A disadvantage of this approach is that it requires significant substrate area around each die for the wire bonds, resulting in a larger substrate. Flip chips using solder bumps require specially designed bond pads on each chip. This precludes the use of off-the-shelf dies on the module. The advantages offered by TAB techniques include pretestibility and very fine pad pitch. Their disadvantages include overall size and added process steps. The thin-film top-side interconnection approach [56] provides high-density interconnection but involves batch processes, which limit its application to high-volume production. The flip-chip approach using a gold bond bypasses the limitations of the other die-attach processes [115,116]. In flip-chip mounting, the chip is flipped over bumps carried by the dies and is aligned with the contact pads. The electrical contact between the chip and the solder bumps is established by bonding the two.

REFERENCES

1. J. S. Hwang, Solder technologies for electronic packaging and assembly, in *Electronic Packaging and Interconnection Handbook*, 3rd ed., C. A. Harper, ed., McGraw-Hill, New York, 2000.
2. C. Cohen and M. T. Shih, Packaging and interconnection of integrated circuits, in *Electronic Packaging and Interconnection Handbook*, 3rd ed., C. A. Harper, ed., McGraw-Hill, New York, 2000.
3. C. Lea, *A Scientific Guide to Surface Mount Technology*, Electrochemical Society, Ayr, Scotland, 1988.
4. M. S. Cole and T. Caufield, Ceramic ball grid array packaging, *Adv. Microelectron.*, Jan./Feb. 1995, p. 18.
5. S. Wischoffer, From road block to routine placement assembly, *Surface Mount Technol.*, Feb. 2000, p. 57.

6. J. F. Pawling, *Surface Mount Assemblies*, Electrochemical Society, Ayr, Scotland, 1987.
7. R. Prasad, Building with missing components, *Surface Mount Technol.*, Feb. 2001, p. 28.
8. J. Lau, *Handbook of Fine Pitch Surface Mount Technology*, Van Nostrand Reinhold, New York, 1983.
9. L. Miller, *Thick-Film Technology and Chip Joining*, Gordon & Breach, New York, 1972.
10. A. Bar-Cohen and A. D. Kraus, *Advances in Thermal Modeling of Electronic Components and Systems*, Vol. I, Hemisphere Publishing, Philadelphia, 1988.
11. M. Hansen, *Constitution of Binary Alloys*, McGraw-Hill, New York, 1958.
12. S. K. Ghandhi, Phase diagram and solid solubility, in *VLSI Fabrication Principles*, Wiley, New York, 1983, pp. 56–58.
13. J. W. Mayer and S. S. Lau, eds., *Electronic Materials Science*, Macmillan, New York, 1990.
14. C. P. Wong, *Polymers for Electronic and Photonic Applications*, Academic Press, San Diego, CA, 1993.
15. J. H. Lee and K. Neville, *Handbook of Epoxy Resins*, McGraw-Hill, New York, 1967.
16. H. Taraseiskey, Adhesives, in *Power Hybrid Circuit Design and Manufacturing*, Marcel Dekker, New York, 1996, Sec. 3.4, p. 83.
17. Ablestik advertisement, *Solid State Technol.*, June 1998, p. 29.
18. W. F. Graham and B. C. Sakiadis, QL: a new thermoplastic ribbon die attach adhesive, *Proc. ISHM*, Minneapolis, MN, 1987, p. 522.
19. K. Gilleo, Direct chip attach using polymer bonding technology, *IEEE Proc.*, 1989, pp. 37–44.
20. H. B. Bakoglu, Advanced packages, in *Circuits, Interconnections, and Packaging for VLSI*, Addison-Wesley, Reading MA, 1990, Chap. 7.
21. G. A. Riley, *Proc. 3rd Workshop Electron. LHC*, CERN/LHCC/97-6.
22. Masterbond, Hackensack, NJ, New Products, *Surface Mount Technol.*, Jan. 2001, p. 150.
23. J. J. Licari and L. R. Enlow, Assembly process, in *Hybrid Microcircuit Handbook*, 2nd ed., Noyes Publications, Park Ridge, NJ, 1998, Chap. 7, (2.4).
24. R. H. Estates, The effect of porosity on mechanical, electrical, and thermal characteristics of conductive die attach adhesives, *Solid State Technol.*, Vol. 6, 1984, pp. 191–197.
25. R. Prasad, Adhesive applications, in *Electronic Processing Handbook*, CRC Press, Boca Raton, FL, 2000.
26. New Products, *Surface Mount Technol.*, Jan. 2001, pp. 250 and 145.
27. R. J. Hunadi, N. Billo, J. Vaccaro, and S. K. Wang, Development of ultra-low extractable ion, thermally conductive die attach adhesives, *Proc. Nepcon West*, 1984, p. 565.
28. W. B. Russel, Effects of interaction between particles on the rheology of dispersions, in *Theory of Dispersed Multiphase Flow*, R. E. Myer, ed., Academic Press, San Diego, CA, 1963.

29. G. C. Bell, Jr., W. F. Graham, C. M. Rossel, and S. T. Joslin, Rheology for high-speed automatic processing: development of a silver/glass die attach adhesive, *Proc. ISHM*, Minneapolis, MN, 1987, p. 542.
30. H. H. Manko, *Soldering Handbook for Printed Circuits and Surface Mounting*, Van Nostrand Reinhold, New York, 1986.
31. M. Pecht, *Soldering Processes and Equipment*, Wiley, New York, 1993.
32. D. Fear, H. Morgan, S. Burcht, and J. H. Lau, *The Mechanics of Solder Alloy Interconnects*, Van Nostrand Reinhold, New York, 1993.
33. P. A. Deane, J. S. Erich, and T. E. Martin, *The Mechanical Strength of Reflow Solder Connection*, ISHM, Minneapolis, MN, 1987.
34. S. W. Zehr and W. A. Backofen, Super plasticity in lead–tin alloys, *Trans. ASM*, Vol. 61, 1968, pp. 300–313.
35. R. K. Yadava and K. A. Padnavan, Fracture of a tin–lead eutectic alloy in region III of superplastic flow, *J. Mater. Sci.*, Vol. 12, 1977, pp. 851–853.
36. H. N. Keller, Temperature aging of external connections condensation soldered to Ti–Pd–Au thin film, *IEEE Trans. Comput. Hybrid Manuf. Technol.*, Vol. 2, 1979, pp. 180–195.
37. J. Lau, *Solder Joint Reliability: Theory and Applications*, Van Nostrand Reinhold, New York, 1991, p. 183.
38. J. McLenagham and J. Belmonte, Solder paste re-flow for through-hole technology, SMTA Convention, Boston, June 12–14, 2000.
39. R. J. K. Wassink, *Soldering in Electronics*, Electrochemical Society, Ayr, Scotland, 1989.
40. Y.-H. Pao, A fracture mechanics approach to thermal fatigue life prediction of solder joints, *IEEE Trans. Comp. Hybrid Manuf. Technol.*, Vol. 15, No. 4, Aug. 1992, p. 559.
41. M. F. Kaminen and C. H. Popelar, *Advanced Fracture Mechanics*, Oxford University Press, Oxford, 1985.
42. C. L. Burber, *Solder: Its Fundamental Usage*, 2nd ed., Kester Solder Co., Des-Plains, IL, 1961.
43. Automated Unlimited Co. advertisement, *Surface Mount Technol.*, Apr. 2001, p. 116.
44. R. Rowland, Hand soldering and rework tools, *Surface Mount Technol.*, Feb. 2001, p. 32.
45. E. A. Fenton, *Soldering Manual*, American Welding Society, New York, 1959.
46. E. A. Wright and W. L. Wolverton, The effect of solder reflow method and joint design on thermal fatigue life of leadless chip carrier solder joints, *Proc. 34th IEEE-EEC*, 1984, p. 149.
47. J. P. Holman, *Heat Transfer*, 4th ed., McGraw-Hill, New York, 1976.
48. J. S. Hwang, *Modern Solder Technology for Competitive Electronics*, IMAPS, Washington, DC, 1995.
49. D. K. Flattery, Infrared reflow for SMT: thermal yield considerations, *Proc. Nepcon West*, 1987; R. J. K. Wassink, IR soldering, in *Soldering in Electronics*, 2nd ed., IMAPS, Washington, DC, 1990.
50. E. R. Patterson, Assembly techniques for packaged components, in *Modular Series in Hybrid Microelectronics*, ISHM, Reston, VA, 1991.

51. J. F. Pawling, *Surface Mount Assembly*, Electrochemical Society, Ayr, Scotland, 1987.
52. G. W. A. Dummer, Basic connection methods in electronics, in *Materials for Conductive and Resistive Functions*, Hayden Book Co., New York, 1970, Chap. 13.
53. J. F. Maguire, Metals joining of electronic circuits, in *Electronic Materials and Processes Handbook*, C. A. Harper and R. M. Sampson, eds., McGraw-Hill, 1994, Chap. 7.
54. J. M. Smith, Metallurgy world wide focus on the use of lead free soldering alloy solder, *Surface Mount Technol.*, June 1999.
55. P. Boka, Global update on lead free solder, *Surface Mount Technol.*, June 1999; J. S. Hwang, *Lead-free Solders for Environmentally Friendly Manufacturing*, Electrochemical Society, London, 2000.
56. M. J. Ruckriegel, The benefits of dual vapor in-line vapor phase soldering, *Hybrid Circuit Technol.*, Nov. 1985, p. 13; Staff report, Vapor phase soldering, *Mater. Eng.*, Feb. 1987.
57. L. M. Lambert, *Soldering Electronic Assemblies*, Marcel Dekker, New York, 1988.
58. L. R. Fox, J. W. Sofia, and M. C. Shine, Investigation of solder fatigue acceleration factor, *IEEE Trans. Comp. Hybrid Manuf. Technol.*, Vol. 8, No. 2, June 1985, p. 275.
59. J. S. Hwang, *Advanced Solder Technology of Competitive Electronics Manufacturing*, McGraw-Hill, New York, 1996.
60. J. H. Lau and D. W. Rice, Solder joint fatigue in surface mount technology: state of the art, *Solid State Technol.*, Oct. 1985, p. 91.
61. D. E. Reimer, The mechanics of solder joint stress relief by leads on the surface mount devices, *Proc. ISHM*, Minneapolis, MN, 1987, p. 603.
62. M. G. Koopman and T. C. Reiley, Thermal mismatch, in *Microelectronic Packaging Handbook*, R. R. Tummala and E. J. Rymaszewski, eds., Van Nostrand Reinhold, New York, 1989.
63. W. Englemaier, Test method considerations for SMT soldered joint reliability, *IEPS Proc.*, 1984.
64. L. H. Van Vlack, Alloys of metals, in *Materials Engineering*, Addison-Wesley, Reading, MA, 1982, Chap. 5, p. 100.
65. Z. Mei and J. W. Morris, Characterization of eutectic Sn–Bi solder joints, *J. Electron. Mater.*, Vol. 21, No. 6, 1992, pp. 599–607.
66. J. S. Hwang and Z. Guo, Effects of Bi contamination on Sn/Pb eutectic solder, *Surface Mount Technol.*, Sept. 2000, p. 91.
67. J. S. Hwang, *Solder Paste in Electronic Packaging*, Van Nostrand Reinhold, New York, 1989, p. 399.
68. J. S. Hwang, *Lead-Free Solders for Environmentally Friendly Manufacturing*, Electrochemical Society, Ayr, Scotland, 2000.
69. P. Vianco, Lead-free interconnect technology, SMTA Convention, Boston, June 12–14, 2000.
70. H. Suganuma and A. Tamanha, Next-generation reflow technology, *Surface Mount Technol.*, Feb. 2001, p. 65.

71. A. E. Riad and F. D. Barlow III, Properties of solders, in *Thin-Film Technology Handbook*, McGraw-Hill, New York, 1998, p. 531.
72. J. S. Hwang, Another strong lead-free candidate the Sn/Ag/Bi/In system, *Surface Mount Technol.*, Sept. 2000, p. 20.
73. C. A. Harper, *Handbook of Microelectronics Packaging*, 2nd ed., McGraw-Hill, New York, 2000.
74. D. P. Seraphin, R. Lasky and C. Y. Li, *Principles of Electronic Packaging*, McGraw-Hill, New York, 1989.
75. M. Pecht, *Handbook of Electronic Package Design*, Marcel Dekker, New York, 1991.
76. K. C. Norris and A. H. Landzberg, Reliability of controlled collapse interconnections, *IBM J. Res. Dev.*, May 1969, pp. 266–271.
77. H. B. Bokoglu, Transmission lines, in *Circuits, Interconnections, and Packaging for VLSI*, Addison-Wesley, Reading, MA, 1990, Chap. 6, pp. 371–373.
78. W. Wein Ming Dai, Multichip routing and placement, *IEEE Spectrum*, Vol. 4, 1992, p. 61.
79. F. W. Kear, *Interconnection Technology in Hybrid Assemblies and MCMs*, Marcel Dekker, New York, 1993.
80. S. P. Murarka, I. V. Verner, and R. T. Gitmann, Overview of IC interconnects, in *Copper: Fundamental Mechanisms for Microelectronic Applications*, Wiley, New York, 2000, Chap. 1.
81. W. S. Young and S. H. Knickerbocker, Multilayer ceramics, in *Interconnects for Electronics*, Marcel Dekker, New York, 1986, pp. 489–526.
82. F. W. Kear, *Hybrid Assemblies and MCM*, Marcel Dekker, New York, 1992.
83. M. Pecht, R. Agarwal, T. Dishough, S. Javadpour, and R. Mahajan, *Electronic Packages and Their Properties*, CRC Press, Boca Raton, FL, 1999; G. Harmon, Metallurgical failure modes of wire bonds, *Proc. 12th Annu. Symp. Reliab. Phys.*, 1974, p. 131.
84. G. Hermon, Wire bonding, in *Microelectronics Materials, Processes, Reliability, and Yield*, IMAPS, Washington, DC, and McGraw-Hill, New York, 1997.
85. J. L. Janssen Mckee, W. D. Toth, and P. M. Fath, The characterization and reliability prediction of a thermocompression wire bonding system, *Proc. ISHM Symp.*, Atlanta, GA, 1986, p. 264.
86. R. D. Jones, Assembly techniques, in *Hybrid Circuit Design*, Marcel Dekker, New York, 1982, Chap. 8, p. 107.
87. R. G. Johnson and R. E. Cote, Thermocompression gold wire bonding to silver bearing conductors, *Proc. ISHM Symp.*, New York, 1980, p. 315.
88. G. R. Madland, ed., *Assembly in Integrated Circuit Engineering*, 5th ed., Integrated Circuit Engineering Corporation, Phoenix, AZ, 1966.
89. S. K. Ghandhi, Deposited films, in *VLSI Fabrication Principle*, Wiley, 1983, Chap. 8; J. S. Fisher and P. M. Hall, Termination of materials for thin-film resistors, *Proc. IEEE*, Vol. 59, No. 10, Oct. 1971, p. 1423.
90. R. D. Jones, *Hybrid Circuit Design and Manufacture*, Marcel Dekker, New York, 1983, p. 108.
91. J. Kristi, Reliability study of wire bonds to silver-plated surfaces, *IEEE Trans. Parts Hybrids Packag.*, Dec. 1977.

92. R. Schmidt, Mechanism of lead failure in thermocompression bonds, *IEEE Trans. Electron. Devices*, Vol. 9, 1962, p. 506.
93. G. R. Madland, ed., *Assembly in Integrated Circuit Engineering*, 5th ed., Integrated Circuit Engineering Corporation, Phoenix, AZ, 1966.
94. V. McGinty, Development of wire bonding, *Electron. Packag. Process.*, Oct. 1995, p. 42.
95. R. C. Jaeger, Thermosonic bonding, in *Introduction to Microelectronics Fabrication*, Vol. V, Addison-Wesley, Reading, MA, 1990, Sec. 8.4.3, p. 157; J. F. Maguire, Metals joining of electronic circuitry, in *Electronic Materials and Processes Handbook*, C. A. Harper and R. M. Sampson, eds., McGraw-Hill, New York, 1994, Chap. 7.
96. A. E. Riad and F. D. Barlow III, Thermosonic bonding, in *Thin-Film Technology Handbook*, McGraw-Hill, New York, 1998, p. 5-20.
97. G. L. Schnable and R. S. Keen, Aluminum metallization: advantages and limitations for integrated circuit applications, *Proc. IEEE*, Vol. 57, 1971, p. 1570.
98. T. Dixon, TAB technology tackles high density interconnections, *Electron. Packag. Prod.*, Dec. 1984, p. 34.
99. F. J. Cheriff and T. E. Salzer, High-reliability ultrasonic wire bonding, *Proc. Int. Symp. Hybrid Microelectron.*, 1974, p. 221.
100. J. H. Lau, *Tape Automated Bonding*, Van Nostrand, New York, 1992.
101. J. Lyman, Tape automated bonding meets VLSI challenge, *Electronics*, Dec. 18, 1980.
102. L. L. Licari and L. R. Enlow, Automated bonding, in *Hybrid Microcircuit Technology Handbook*, 2nd ed., Noyes Publications, Park Ridge, NJ, 1998, Chap. 3.
103. M. F. Bregman and C. A. Kovac, Plastic packaging for VLSI-based computers, *Solid State Technol.*, June 1988, p. 77.
104. R. L. Gadenhead, Assembly of hybrid microelectronics, in *Hybrid Microelectronics Handbook*, 2nd ed., J. E. Sergent and C. A. Harper, eds., McGraw-Hill, New York, 1995, Chap. 6.
105. M. F. Bregman and C. A. Kovac, Plastic packaging for VLSI based computers, *Solid State Technol.*, June 1988, p. 75; P. Burggraaf, Tab for high I/O and high speed, *Semcond. Int.*, June 1988, pp. 72–77.
106. R. C. McFarland and R. H. Gephard, Automatic bonders simplify LSI computer manufacturing, *Electron. Packag. Prod.*, Feb. 1984; Tames high density chip I/Os, *Electron. Packag. Prod.*, Dec. 1988, pp. 42–45.
107. G. Silverberg, Single-point TAB (SPT): a versatile tool for TAB bonding, *Proc. ISHM*, Minneapolis, MN, 1987, p. 449.
108. B. M. Romenesk, H. K. Charles, Jr., G. V. Clatterbaugh, and J. Weiner, Thick-film bondability: geometrical and morphological influences, *Proc. ISHM*, 1985, p. 408.
109. D. A. Behm and B. Perizek, *Beam Tape Carriers: A New Design Guide*, ISHM, Anaheim, CA, 1985, p. 387; S. Y. Poh and T. L. Michalka, TAB tapes, *IEEE Trans. Comp. Hybrid Manuf. Technol.*, Vol. 15, No. 4, Aug. 1992, pp. 524–541.
110. P. Hoffman, Tape-automated bonding implementation and trends, *Solid State Technol.*, June 1988, p. 85.

111. C. Banas and R. Webb, Laser micro-materials processing, *Proc. IEEE*, Vol. 70, No. 6, June 1982, pp. 556–565.

112. L. Banas and S. Engel, Kilowatt welding with a laser, *Laser Focus*, Feb. 1976.

113. C. E. Goodman and M. P. Metroka, A novel multichip approach using gold ball flip chip bonding, *IEEE Trans. Comp. Hybrid Manuf. Technol.*, Vol. 15, No. 4, 1992, p. 457.

114. J. C. Blakenhorn, *BGA, COB, and Flip Chip PCB Design*, IMAPS, Washington, DC, 1988.

115. K. Gilleo, Flip chip technology, in *Electronic Packaging and Interconnection Handbook*, 3rd ed., C. Harper, ed., McGraw-Hill, New York, 2000, Sec. 1.2-1.54.

116. A. E. Riad and F. D. Barlow III, Flip chip bonding attachment technology, in *Thin-Film Technology Handbook*, McGraw-Hill, New York, 1997, Chap. 5.

RECOMMENDED READING

Al, J. H., *Solder Joint Reliability: Theory and Applications*, Van Nostrand Reinhold, New York, 1991.

Frear, D., H. Morgan, S. Burchett, and J. H. Lau, *The Mechanics of Solder Alloy Interconnections*, Van Nostrand Reinhold, New York, 1963.

Harper, C. A., *Electronic Packaging and Interconnection Handbook*, 3rd ed., McGraw-Hill, New York, 2000.

Hwang, J. S., *Solder Paste in Electronics Packaging*, Van Nostrand Reinhold, New York, 1989.

Lau, J. H., *Flip Chip Technologies*, McGraw-Hill, New York, 1996.

Lau, J. H., *Handbook of Tape-Automated Bonding*, Van Nostrand Reinhold, New York, 1992.

Lau, J., C. P. Wong, J. Prince, and W. Nakayana, *Electronic Packaging, Design, Materials Process, and Reliability*, McGraw-Hill, New York, 1998.

Lee, J. H., and K. Neville, *Handbook of Epoxy Resins*, McGraw-Hill, New York, 1967.

Manko, H. H., *Soldering Handbook for Printed Circuits and Surface Mounting*, Van Nostrand Reinhold, New York, 1986.

Pecht, M., *Soldering Process and Equipment*, Wiley, New York, 1993.

Tummala, R. R., E. Ramaszewski, and A. Koperstein, *Handbook on Electronic Packaging*, Chapman & Hall, London, 1997.

Wassink, R. J. K., *Soldering in Electronics*, Electrochemical Society, Ayr, Scotland, 1989.

9

ADJUSTMENT OF PASSIVE COMPONENTS

Growing technological requirements and widespread acceptance of sophisticated electronic devices have created an unprecedented demand for high-density multilayer hybrid circuits [1–3]. In high-density multilayer hybrid circuits, the number of circuit elements per unit area has increased enormously, and the size of the circuit elements has decreased at the same rate [4]. As a result, there is increased emphasis on the reliability and tolerances of the ever-decreasing size of devices [5]. As circuit dimensions are decreasing, the demand for tolerances above ±0.01% is increasing. In modern electronic circuitry, the majority of circuits require tolerances better than ±0.01%. In fact, analog-to-digital or digital-to-analog converters require tolerances of better than 0.01% [6]. The most important criterion of thick- and thin-film hybrid circuits is that one can adjust the value of the resistor and capacitor deposited by more than 0.01% by trimming even before they are in the process of final measurement. This is not possible with ICs or PCBs. This has given hybrid circuit manufacturers an additional tool to cope with the overall tolerances of the electronic circuitry. Thus the demand for increased circuit sophistication and better quality have been handled more efficiently using modern hybrid circuit assembly [7].

The advent of trimming as a manufacturing tool has permitted sweeping improvements in manufacturing techniques and in the overall efficiency of production in hybrid microelectronic circuits. Several trimming methods have been developed to adjust, primarily, resistors, as capacitors and inductors are generally selected from standard values from different vendors [10]. These include airbrasive, laser [8,9], ultrasonic, thermal, and high-voltage pulse trim-

Handbook of Thick- and Thin-Film Hybrid Microelectronics, By Tapan K. Gupta
ISBN 0-471-27229-9

ming [11]. The majority of manufacturing problems with trimming technology are due to a lack of good circuit design, poor process control, and poor understanding of the software and hardware systems used by the trimming machine. The relative merits of various trimming methods are frequently argued, and many people feel that laser and airbrasive trimming are competitors [12,13]. Another method of adjusting resistors and capacitors is high-frequency electric discharge. This method has not yet found a place in the hybrid industry, but it is capable of either increasing or decreasing the value of resistors and capacitors [14].

To handle modern hybrid circuit assembly more efficiently, we must consider some of the basic rules of trimming [15]: (1) The highest-quality trim should have the longest trim length and minimum penetration; (2) trimming should be done at the center of the element to be trimmed, without heating the terminated conductor; (3) tophat trim should have uniform resistor leg widths; (4) the trimming link should be placed close to the source terminal probe of the bridge network, to minimize the development of static charge; and (5) care must be taken in the selection of thick- and thin-film materials and trimming speed, as the ingredients of a particular thick-film ink or thin-film material and the trimming speed have both been found to affect the percentage drift value of the resistors.

Fired resistors generally differ from their design values by +5 to 15%. However, in the 75% design rule, resistors are intentionally designed and deposited so as to give printed resistor values 25% below their final values. After firing, the resistor shrinks and the resistor dimensions change. The amount of shrinkage depends on many factors, including firing temperature, ink rheology, and firing time. Fired resistors are generally expected to shrink to about 40% of their wet dimension values. The resistors are measured and the final adjustments to the values desired are done through trimming.

9.1 AIRBRASIVE TRIMMING

Airbrasive trimming is the most economical method in use for trimming resistors, inductors, and capacitors [16]. In airbrasive trimming, 300-mesh alundum powder is forced through a 0.14-in. (0.36-cm)-diameter nozzle with high velocity and is blown on the deposited passive circuit element to be trimmed [17]. This brings some changes in the dimension of the circuit element and thus changes its value. The amount of material removed from an element in a given period of time depends on the gas pressure, nozzle configuration, hardness of the device material, and the material's thickness. It generally takes 3 ms to halt and reverse the motor that drives the trim nozzle. It takes 20 ms for the stop valve to shut off the supply of abrasive. Sometime an additional chamber is added to the system to control the percentage flow of the alumina particles precisely. During abrasion, probes connected to a bridge circuit monitor the value of the discrete circuit element.

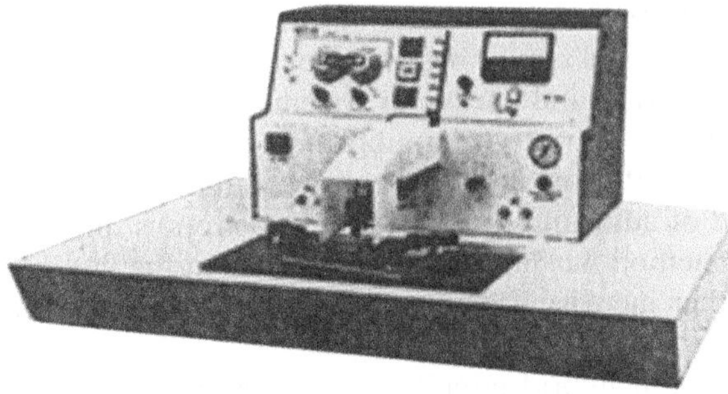

Figure 9.1 Airbrasive resistor trimmer, showing cutting nozzle, substrate holder test probe, and feedback controller [8].

Commercial airbrasive trimmers are expected to trim resistors and capacitors more than 4000 trims per hour with an accuracy of +0.1%. Figure 9.1 shows an airbrasive resistor trimmer with a feedback controller. Resistor drift after airbrasive trimming is less than that of a trimmed resistor after high-speed laser trimming. Airbrasive trimming permits the use of a variety of media for different end results. Even very thin substrates (0.01 in.) present no problem for airbrasive trimming. The airbrasive trimming process is relatively inexpensive, and the machine is capable of trimming up to 700 to 900 resistors per hour, depending on the operator's efficiency. Although airbrasive trimming is more economical than laser trimming, it is considered obsolescent because it (1) generates a dust fog, which contaminates the environment; (2) produces electrostatic charges due to particle motion; (3) is very difficult to conduct in a clean room (class 100 to class 10); (4) leaves an exposed edge, which is susceptible to the environment; (5) is unable to trim areas bellow 0.002 to 0.003 in. (0.005 to 0.008 cm) due to minimum orifices; and (6) is slower than laser trimming. However, dramatic improvements are being made in precision and speed using airbrasive trimming [18]. In recent years, people have begun to realize that it is not possible to perform all trimming functions satisfactorily using a laser. Some operations can be done very easily with airbrasive trimming that are impossible to do with a laser system [12,19].

9.2 LASER TRIMMING

Lasers have four unique properties that distinguish them from ordinary light sources [20,22]: monochromaticity, spatial and temporal coherence, high directionality, and high brightness. Spectral broadening is another special characteristic observed in lasers. Depending on the type of laser, the broadening may be Doppler, thermal, collisional, or natural. *Doppler broadening* occurs due to the relative motion of a source and observer. Since the atoms are in random motion, an observer will measure a range of frequency, depending on to the magnitude and direction of the component velocity of the atom. *Collision broadening* occurs when the ions in an active gas medium collide with a phonon (i.e., quantized lattice vibrations). As a result, the phase of the wavetrain associated with the photon is suddenly altered. *Natural broadening* is the act of an electron emitting energy in the form of a photon, leading to an exponential damping of the amplitude of the wavetrain. The effect of this is similar to collision broadening, in that it effectively shortens the wavetrain and produces a broadened spectral line [23].

Thermal broadening: Local variations of temperature lead to in homogeneous broadening of the emission and absorption line shapes. This nature of broadening is important in several respect and is associated with Doppler shift due to unresolved line-of-sight motions

$$T_{\text{eff}} = T_i + C\frac{m_i}{2k}\langle V_{\text{LOS}}^2\rangle,$$

where m_i mass of the ion, T_i is the temperature of the line forming ion, k is Boltzmann constant, V_{LOS} is the line of sight velocity, and $\frac{2}{3} < C < 1$. Thermal broadening contributes to the intrinsic Doppler broadening of spectral lines.

In general, Doppler broadening dominates in Xe ion lasers, collision broadening in carbon dioxide gas lasers [24], and thermal broadening in solid-state lasers. One of the important parameters to consider when processing materials with a laser is the pulse width. The longer the pulse width, the deeper the penetration will be. Pulses can be *Q*-switched where the pulses are a few tens of nanoseconds long and the pulses have high peak powers [23]. However, *plasma shielding* can occur when *Q*-switched laser pulses are used to process materials [24]. The rate of travel of the beam and the *Q*-switched repetition rate must be adjusted such that sufficient overlap in the pulses exists to produce a clean kerf without excessive local heating [25]. A *Q*-switched pulse 44 ns long with 703 W/in^2 (109 W/cm^2) power density will penetrate copper 0.09 mil (2.2 μm); a pulse 600 μs long with 1.55 W/in^2 (10 W/cm^2) power density will produce 400 times as much generation [26].

Laser trimming has emerged as a practical tool [27,28] for use on the passive circuit elements of a hybrid microcircuit. The laser trimming system is used as a clean, noncontacting tool. It has allowed great advances in automation and process control, resulting in improved total productive capacity [29,30]. Effective high-volume laser trimming of a hybrid microelectronic circuit depends on good coordination interaction of several factors [31], among them [32] (1) the type of material being trimmed (e.g., thick- and thin-film circuit elements); (2) the geometry of the circuit, including the number of circuit elements and the size and layout of each; (3) the type of laser used and its mode of operation; and (4) the speed of the trimming process. The optimization of each of these factors becomes increasingly important as manufacturers strive for higher output, better quality product, and increasing circuit sophistication [18]. In recent years, the evolution of hybrid circuit element adjustment by laser scribing has

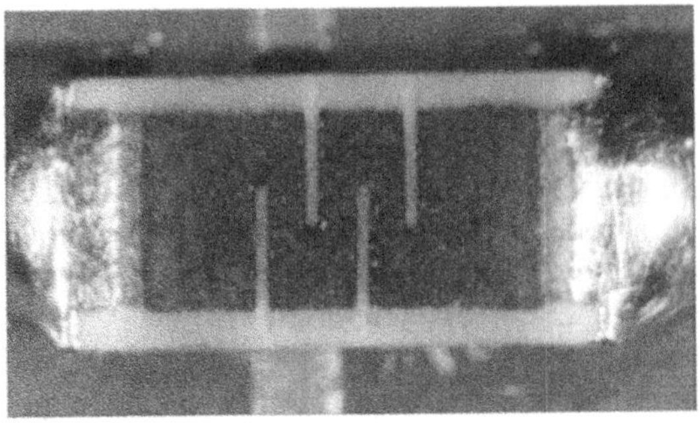

Figure 9.2 Laser-trimmed M-cut to achieve a high resistance change in a resistor [33]. (Courtesy of LS Laser Systems, Munich.)

greatly increased the speed and precision of hybrid circuit element trimming over those of conventional airbrasive trimming. The commonly used laser systems are carbon dioxide, xenon, and yttrium–aluminum–garnet (YAG) neodymium-doped. A laser-trimmed M-cut resistor is shown in Figure 9.2.

As products such as cellular phones, personal computers, and medical devices continue to shrink, the use of lasers to drill microvias on multilayer circuit boards will inevitably grow. The CO_2 laser is the most popular for via drilling, but it has some shortcomings: Hole sizes are limited and it cannot drill through copper. A frequency-tripled Nd:YAG can drill through copper, but it is difficult to drill a hole less than 10 μm in diameter. An option is the excimer laser. Rather than drilling one hole at a time directly, as with a CO_2 or Nd:YAG laser, the excimer's larger beam is capable of doing the job when the desired hole pattern is projected onto a substrate through near-field mask imaging [33].

Excimer lasers are gas lasers that emit pulses of light ($\Delta\tau \sim 20$ ns) in the ultraviolet (UV) region. The two most commonly used excimer lasers are krypton fluoride and argon fluoride, which emit light at wavelengths of 248 and 198 nm, respectively. These short pulses of UV light can also be used for the efficient ablation of material to produce high-quality microstructures [34]. Although users must decide which laser is best suited for a particular drilling application, users and manufacturers alike contend that the technologies are complementary [35].

9.2.1 Carbon Dioxide Laser [36]

With the almost unstoppable thirst for smaller and lighter electronic equipment, the demand for multilayer hybrid microcircuits is on the rise. To meet production requirements, many companies are exploring alternative manufacturing techniques. Drilling tiny interconnection holes (vias) through the boards of electronic multilayer hybrid microcircuits is one alternative, but there are some fundamental problems. The researchers at Matsushita Electric Indus-

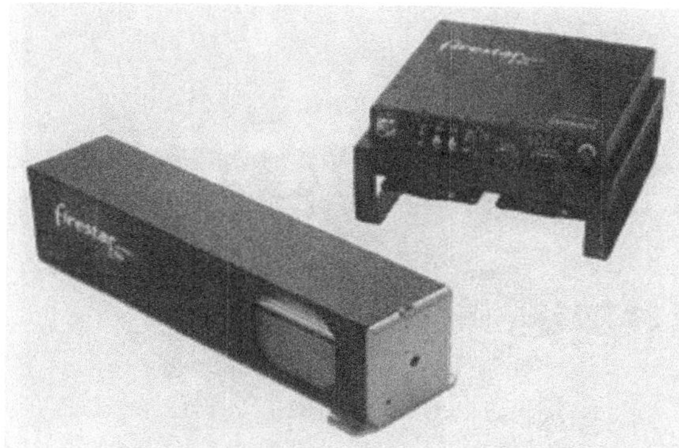

Figure 9.3 Synrad's T-60 carbon dioxide laser and laser power supply. (Courtesy of Synrad, Mukliteo, Washington.)

trial Co. in Japan have come out with the novel idea of redesigning the optical system of a CO_2 laser to drill holes 15.8 mil (400 μm) in diameter in a 39-mil (1-mm)-thick glass epoxy PCB [37]. A recently developed commercial CO_2 laser can cut a hole as small as 3 mils (75 μm).

Carbon dioxide laser systems are used widely in industry [38]. The energy emitted by these systems is in the far-infrared region (wavelength of 10.6 μm), and the ray is not absorbed by thick-film materials. Xenon lasers are fine for thin films but not powerful enough to trim thick-film resistors. Furthermore, the power stability of a xenon laser is inferior to that of CO_2 and YAG laser systems. However, xenon lasers do have the advantage of simplicity, low cost, and operation on standard 117 V ac with no cooling system. Whereas other gas lasers have efficiencies of 0.1% or less, a CO_2 laser may have an efficiency of more than 30%. Because of this high efficiency, it is relatively easy to obtain continuous-wave outputs of 100 W for a laser 1 m long. The power output of a CO_2 laser is approximately proportional to the tube length [39].

Figure 9.3 shows a CO_2 laser manufactured by Synrad. The figure on the right-hand side shows the power supply for the laser. The 100-W Synrad CO_2 laser can make holes of 50 and 100 μm in a 12-mil ceramic plate with a 2.5-in. positive meniscus focusing lens, with 40-psi compressed air as an assist gas. For a 50-μm hole, a laser pulse of 650 Hz is used with an average power of 14.3 W; for a 100-μm-diameter hole, the pulse frequency is adjusted to 2.5 Hz with a power of 55.9 W.

9.2.2 Yttrium–Aluminum–Garnet Laser

YAG [40] lasers operate in the infrared region but can provide coherent green light in the green region with complicated frequency-doubling techniques. Figure 9.4 shows Alessi's LY1 YAG laser model, which combines cutting and probing capabilities. If resistors on ICs with nitride passivation are to be

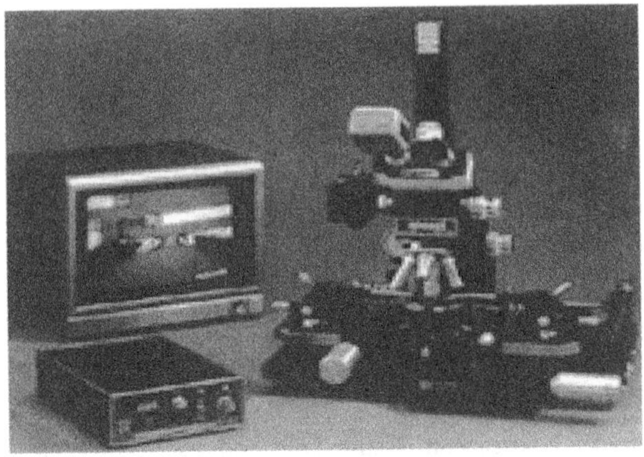

Figure 9.4 YAG laser system [41]. (Courtesy of Cascade Microtech, Beaverton, Oregon.)

trimmed, a frequency doubler may be needed, depending on the oxide thickness, which may be just the right distance to cause reflective cancellation at the IR wavelength. Within the YAG laser technology, pulses are produced either by flash pumping or by continuous pumping with *Q*-switching. Pumping with a xenon-filled flash lamp produces laser pulses of approximately 100 s. *Q*-switching, with either a rotating mirror or acousto-optic transmission of continuously pumped lasers, results in pulses of 100 ns.

Pulse characteristics of a typical flash-pumped system and a continuously pumped switched system are presented in Table 9.1. The table shows that enormous power density is available from a well-focused *Q*-switched neodymium-doped YAG laser even when operating at low pulse energy. This aspect, combined with the fast repetition rate available with *Q*-switching, is the significant feature of a successful laser trimming system. Because of the large pulse rate of a *Q*-switched YAG laser, relatively fast mechanical movements

TABLE 9.1 Comparison of a Flash-Pumped Laser and a Continuously Pumped *Q*-Switched Neodymium-Doped YAG Laser

Pulse Characteristic	Flash Pumped	Continuously Pumped
Pulse rate (s^{-1})	30–60	800–1200
Pulse energy (mJ)	17	1
Pulse width	100 μs	120 ns
Peak pulse power (W)	170	8000
Average power (W)	0.5	0.5
Spot size (cm)	0.02	0.005
Energy density (J/cm^2)	52.4	49.35
Power density (W/cm^2)	5.24×10^5	3.94×10^8

Source: From [42].

TABLE 9.2 Speed and Energy Relationship for Three YAG Laser Systems (Neodymium-Doped)

Data for Thick Film	Flash Pumped	*Q*-Switched Rotating Etalon	*Q*-Switched Acousto-optic
Pulse rate (s^{-1})	30–60	1100	4000
Pulse energy (mJ)	17	1	0.1
Trim rate (cm/s)	0.085	5	0.762
Spacing (cm)	0.00125	0.0046	0.00019
Linear energy density	13.4	0.2	0.66

Source: From [42].

can obtain high rates of trimming. The primary requirement for production of a linear kerf is that craters blasted by single laser pulses should overlap by at least 50% (linear direction). The minimum allowable trimming speed is given by

$$\text{trimming speed} = \text{crater spacing} \times \text{repetition rate}$$

With a repetition rate of 1000 pulses per second and a final spot of 0.05 cm, the trimming speed may reach 1 in. (2.54 cm) per second. Increasing either the crater size or the repetition rate at 4000 pulses per second should be matched by a mechanism capable of trimming at 5 in. (12.5 cm) per second. Typical data for YAG laser set at three different modes are shown in Table 9.2.

Table 9.3 shows the characteristics of three different types of laser systems frequently used in research and in industrial applications. Laser systems available for noncontact removal of materials without damage to the underlying layers are in common use in the field of thick- and thin-film hybrid circuits for circuit trimming, etching, and marking in plastics and ceramics, drilling hard materials, and in medical science for sophisticated surgery. Laser processing provides a valuable and unique capability for selective high-precision etching. Proper integration of motion control, beam delivery, and focal point matching make this technique a cost-effective way to trim film.

9.3 LASER TRIMMING SYSTEMS

On the basis of the hardware arrangements, laser trimming systems can be classified into two categories. The first is a two-resistor probe system with fixed laser beam in which the substrate can be moved along the *X* or *Y* direction. The second is capable of sweeping the laser beam over the fixed substrate and is provided with multiple probes. Of the systems surveyed, most fall into the first category. The first consumes much of the operator's time in the course of probing one resistor with another. Thus the overall yield is poor, due to long operation time. The second has a sweeping beam, *Q*-switched acousto-optic

TABLE 9.3 Characteristics of Different Laser Systems

Type of Laser	Medium	Wavelength	Materials/Characteristics	Applications	Advantages and Disadvantages
Gas	Excimer	193–351 nm (a) F_2: working wavelength, 157 nm; average power, <5 W (b) ArF: can work in 193 nm; average power, 30 W (c) KrCl: works in 222 nm; average power, 30 W (d) KrF: works in 248 nm; average power, 50–200 W (e) XeCl: works in 308 nm, average power, 50–200 W (f) XeF: works in 351 nm; average power, <50 W	Excimer laser system was introduced in the year 1970 and in 1990 became compatible with other laser systems. The excimer laser gas can be F_2, ArF, KrCl, KrF, XeCl, or XeF. Among these, KrF is used commonly in industry; Lambda Physik, Germany, and Lamonics, Inc. Canada, are the two main manufacturing companies for excimer lasers. The pulse frequencies for KrF lasers can vary from 300 to 600 mJ, and the maximum average power can go from 30 to 80 W. The beam divergence (mrad) can be either 4.5×1.5 or 1×3, and the repetition pulse rate can vary from 0 to 50 and 0 to 200 pulses/s. Attainable aspect ratio is generally >100:1.	Cutting and drilling of metals and ceramics where the feature sizes and resolution are important	Ablation of the target material can be accomplished without heat and a wavelength of 193–351 nm permits high resolution. Due to the shallow absorption depth, tighter control of feature sizes is possible. System maintenance is costly, and the gases used are mostly toxic.

CO_2 (a) CW (b) Gate pulsed (c) Enhanced pulsed (d) TEA	Uses a gas mixture of $CO_2:N_2:He$. CO_2 is the lasing medium, N_2 the energy transfer medium, and He enhances the population inversion. Penetration depth is normally 5–10 μm, feature resolution 50 μm; as the processes involve thermal interaction, inert gas is used to limit oxidation. The attainable aspect ratio is ~100:1.	(a) *CW:* high-speed profile cutting, cladding, seam welding, and engraving (b) *Gate pulsed:* spot welding and for cutting and drilling metals (c) *Enhanced pulsed:* cutting and drilling in IR refractive materials (d) *TEA:* via drilling, stripping wires, and marking on materials that are thermally insulated	CO_2 laser is inexpensive and can deliver high-power output in a very efficient way. Since the material interaction is through thermal overload and vaporization, operation is not safe, especially if the material used for the processing is toxic. Most of the time, due to thermal heat, oxidation results and ultimately more power is required to achieve the initial ablation rate. Penetration depth is very high and control of the feature size is difficult.

(Continued)

TABLE 9.3 *(Continued)*

Type of Laser	Medium	Wavelength	Materials/Characteristics	Applications	Advantages and Disadvantages
Solid state	Nd:YAG	1.064 μm fundamental, 532 nm for doubled, 355 nm for tripled, and 266 nm for quadrupled	Typical solid-state lasers are pumped electronically and are easy to pump. Emission wavelengths are dependent on the lattice structures of the host materials, and the frequency of Nd laser can be doubled, tripled, or quadrupled and can respond to *Q*-switching. The attainable aspect ratio is 100:1.	(a) *CW:* profile cutting, engraving, cladding, and seam welding (b) *CW pumped, Q-switched:* spot welding, drilling and cutting metals and ceramics (c) *Flashed pumped, pulsed:* spot welding, cutting and drilling metals and ceramics	
	Nd:YLF	1.047 μm; its pulse energy is greater than that of a similarly constructed Nd:YAG laser.			Harmonic generation is possible through an anisotropic medium, which can produce different wavelengths. However, output of the shorter-wavelength component is typically reduced to half the energy of the incident radiation or sometimes, less than half.

Nd:YAG (second harmonic)	532 nm
Nd:YAG (third harmonic)	355 nm
Nd:YAG (fourth harmonic)	266 nm

system. This system can be fully or semiautomated. Most manufacturing models are fully automated, and the speed, accuracy, and yield are far better than with the semiautomated system. The semiautomated version is popular in R&D laboratories because its cost is less than that of the automated version. An automated variation of the flash pump system employs steerable probes and overall control of the trimming system.

The *Q*-switched rotating etalon system achieves a laser beam sweep by mounting the entire laser head on an X–Y table. Because thick-film trimming requires automatic control of trim movement, most operators prefer motor-driven versions of the X–Y table to the micrometer-driven manual type. Instruments for the measurement and control of laser trimming may be used independently, with an operator serving as a feedback control element while manipulating micrometer heads for X–Y movements. Small process control programmable computers are frequently employed to complete the control system. To achieve minimal cracking, linear energy density deposited by the laser should be less than 3.22 J/in^2 (0.5 J/cm^2). Since the linear speed must be several centimeters per second to limit the energy density deposited, the use of L scribing is desirable. These requirements prescribe an X–Y drive with low inertia and with linear precision within 0.013 in./in. (0.005 cm/cm).

The most desirable resistor response to laser scribing appears to be an explosive fragmentation of resistor material in the kerf. The scribing mechanism is associated with the enormous power densities of the *Q*-switched laser pulse [43]. Focusing the laser beam to 0.003- to 0.005-cm spot diameters results in power densities exceeding 10^8 W/cm^2.

9.3.1 Trimming Procedure [44]

Optical System The beam from the laser, which is located on the bottom of the chassis, is routed up to the X–Y stages on the surface of the substrate carrying the element to be trimmed with a series of mirrors. Attached to the X–Y stage is a video camera and the optics head, which contains the necessary lens, filters, and beam splitters. There are three interchangeable optics heads. The video camera, which is designed by its position on the X–Y stage, is designed to be adjustment-free after changing heads. The spot size is somewhat less dependent on the aperture size used on the laser rail. The aperture is used to chop out the center of the beam emerging from the cavity, which has a Gaussian power distribution curve or TEM_{00} mode, highest in the center and trailing off radially outward. A TEM_{00} mode laser beam can be considered to be *in focus* when the beam has a minimum circular cross section. An excessively large aperture may limit spot size with a toroidal power distribution that has a low power spot in the center of the beam. On the other hand, operating with a small aperture make it difficult to achieve the required spot.

Positioning Generally, there are two stepper-motor-driven X–Y stages, one for fine positioning of the laser beam in 0.1-mil (2.5-μm) increments and a coarse stage in 1.0-mil (25-μm) steps. The fine stage contains the mirrors, which position the beam to the device, while the coarse stage moves the device under the beam. For fine trimming, the fine stage operates while the laser is on, and the coarse stage is used to go quickly to the starting-point position for each resistor trim and for loading and unloading devices from the test socket on the trim card. To achieve 0.1-mil (2.5-μm) resolution, the fine stage uses stepper motors with 200 positions or 1.8° resolution combined with a 50-pitch (50 threads per inch) lead screw.

$$\frac{1}{50}\ \text{inch/thread} \times \frac{1}{200}\ \text{rev/step} = \frac{1}{1000}\ \text{in.} = 0.1\ \text{mil}$$

The coarse X–Y stage also serves as a stepper to index across a plate or wafer. Mounted on the coarse stage is a vertical Z stage with 2 in. of travel, which is lowering the device in 0.2-mil (5.1-μm) steps using a 25-pitch lead screw. The advantage of the Z stage is the use of software autofocusing to compensate for surfaces, which are not flat or are tilted and is always a problem when trimming packaged units in a test socket.

Computer Control A desktop computer is used to control the laser trim system using an IEEE interface, which is a good real-time controller with fast I/O speed and high-speed computation with a powerful engineering-oriented instruction set. For manual movement of laser positioning, a continuous rotation knob is located on the keyboard whose function is programmable and can be changed by five dual-function soft keys. One soft key toggles between the fine- and coarse-stage controllers. The knob alone causes movement in the X direction, and the knob together with the shift key depressed actuates the Y motor. Another soft key allows Z motion.

9.3.2 Design Criteria for Resistor Trimming

Resistor trimming by laser is the most advanced and commonly adopted technology for both thick- and thin-film passive circuits. The most challenging problem of the trimming technology is to maintain the tight tolerance of the passive circuit elements' values [45]. When more than one cut is necessary to get the desired value from the trimmed passive element, the probability of exceeding the desired value is higher. To keep the desired passive circuit element's value within the permissible tolerance, great care must be taken at the initial stage. A greater number of cuts also increases the chance of increasing the instability of the trimmed circuit elements. Because of manufacturing uncertainties and tolerance requirements, a careful analysis must be made for each passive element to be trimmed. Unsatisfactory trimming might be due to an extremely low initial value of the passive circuit element and/or if the trim programming is in error.

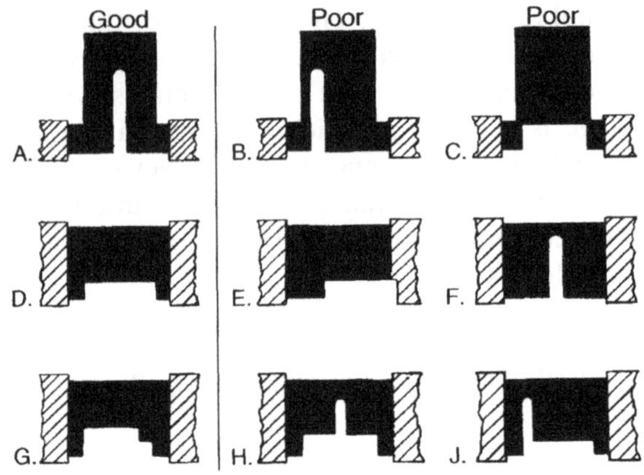

Figure 9.5 Common trimming designs for resistors [12].

When a resistor has to be trimmed, the aspect ratio (L/W) must be given important consideration because the resistor geometry is an important factor for acceptable trimming. By utilizing the computational ability of the computer, trimming can be done on a percentage basis rather than a dimension basis. If two L cuts are to be made, the first cut will be a percentage and the last cut should be absolute. The maximum allowable percentage without violating power requirements can be determined from the dynamic characteristic curve for a particular resistor geometry. Once this is known, the proper percentage can be programmed and the depth of cut will vary as a function of the initial resistor value. This method provides very absolute control over the depth of cut and essentially eliminates all problems associated with mechanical tolerances. Laser trimming of thick-film resistors [46] requires a better thermal match to the substrate than airabrasive trimming. The high thermal gradients of laser trimming induce stresses. When these stresses are released due to environmental changes, resistances are seen to change their values. The absolute *noise index* does not by itself indicate stability [47,48], but the difference in noise between abrasive and laser trim does. Higher laser trim noise indicates poorer stability [49,50]. Figure 9.5 shows some common trimming designs for resistors.

Use of a given approach, such as a single plunge cut for all resistors, could result in a drastic reduction in yield. A better practical approach for laser trimming is the application of two cuts instead of one. The reason is that sometimes, to compensate for the difference between the final (nominal) and initial resistor values, a single cut may be too deep, which can result in resistor instability. To avoid such instability, there should be two cuts instead of one. The first cut may be an L-cut, which will compensate 80% of the total value of the resistor to be restored, and the remaining 20% will be compensated by the second cut [51,52]. Figure 9.6 shows a laser-trimmed thick-film resistor. Even if perfectly centered in the resistor, the L-cut will result in a net resistor change of less than 100%. A long L-cut will expose an excessive amount of *raw* edge, and

Figure 9.6 Laser-trimmed thick-film resistor [13].

microcracking is seen to develop along the edge of the resistor. The first cut must always be started at a fixed distance from the resistor termination, to compensate for all possible tolerances present. The second trim should never be cut into the opposite termination, since it could result in *severing* the resistor completely. Figure 9.7 shows a laser-trimmed thin-film resistor.

Resistance change can be defined as [13]

$$\Delta R_c = \frac{R_1 - R_0}{R_0} \tag{9.1}$$

where, ΔR_c is the change in resistance value, R_1 the value after it is trimmed, and R_0 the untrimmed value. The sensitivity can be defined as

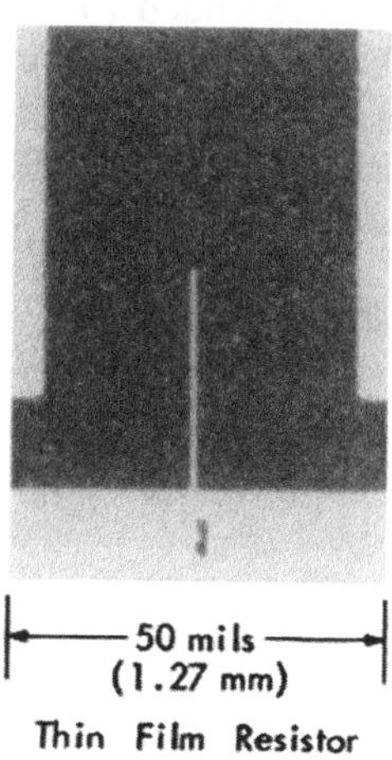

Figure 9.7 Laser-trimmed thin-film resistor [13].

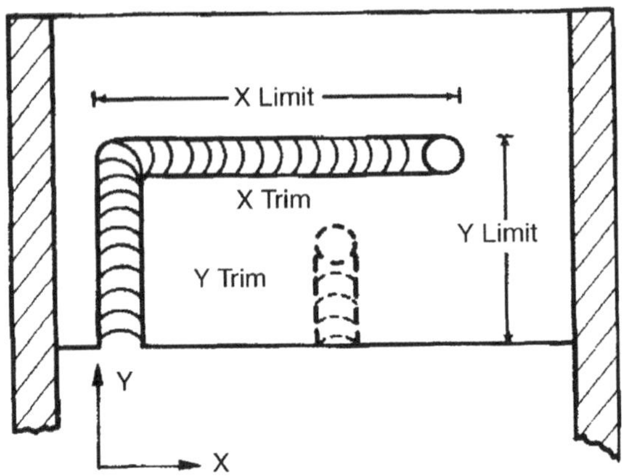

Figure 9.8 Constraint of a Y-cut trim in a resistor with a second cut in the Y-direction [22].

$$\frac{\Delta R}{\Delta l} = \frac{R_{n+1} - R_n}{R_n} \frac{l'}{l_{n+1} - l_n} \tag{9.2}$$

where ΔR is the incremental resistance change in proportional parts of the resistance before the change, Δl the incremental cut length in proportional parts of the reference length w, R_{n+1} the resistance after the $(n+1)$th incremental cut, l' the reference length (width of the resistor where the leads are connected), l_{n+1} the length after the $(n+1)$th incremental cut, and l_n the length before the $(n+1)$th incremental cut.

L-shaped cuts, parallel cuts, or serpentine cuts are generally used for multiple-aspect-ratio (L/W) resistors. The most desirable type of cut for a multiple square resistor depends on the size of the resistor, type of resistor material, and deviation from the final value. For an L-shaped cut the change in resistor value is not much along the Y direction cut (Figure 9.8). However, the cut along the X-direction is appreciable. If the cut along the X-direction is extended to the edge, the slope of the curve would have decreased to zero. The small vernier cut along the Y-direction has a negligible change in resistance value. The percentage $(\Delta R/\Delta l)$ along the length of cut is almost linear with the Y-cut, while hyperbolic changes occurs during a cut along the X-direction. The sensitivity is almost zero with a vernier cut.

An empirical formula has been developed for a Y-cut [51]:

$$\text{resistance change} = \left(\frac{l}{w}\right)^2 \frac{A+B}{[1-(l/w)]^{1/2}} \tag{9.3}$$

where l is the length of cut, w the width of cut, and A and B are empirically determined constants. The change in resistance with a laser cut along the Y-direction will be different if the aspect ratio of the resistor is 4 or more

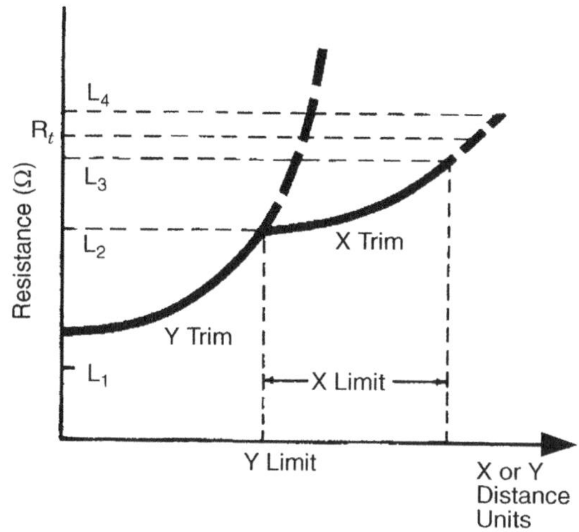

Figure 9.9 Change in resistance value with trimming along X–Y. R_t, target resistance; L_1, blank low limit; L_2, Y trim proportion of range; L_3, low tolerance limit; L_4, blank high tolerance limit [42].

(Figure 9.9). In the case of a rectangular resistor with $L/W = \frac{1}{4}$, the resistor change can be fitted with a mathematical model as follows:

$$\text{resistance change} = \left(\frac{l}{w}\right)^2 \frac{A+B}{[1-(l/w)]^{1/2}} \tag{9.4}$$

Figure 9.10 shows scanning electron micrographs of three L-cut trimmed resistors with three different compositions. The length, width, and thickness of the laser cut resistor will depend on several things, including the composition of the resistor material. When the aspect ratio is more than 5, tophat resistor design is preferable (Figure 9.11). Going from the top to the bottom has either too little change or too much sensitivity, with only a very narrow range for easy trimming. Frequently, a second cut is used for the final adjustment. There are two methods often used for tophat resistor configuration. The first cut may be along the top and then toward the downside of the hat. The gap between the two cuts (upward and downward) should leave enough space (at least $2w/3$). The tophat trim is sensitive to location across the tophat. If the cut gets too close to one edge, it leaves a long high-resistance path, which increases sensitivity and produces excessive current crowding.

When more than one cut is made, care must be taken so that the depth of the trim and spacing between trims do not violate the rigid power capability of the resistor. More than one cut, however, decreases the stability of the resistor. If it is necessary to have two cuts, the first cut is made deep and the second cut

Figure 9.10 Scanning electron micrographs of three L-cut trimmed resistors with three different compositions [42].

is made shallow in the shadow of the first cut. Figure 9.11 shows several trimming characteristics for a tophat design. Going from the bottom of the tophat to the top gives a long-range, fairly constant sensitivity at a level which is usually desirable for fast, accurate trimming. Going from the top of the tophat down, one can see either too little change or too much sensitivity, with only a very narrow range for easy trimming. So a second cut is recommended for final adjustment of the resistor value. This cut should be at a lower sensitivity, which can be obtained by a cut across the *brim* of the hat and cutting up on one side for the final trim. The second cut can make appreciable changes at reasonable sensitivity values. The tophat trims are sensitive to location across the top of the hat.

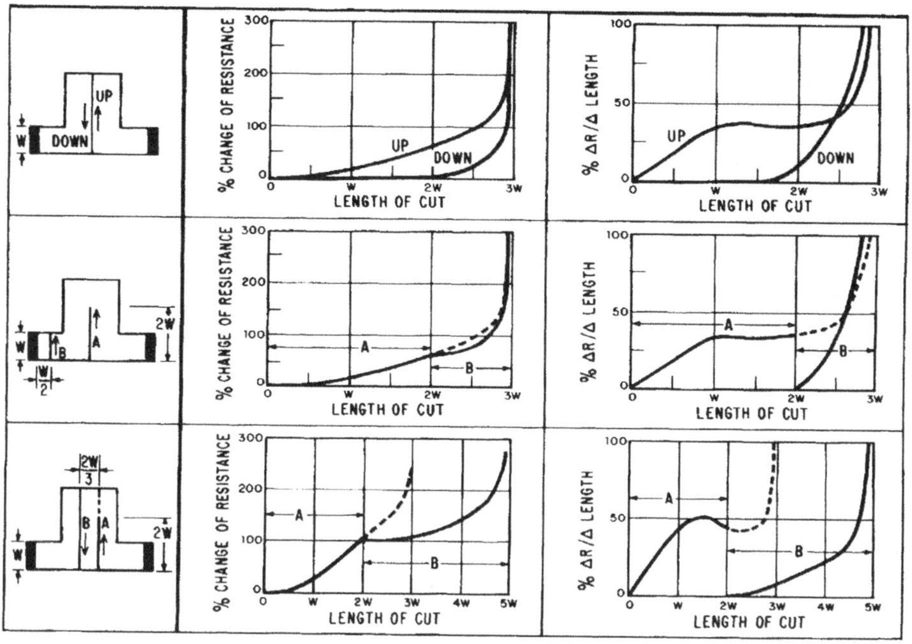

Figure 9.11 Tophat resistor trims [13].

9.4 DEFINITIONS

- *Average power:* time-averaged power of the individual pulses generated by the laser, which is the power measured using the thermoelectric power.
- *Peak power:* maximum power in an individual laser pulse. Since peak power is not easily measured, it is usually delivered from average power measurements by the formula [32]

$$\text{peak power} = \frac{\text{average power}}{\text{pulse width} \times \text{repetition rate}}$$

where power is in watts, pulse width is in seconds, and repetition rate is in hertz.

- *Repetition rate:* time rate at which laser pulses are generated.
- *Trimming speed:* speed at which the laser beam is moved over the work.
- *Overlap:* irradiated area common to two successive laser pulses. Percent overlap is given by

$$L = \frac{K(S/F)}{K \times 100}$$

where L is the percent overlap, K the kerf width, S the table speed, and F the pulse repetition rate.

- *Kerf edge:* characteristics that the kerf wall and edge kerf make at the surface of the work.
- *Kerf cleanliness:* amount of detritus within the kerf.

A second factor to consider in selecting trimming parameters is the change in resistance to be expected for each *laser pulse*, or *bite size*, for want of better term. This change must be small enough to allow the resistance to be trimmed to within the desired tolerance. A rule of thumb is that each pulse should change the resistor value no more than one-tenth of the total tolerance band. Bite size can be estimated by [27]

$$D = \frac{2R(100 - L)}{100} \tag{9.5}$$

where D is the amount of material removed per pulse, R the radius of the spot, and L the percent overlap.

REFERENCES

1. G. K. Rao, *Multilayer Interconnect Technology*, McGraw-Hill, New York, 1993.
2. D. Baudrand, Multilayer PWBS, in *Electronic Materials and Processes Handbook*, C. A. Harper and R. N. Sampson, eds., McGraw-Hill, New York, 1994, Chap. 9.
3. M. Huschka, Multilayer construction, in *Multilayer Bonding Guide*, IMAPS, Washington, DC, 1990.
4. C. Cohen and M. T. Shih, Packaging and interconnection of integrated circuits, in *Electronic Packaging and Interconnection Handbook*, 3rd ed., C. A. Harper, ed., McGraw-Hill, New York, 2000, Chap. 7.
5. M. Pecht, *Integrated Circuits, Hybrid, and Multichip Module Package Design Guidelines: A Focus on Reliability*, Wiley, New York, 1994, Chap. 1.
6. R. D. Jones, *Hybrid Circuit Design and Manufacture*, Marcel Dekker, New York, 1982.
7. C. W. White and P. S. Percy, *Laser and Electronic Beam Processing of Materials*, Academic Press, San Diego, CA, 1980; M. Pecht, J. Evans, and J. Evans, *Quality Conformance and Qualification of Microelectronic Packages and Interconnects*, Wiley, New York, 1994.
8. R. A. Rikoski, Laser resistor trimming, in *Hybrid Microelectronics Circuits: Thick Film*, Wiley, 1980, p. 38.
9. D. C. O'Shea, W. R. Callen, and W. T. Rodes, *Introduction to Lasers and Their Application*, Addison-Wesley, Reading, MA, 1997.
10. S. S. Charschan, *Lasers in Industry*, Van Nostrand Reinhold, New York, 1972.
11. R. Glang, K. Jaeckel, W. Perkins, and L. I. Maissel, in Thin Film resistor, p. 18-31,

in Handbook of Thin Film Technology, L. I. Maissel and R. Glang ed., McGraw-Hill, New York, 1983. *IEEE Spectrum*, Vol. 6, 1969, p. 71.

12. H. G. Weightman, Abrasive trimming: the important forgotten, *Proc. Int. Symp. Hybrid Microelectron.*, 1977, p. 25.
13. E. J. Swenson and G. D. Vincent, Laser trimming analysis using a resistive sheet analogy, *Proc. Int. Symp. Hybrid Microelectron.*, 1972, p. 1-A-5-1.
14. Y. Taketa and M. Haradome, Thick-film resistor adjustment by high-voltage discharge, *Proc. Int. Symp. Hybrid Microelectron.*, 1972, p. 1-A-4-1.
15. R. E. Cote, R. C. Headley, J. T. Herman, and A. Howe, Factors affecting laser trim stability of thick-film resistors, *Proc. Int. Symp. Hybrid Microelectron.*, 1976, pp. 128–132.
16. J. J. Licari and E. R. Enlow, Abrasive trimming, in *Hybrid Microcircuit Technology Handbook*, 2nd ed., Noyes Publications, Park Ridge, NJ, 1998.
17. C. Ingulli, Adjusting deposited thick-film resistors, *Proc. Int. Symp. Hybrid Microelectron.*, 1967, p. 179.
18. D. L. Bunch, New multistation automatic airbrasive trimming system, *Proc. Int. Symp. Hybrid Microelectron.*, Princeton, NJ, May 26, 1971.
19. J. F. Ready, *Effects of High-Power Laser Radiation*, Academic Press, San Diego, CA, 1971.
20. A. Yariv, *Introduction to Optical Electronics*, Holt, Rinehart & Winston, New York, 1971.
21. F. T. Arecchi and E. O. Schulzdubois, *Laser Handbook*, Vol. I, Part A, North-Holland, Amsterdam, 1972.
22. W. V. Smith and P. P. Sorokiz, *Lasers*, McGraw-Hill, New York, 1966.
23. J. Wilson and J. F. B. Hawkes, *Optoelectronics: An Introduction*, Prentice Hall, Upper Saddle River, NJ, 1983, p. 170.
24. A. Khachatourians, New applications of a 30 W (MM) Nd^{+3} YAG laser: GaAs via drilling and ITO etching, *Proc. Int. Symp. Hybrid Microelectron.*, 1986, p. 100.
25. W. W. Duley, *Laser Processing and Analysis of Materials*, Plenum Press, New York, 1983.
26. P. F. Parks and W. I. Austin, Electro-optic *Q*-switches for laser resistor trimmers, *Proc. Int. Symp. Hybrid Microelectron.*, 1976, p. 25.
27. G. Hurtig and J. Swenson, eds., Overview of laser functional trimming techniques, *Proc. Electron. Comput. Conf.*, 1973.
28. E. A. Metzbower, Applications of lasers in materials processing, *ASM Conf. Proc.*, Washington, DC, Apr. 18–20, 1979.
29. D. R. Herriot, Applications of laser light, *Sci. Am.*, Vol. 219, No. 3, Sept. 1968, pp. 141–156.
30. J. E. Harry, *Industrial Lasers and Their Applications*, McGraw-Hill, New York, 1974.
31. R. C. Headley, Laser trimming is an art that must be learned, *Electronics*, June 1973, pp. 121–125.
32. T. Richardson and P. Perks, A method of obtaining optimum trimming performance from a laser system, *Proc. Int. Symp. Hybrid Microelectron.*, 1974, pp. 398–401.

33. A. J. Hand, Lasers squeeze into tighter board assemblies, *Photon. Spectra*, July 1999, p. 96.
34. N. Rizvi, Excimer lasers machine novel microstructures, *Laser Focus World*, Nov. 2000, p. 75.
35. K. Mukherjee and J. Mazumdar, eds., *Lasers in Metallurgy*, Materials Society of AMIE, Warrendale, PA, 1981.
36. C. K. N. Patel, High-power carbon dioxide laser, *Sci. Am.*, Vol. 219, No. 2, Aug. 1968.
37. P. N. Powell, Uniform beam drills better holes, *Laser Focus*, Nov. 2000.
38. M. J. Beesley, Carbon dioxide laser, in *Lasers and Their Applications*, Taylor & Francis, London, 1976, Chap. 7, p. 118.
39. J. Wilson and J. F. B. Hawkes, Molecular lasers: the CO_2 laser, in *Optoelectronics: An Introduction*, 2nd ed., Prentice Hall, Upper Saddle River, NJ, 1999.
40. M. J. Beesley, YAG (Nb-doped) laser, in *Lasers and Their Applications*, Taylor & Francis, London, 1976, Chap. 6, p. 99.
41. A. Chestnut, Product showcase "top five," *Semicond. Int.*, June 1988, p. 152.
42. R. T. Howard and R. Allen, Characterization of a laser-trimmed thick-film resistor by a scanning electron microscope, *Proc. Int. Symp. Hybrid Microelectron.*, 1971, p. 4-1-1.
43. R. P. Knapp and S. J. Wendel, A laser system for active trimming of precision hybrid devices, *Proc. Int. Symp. Hybrid Microelectron.*, 1986, p. 85; L. Braun and D. E. Lood, Precision thin-film cermet resistors for ICs, *Proc. IEEE*, Vol. 54, Nov. 1966, pp. 1521–1527.
44. B. G. Streetman, Lasers, in *Solid-State Electronic Devices*, 2nd ed., Prentice Hall, Upper Saddle River, NJ, 1980, Chap. 10.
45. A. K. Kakar and A. Polenske, On laser trimming of thick- and thin-film resistors, *Proc. Int. Symp. Hybrid Microelectron.*, 1974, pp. 110–118.
46. N. S. Span, R. Headley, G. D. Vincent, and E. J. Swenson, The effects of laser trimming on birox and 1100 series thick film composition, *Proc. Int. Symp. Hybrid Microelectron.*, 1971, p. 4-2-1.
47. K. L. Haynes, V. W. Ruwe, and R. E. Shannon, Experimental characterization of thick-film resistor stability and performance, *Proc. Int. Symp. Hybrid Microelectron.*, 1977, p. 76.
48. J. Houge and L. K. Keys, Long-term stability characterization of thick-film resistors trimmed down in resistance by RF technique, *Proc. Int. Symp. Hybrid Microelectron.*, 1987, p. 123.
49. T. M. Chen, J. G. Rhee, and B. A. Shojai, The effects of trimming on current noise of thick-film resistors, *Solid State Technol.*, June 1977, pp. 49–53.
50. C. Y. Kuo, The effects of trimming method on noise in thick-film resistors, *Proc. 21st Electron. Comput. Conf.*, 1971.
51. E. J. Swanson, G. D. Vincent, and J. Riley, Laser trimming analysis using a resistive sheet analogy, *Proc. Int. Symp. Hybrid Microelectron.*, 1972, p. 1-4-51.
52. J. J. Licari and L. R. Enlow, Types of resistor trims, in *Hybrid Microcircuit Technology Handbook*, 2nd ed., Noyes Publications, Park Ridge, NJ, 1998, Chap. 4.

RECOMMENDED READING

Chandrasekhar, S., *Radiative Transfer*, Dover Publications, Mineola, NY, 1960.

Jacobs, S., M. Sargent III, and M. O. Scully, *High-Energy Lasers and Their Applications*, Addison-Wesley, Reading, MA, 1974.

Sargent, M., III, M. O. Scully, and W. E. Lamb, Jr., *Laser Physics*, Addison-Wesley, Reading, MA, 1974.

10

PACKAGING AND THERMAL CONSIDERATIONS

There is an increased awareness in the industry that the assembly/packaging operation is becoming a differentiator in product development [1–3]. Package design and fabrication are important to system applications. It is no longer just a means of protecting circuits but also a way for the systems designer to ensure form, fit, and function for today's products—from consumer products to high-end workstations [4,5]. After active and passive components are attached to a screened and fired substrate, the resulting microcircuit is ready for mounting with a package [6–8]. The package protects the circuit from its environment, supports the electronics mechanically, provides a heat transfer path from the circuit substrate to the air, and allows external leads to be attached to relatively strong package leads or pins rather than to fragile substrate lands. In fact, some are saying that it is not the chips but the circuit interconnections that will dominate performance and cost [9,10]. Dramatic changes are under way in the computer, telecommunication, automotive, and consumer electronics industries. In view of the drastic changes in these industries, packaging becomes the best vehicle to address the needs of these industries.

According to one observer [11], highly integrated, low-cost, thin, light-weight, reliable, high-density packaging can lead to very cost-effective consumer product integrating the functions of a cellular video telephone, the power of a miniframe computer, and combining a wireless fax modem with voice and handwriting recognition into a single palm-sized machine selling at consumer prices. The definition of high-density packaging varies from one manufacturer to another. Some define it by the I/O or interconnect pitch; others define it as

Handbook of Thick- and Thin-Film Hybrid Microelectronics, By Tapan K. Gupta
ISBN 0-471-27229-9

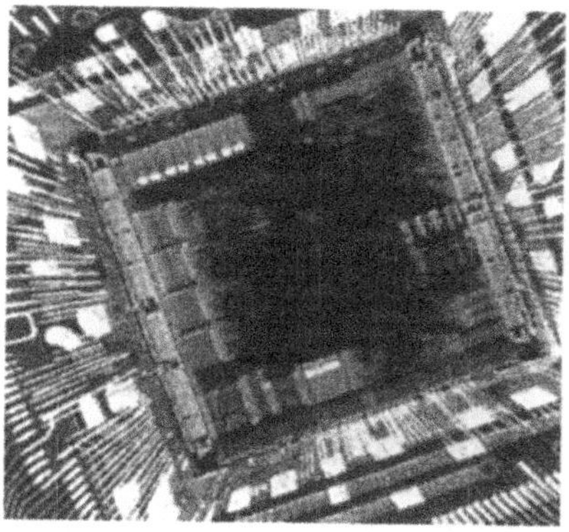

Figure 10.1 High-density packaging system [3]. (Courtesy of Abpac, Phoenix, Arizona, as published in *Semiconductor International*, Vol. 23, no. 2, 2000.)

any package that has to be codesigned with a chip to perform as required. Regardless of definition, it is understood that high-density packages are a necessity in high-performance circuits and systems. The wiring on a high-density package is very dense, and the interconnections should have low capacitance and impedance.

A high-density package (Figure 10.1) provides a stable power supply level even when the circuits are operating at high speeds and many output drivers are switched simultaneously. In addition, the thermal properties of the package and the semiconductor (silicon) chips seating on the alumina match well [12,13]. This avoids stress-induced cracks and failures when the parts expand and contract as the temperature varies. The package can also remove large amounts of heat generated by the circuit. The reliability of the package should be perfect and cost should be minimal. A *perfect* package is defined as a compact unit that supplies the chips with a large number of signals and power connections and which has very low capacitance, inductance, and resistance values.

10.1 PACKAGING MATERIALS [14]

Silicon carbide (SiC), aluminum nitride (AlN), beryllia (BeO), and alumina (Al_2O_3) are some of the ceramics used in electronic packaging. In comparison with other ceramics, SiC has a thermal expansion coefficient closer to that of silicon, and as a result, less stress is generated between the dice and the substrate at elevated temperatures. Its high dielectric constant, however, makes it undesirable as a substrate to carry interconnections. Alumina and beryllia have properties similar to those of silicon carbide. However, alumina is the first choice for a hybrid microcircuit substrate. Although its thermal conductivity is not as good as that of SiC or BeO, it is still better than plastic packages or PCBs. Gallium arsenide and germanium have thermal expansion coefficients closer to those of alumina and beryllia [15]. However, germanium is not suit-

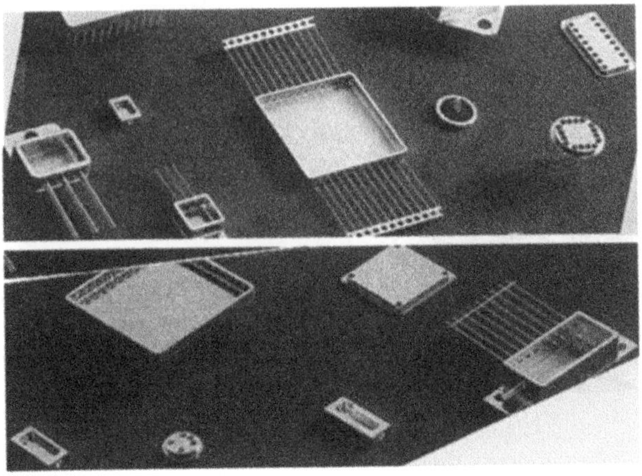

Figure 10.2 Common hybrid packages [17]. (Courtesy of Advanced Technology Group, Rockway, New Jersey.)

able above 100°C. It goes without saying that thermal mismatch between a substrate and the chips attached to it should be the primary concern for the packaging engineer [16]. Figure 10.2 presents some common packaging units used in hybrid microelectronic circuits.

Thermal mismatch problems can be solved by incorporating a metal core inside a ceramic substrate or by using a metal-core board as a base and mounting a regular multilayer ceramic board on top [18]. The two popular structures that solve the coefficient of thermal expansion mismatch problem are copper-clad invar and copper-clad molybdenum. Molybdenum has a coefficient of thermal expansion close to that of alumina, and it is very rigid. The coefficient of thermal expansion of copper-clad invar can be adjusted to be exactly the same as the ceramic package mounted on top of it, or the same as silicon if chips are mounted directly on the board. Even the temperature difference between the substrate and the dies can be compensated by making the temperature coefficient of thermal expansion of the copper–invar–copper substrate slightly greater than that of silicon in order to match their thermal expansion exactly [19]. The advantage of copper-clad molybdenum over copper-clad invar is its higher thermal conductivity and rigidity [20].

First let us consider the lead configuration. In an all-metal package, leads exist through holes in the sidewall. They are insulated from the package shell by individual glass beads surrounding each lead. If a user has applications for several lead arrangements within the same basic package size, new stampings are required for each arrangement. It is apparent that such a user should consider a glass-to-metal sealed package. An advantage of a glass-sealed package over metal packages, for example, is the ability to change the base material. The base can be either metal or ceramic. Metal is the preferred choice when a ground plane and/or shielding is required. The metal generally is an iron–nickel–cobalt alloy conforming to ASTM F-15. This alloy provides a coefficient of thermal expansion closely approximating that of the glass used in package sidewalls. The other basic material is ceramic, which is the choice when insulation is desired in the base. Beryllia has a high thermal conductivity and should be used for high-power applications requiring heat sink [21]. Due to beryllia's cost and toxicity, alumina is more economical and safer to handle. These ceramics also match the thermal coefficient of expansion of available glasses, so that the packages can be operated over the temperature range −65 to 200°C without failure due to expansion mismatch. With proper design and process control, glass-to-metal micropackages can provide a hermetic seal to a level of 1×10^{-8} mL/s maximum leakage. This is important for containing a neutral atmosphere such as nitrogen, argon, or helium. Conversely, the package seal keeps out moisture. Hermetic package [22] sealing is probably the single most important step in the many process steps required for producing hybrid microelectronic circuits. The majority of these circuits are used in space, military, and commercial applications which need 100% hermetic seal testing per the MIL-STD-883 specifications for substrates. Several alternatives are available, including plastic, metal, ceramic-to-metal, and glass-to-metal packages. In broad terms, hermetic sealing is any method whereby a metal or glass package is closed by fusion, diffusion, or solder to form a gas- or airtight interior cavity [23]. Two common failure modes attributed to an undesirable package atmosphere may be involved. One disturbs the basic electronic functions of a junction device. A prime example is charge spreading, which can lead to inversion layer formation with degradation of junction performance. Surface leakage may be an extension of this. The other basic mode exerts itself by chemically changing the physical components of a microcircuit. It is generally accepted that moisture is the major contributor, although some think that oxygen can be detrimental. Hydrogen is often one of the major impurities found in semiconductor packages sealed in dry nitrogen. Chlorine is the next important gas, which corrodes aluminum leads and destroys the native lead oxides. Manufacturers are confronted with the decision of how to package to protect aluminum [20]. Table 10.1 represents some of the packaging material characteristics.

10.2 PACKAGING SYSTEMS

Advances in surface-mounted components have led to a shift from through-hole mounting to surface mounting. All packaged components typically use conventional transfer molding to encapsulate a device. Typical packages include DIPs, small-outline packages, plastic-leaded chip carriers, small-outline J-leaded packages, and leaded chip carriers [23–25]. These packages are manufactured prior to the board assembly process, and to mold these packages to less than 40 mil (1 mm) thickness is difficult. Chip-on-board (COB) technology has become very popular, because of its thinness and light weight and its promise for use as high-density packaging. The difficulty with COB technology

TABLE 10.1 Characteristics of Packaging Materials

Material	Thermal Conductivity (W/cm·K)	Coefficient of Thermal Expansion ($\times 10^{-6}$ K)	Dielectric Constant, K	Electrical Resistivity ($\mu\Omega$·cm)
Gold (Au)	2.96	14.2	—	2.35
Silver (Ag)	4.3	18.9	—	1.59
Copper (Cu)	4.0	17.0	—	1.67
Aluminum (Al)	2.3	24.0	—	2.65
Tungsten (W)	1.7	4.0	—	5.65
Molybdenum (Mo)	1.4	5.0	—	5.0
Kovar A	0.7	0.5	—	4.9
Nichrome (Ni–Cr)	0.13	1.8	—	10.8
Phosphor bronze	0.71	1.8	—	11.5
Silicon (Si)	1.5	2.5	11.8	—
Germanium (Ge)	0.7	5.7	16.0	—
Gallium arsenide (GaAs)	0.5	5.8	10.9	—
Silicon carbide (SiC)	2.2	3.7	—	42.0
Beryllia (BeO)	2.0	6.0	6.7	—
Alumina (Al_2O_3)	0.3	6.0	9.5	—
Silicon dioxide (SiO_2)	0.01	0.5	3.9	—
Polyimide	0.004	—	—	—
Epoxy glass	0.003	15.0	5.0	—

Source: From [1]; © reprinted by permission of Pearson Education, Inc.

is the sealing resin. There is an ongoing research to replace this resin with a screen-printable epoxy [26]. Table 10.2 presents some electronic packaging requirements.

10.2.1 TO Packages

TO packages are one of the most common package units used in hybrid microcircuits. Basic packages are the TO-5, TO-8, and TO-18 cans; butterfly, dihe-

TABLE 10.2 Electronic Packaging Requirements

Speed: high speed with short propagation delays and high transmission frequency
Pin count: maximum I/O pin count per device
Wireability: dense wiring
Size: compact size
Noise: low noise coupling among wires with high-quality transmission
Thermal and mechanical reliability: high heat dissipation rate and good match between thermal expansion coefficients of the die and chip carriers
Test and cost: should be easy to test, fix, and change; cost to perform these operations should be minimal

Figure 10.3 TO package [28]. (Courtesy of Integrated Circuit Engineering Corporation, Phoenix, Arizona.)

dral, platform, and cold-weld cases; custom flatpacks with kovar or ceramic-to-glass seals; and plastic types [27]. TO-5 and TO-8 cans are 350 mils (8.75 mm) in diameter with 6, 8, 10, or 12 leads. The TO-5 can dissipate 300 mW at 130°C in free air. TO-18 cans are similar to but slightly larger than the TO-5 size. TO-99 is a low-profile version of the TO-5, which has been used to package hybrids for over 20 years. The original packages housed thick-film resistors, capacitors, and chip and wire semiconductors on multiple levels of ceramic substrates. With increased use of small thin-film chip resistors, circuits of moderate complexity are being fabricated on a single level and packaged in low-profile TO-99s. The material cost is lowest in this construction, due to the volume use of headers and covers by the semiconductor industry. The yields are good despite high circuit density. Figure 10.3 shows a typical TO package used in hybrid microcircuits.

10.2.2 Flat Case Packages

Flatpack cases are packages that hold substrates as small as 150 mil (3.8 mm) square or as large as 1.25 by 1.35 in. (3.18 by 3.43 cm), dissipating 250 mW to 2.5 W, respectively [29]. The 14-pin DIP is a very popular rectangular flat package with two rows of pins on its two sides. Here a die is bonded on the lead frame, and in the next step, chip I/O and power/ground pads are wire bonded to the lead frame and the package is molded in plastic. The packaging density of DIPs, however, is degraded by many factors. For example, DIPs can get excessively large at pin counts greater than 64, and through holes limit the board packing density by blocking the lines that might have been routed below them, making it impossible to mount chips on both sides of the board in a way that would increase board density. This results in long interconnections at the board level, which combined with large capacitance and inductance of DIP pins, degrade system speed. The 14-pin DIP evolved from the 8-pin TO-5 as the most popular hermetic package for integrated circuits [24]. The highest-volume glass ceramic package requires high-temperature sealing [30]. The side-braze

Figure 10.4 Flat case package.

package is coming down in cost as more use is made in packaging precision devices. The special 145-pin side braze increases the area and is effective in a hybrid package. Figure 10.4 shows the top view of a caped flat case package.

Dual-Inline Packages A DIP is a rectangular package with two rows of pins on its two sides. DIPs have been the traditional low-cost packaging choice and remains the most widely used package type. Because of the low thermal conductivity of the plastic molding, DIPs have moderate heat removal capacity [29]. During board assembly, DIPs are mounted in holes drilled through the substrate and soldered down (Figure 10.5). Special solder solution that adheres to the metal pins, copper-plated walls of the holes, and contact areas surrounding the holes is used, adopting wave soldering techniques. The packaging density of DIPs is, however, degraded by many factors. Long interconnections result in high capacitance and inductance values, which degrade the speed of the system [31–33].

Pin Grid Arrays [23] A pin grid array (PGA) usually utilizes a multilayer ceramic substrate, which has a high dielectric constant (ca. 9.5). It has leads on

Figure 10.5 Dual-inline package [28]. (Courtesy of Integrated Circuit Engineering Corporation, Phoenix, Arizona.)

its entire surface rather than on its periphery, and this has provided PGAs with larger pin counts (300 or more) compared to DIPs, which can have a maximum of 64. PGAs can be available in a cavity-up or cavity-down version. In the latter case, a die is mounted on the same side as the pins facing toward the substrate, and a heat sink can be mounted on its back side. When the cavity and the pins are mounted on the same side, the total number of pins is reduced. The mounting and wire bonding of the dies become difficult because of the existence of pins next to the cavity. In cavity-up PGAs, pins and chip are on opposite sides of the square package. This type of PGA is easy to handle and offers a dense pin array on its entire bottom surface. Power dissipation is less than in the cavity-down version, however, because the back side of the die faces the board and cannot be bounded directly to the heat sink [12].

An alternative to ceramic substrate, epoxy fiberglass board is used for PGAs. These packages are cheap, smaller parasitic lead inductance and capacitance but cannot dissipate as much power as ceramic substrates. To improve heat dissipation in PGAs, through-holes are drilled in the substrate and are filled with metals. These metal fillers are ultimately connected to the sink.

Ribbon-leaded flat packages have not been as popular with hybrid manufacturers as with package suppliers. The shorted lead frames protect the leads until the first electrical test, when expensive carriers and sockets are required. The leads are impossible to bond ultrasonically, due to the viewing angle required. Platform packages are much more efficient in hybrid packages. For one reason, major suppliers of glass flatpacks have not made a glass platform; the smallest houses have not been able to break through the volumes required for low-cost packages through industry standardization. Gold-plated flat platform packages are readily available, but due to fluctuation in the gold market, they are not often used.

Flatpack Plastic Package Flatpack ceramic chip carriers are used in surface-mounting carriers with more than 100 terminals and lead spacing as small as 25.35 mils (0.65 mm).

10.2.3 Chip Carriers

The chip carrier concept has become very popular in industry, and substantial production programs have been adopted. The use of chip carriers is expected to improve circuit density at least fourfold compared to DIPs. Alumina and beryllium oxide chip carriers have brought outstanding thermal conductivity improvement.

Leadless Inverted Devices Leadless inverted devices (LIDs) carry packages from diodes, transistors, and ICs. The semiconductor die is attached in the trough of the LID by either eutectic bonding or soldering. Leads from the die are wire bonded to the legs of the LID. The die is passivated and the LID is limited. However, LIDs are not widely used in commercial hybrid circuits.

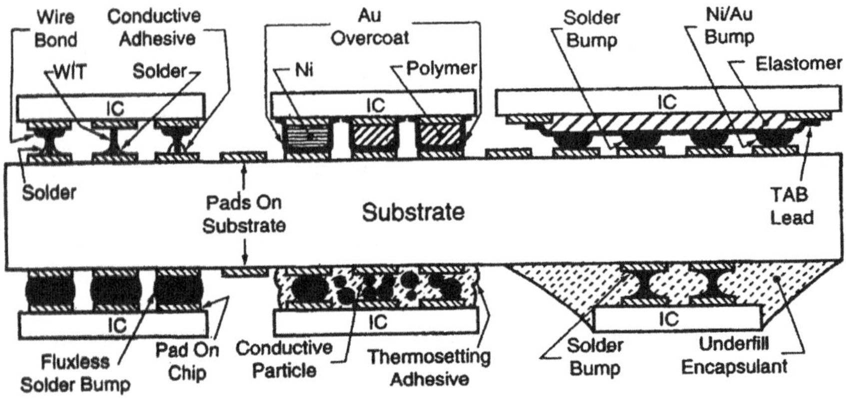

Figure 10.6 Flip-chip technologies [34]. (Reproduced with permission of The McGraw-Hill Companies.)

Flip Chips [34] In flip-chip attachment (Figure 10.6), a chip is passivated and flipped in order to attach the chip to the substrate. This attachment may be tape automated, fluxless solder bumps, wire interconnects, isotropic and anisotropic conductive adhesives, metal bumps, compliant bumps, and pressure contacts [34]. The passivated chip carries small solder bumps on its termination pads and are attached to the metallized surface of the substrate. Solder fluxes are used to keep the chip in its place [35,36]. The process is economical and easy to carry out, and the resulting smaller package size provides a higher package density [37].

Ball Grid Arrays Starting in the late 1980s and through 1990, surface-mounting technology migrated from perimeter-leaded packages to area array interconnections [ball grid arrays (BGAs)] and increased the number of interconnections in a power law fashion [34,38]. Several types of BGA packages are available in the market [e.g., ceramic BGA (CBGA), plastic BGA (PBGA), tape BGA (TBGA)]. The main difference in the construction of these three BGAs is in the substrate material. In ceramic BGA, cofired alumina is used as substrate, and PBGA utilizes PCB technology for the substrate and transfer molding encapsulation. Cofired ceramic substrate has excellent inherent electrical and thermal characteristics, including built-in reference and power planes and low inductive power connections with multipower paths. Ceramic BGAs can support all types of chip interconnections (wire bond, flip chip, and/or TAB), plus a variety of lead sealing or encapsulation techniques. Flip-chip CBGA packages are typically not hermetic and may be capless or capped, but compared to plastic BGA, they are expensive. Figure 10.7 shows a typical ceramic BGA package. PBGA technology has been gaining industry-wide interest and commitment as potentially the lowest-cost package for high-I/O devices and even for some lower pin count applications [39,40]. Figure 10.8 is a

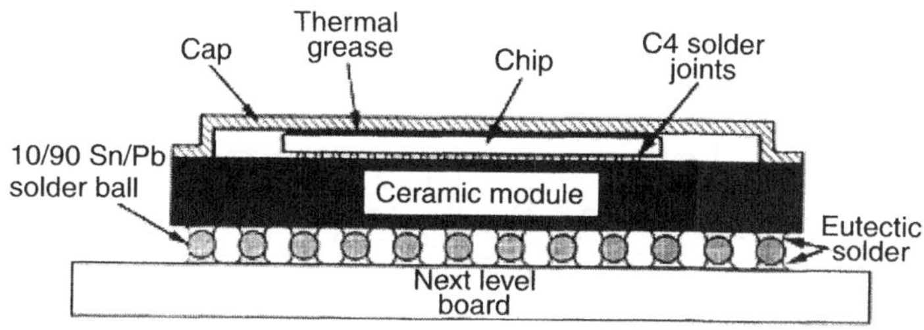

Figure 10.7 Ceramic ball grid array package [41].

cross-sectional view of a PBGA packaging system. The major advantage of PBGA is the ease of achieving low solder joint defect levels using existing SMT equipment and processes. Six sigma (6σ) assembly yields are achievable much more quickly with PBGA than with plastic quad flatpacks, saving several years of process improvement.

In TBGAs a tape automated bonding tape frame puts the chip in contact with the next-level board. However, all BGA packages can be attached to industry standard FR4 cards using standard surface-mounting assembly processes. The tape frame is generally a two-conductor-layer tape with a 2-mil (50-μm)-thick dielectric layer sandwiched between. Usually, one conductor layer contains the signal traces, and the other contains the ground plane. High-melting-point solder balls are partially reflowed into each plated through-hole. Figure 10.9 shows a typical TBGA packaging system.

Multichip Packages Advances in semiconductor technology continue to result in increased speed and complexity of electronic circuits. In conventional hybrid microcircuit packaging technology, individual chips are mounted on the single substrate and connected by wires. The parasitic reactance on the bond wires and bonding pads degrade the electrical performance of the circuit. To increase packaging density with minimal compromise with performance, multichip packaging (MCP) systems without bond wires for interchip interconnections have been developed. The circuit is reported to function normally without

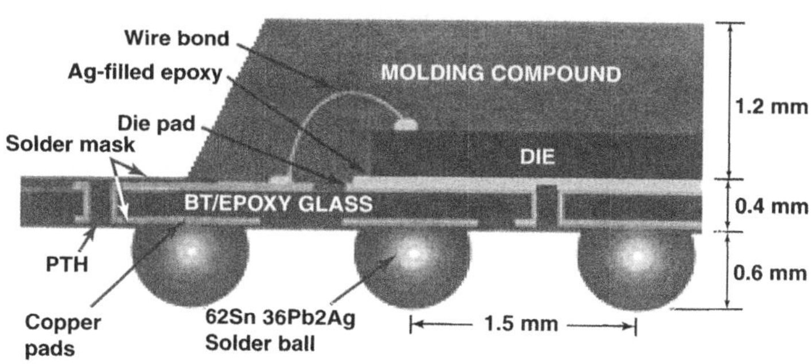

Figure 10.8 Cross section of a PBGA [39].

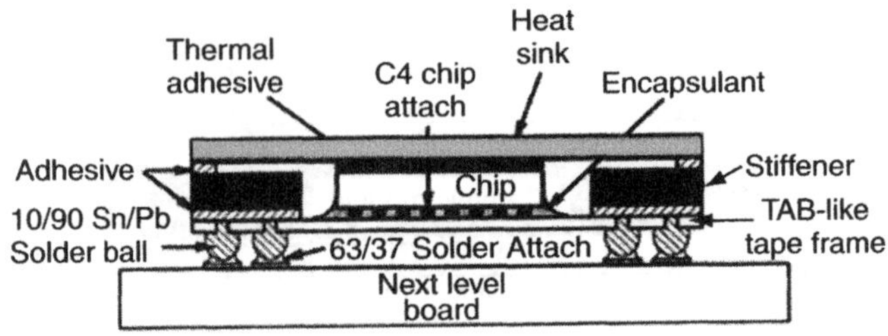

Figure 10.9 Tape ball grid array [41].

degradation in speed and performance [42]. According to some experts, the hermetic sealing process may become a roadblock of MCP applications, especially for military and space applications [43].

Multichip packages are referred to as *complex ICs* because the component dies logically belong together but are not realized as a single die. In high-density applications, multiple unpackaged dies are placed in one module to save space. The term *multichip package* is starting to emerge, underscoring the difference between the newer designs from the integrated concept of a high-reliability multichip module [1]. According to an observer [44], the maximum die-size system is 0.58 by 0.58 in^2 (15 by 15 mm^2). For larger areas the package yield is very poor. Alpine microsystems have a silicon substrate interconnect system for multichip packages.

Plastic-Leaded Chip Carriers A plastic-leaded chip carrier (PLCC) unit may be either gullwing or J-leaded. The pin count sometimes exceeds 124 with 50-mil (1.27-mm) spacing. A J-leaded PLCC chip carrier pack is denser and more suitable for automation than is a gullwing leaded carrier. Another advantage of J-leaded PLCCs is that they can be inserted into sockets.

Leadless Ceramic Chip Carriers Leadless ceramic chip carriers (LCCCs) also utilize multilayer ceramic technology as pin grid arrays. In LCCCs the arrays of pins are not brazed on the substrate. The conductors are bonded to chip pads that are left exposed around the package periphery to provide contacts for surface mounting. The spacing of the leads is generally 30 mils (0.75 mm) and there are 124 terminals. For military and space applications LCCCs are hermetically sealed. Because of the ceramic substrate, LCCCs do not have any mismatch problem regarding thermal expansion when they are mounted on thick- and thin-film hybrid circuits using alumina substrate. However, the ceramic substrates are not cost-effective.

10.2.4 Small-Outline Packages

The small-outline package (SOP; Figure 10.10) family offers several features that reduce the total volume by almost one-third to one-half of the volume

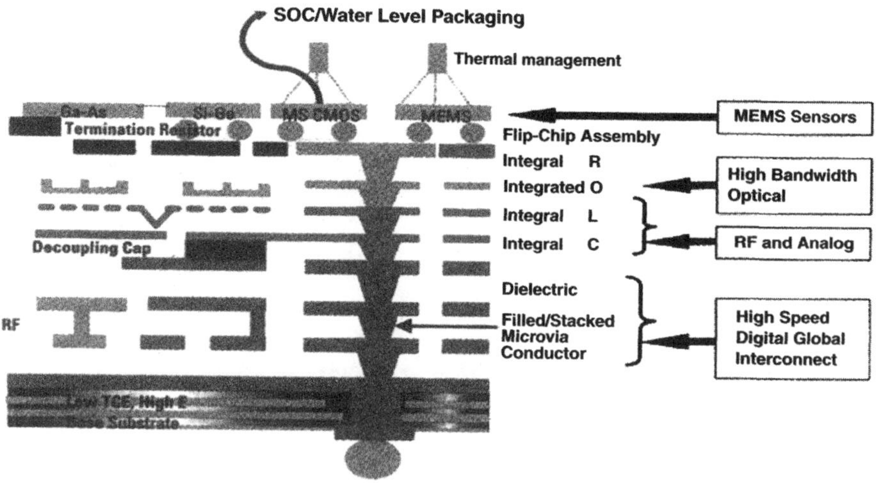

Figure 10.10 Small-outline package system in a single-layer integrated module [47].

of earlier component packages. Industry-standard SOP is available in three standard packages: (1) thin small-outline packages (TSOPs), (2) plastic small-outline packages (PSOPs), and (3) shrink small-outline packages (SSOPs). All three package types offer X8 or X16 organization options. The X8 option reduces component count, power consumption, and board space. The 44-lead PSOP (X16) device replaces two X8 32-lead PLCC devices, decreasing required board space by approximately 30%. Furthermore, SOP pin counts provide space for a single-pin upgrade path for 2 to 8 megabits in a consistent package. The TSOP thickness of 48 mils (1.20 mm) is one-third to one-half that of small-outline J-leads (SOJ) and PLCCs. The two-sided pinouts and standard/reverse pinout offerings of TSOP minimize both board area and circuit board layers consumed by memory. The 112-mil (2.80-mm) thickness of PSOP and 72-mil (1.80-mm) thickness of SSOP are less than half that of standard plastic DIPs. PSOPs and SSOPs have 51-mil (1.27-mm) and 32-mil (0.8-mm) pitch gullwing leads located on the two short sides of the package. Leads are on two sides of the package and can be placed much closer together and in closer proximity to the bound components [44–46].

Small-Outline IC Package A small-outline IC package is a plastic package with gullwing leads spaced equally at a distance of 50 mils (1.25 mm). These packages are soldered to the pads on the surface and do not require holes.

10.2.5 Systems on a Chip

As demand continues to grow for more functionality in electronic systems, more effort goes into developing system-on-a-chip (SOC) designs and devices. Systems in a package (SIPs) are expected to be holding a significant market

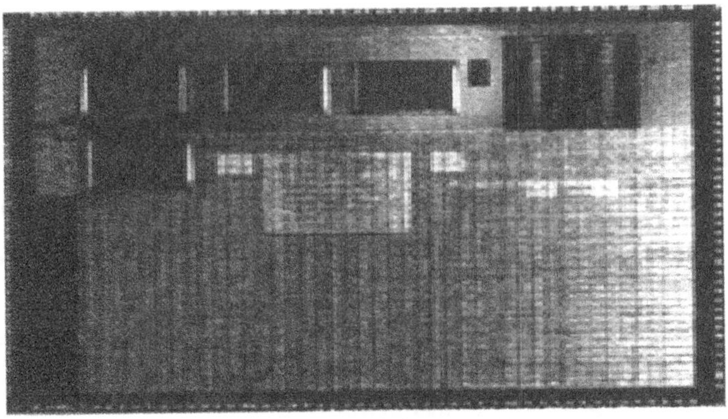

Figure 10.11 System on a chip [49]. (Courtesy of LSI Logic, Milpitas, California.)

share in 2003, offering a favorable combination of cost, speed, and density [48]. Depending on the particular application, SOC can be the best solution. The *1999 ITRS Road Map* has a chapter devoted to SOC and proposes a cost analysis on the basis of yield [5]. The yield of an SOC die is not based on die size. Creation of an SOC involves a combination of many different device types, each with its own yield and economic dynamics. Although there can be good reasons to use an SOC design, there also are good reasons for seeking alternatives. One reason offered for the increased use of SOC is that on-chip integration allows faster interconnection between circuit components (Figure 10.11). Although SOC integration will be the best choice in some situations, it definitely will not be best in all system integration situations. Many factors come into play in determining whether SOC is a suitable choice, and many more options exist for packaging multichip systems. It is expected that SOP technology will be more adaptable from the point of view of performance and cost, and it will reduce packaging system size 1000-fold [49,50]. An SOP is packaging that provides all the functions of tomorrow's systems needs.

10.2.6 Chip-Scale Packages

The demand for smaller, lighter, and thinner consumer products has caused a significant increase in the use of chip-scale packages (CSPs) [51]. CSP formats can be categorized as either (BGA) or leaded (non-BGA). The ball grid version of CSPs uses rigid substrates as interposers in an attempt to push BGA component technology into fine pitch/small footprint applications. These devices typically use wire-bonded dies mounted to a rigid ceramic or organic BGA interposer [23]. Several flip-chip die versions are being developed as well in chip-scale packages and qualify as high-density packaging. On the other hand, the lead-on-chip (LOC) version of CSP uses mold resin and support lead frames as package constructing materials, which allows the CSP to retain the

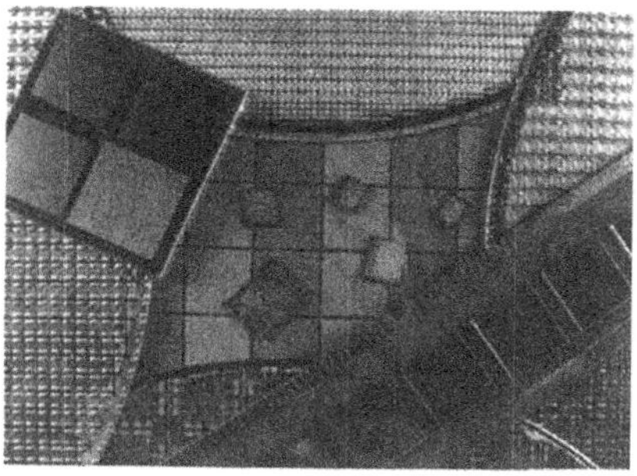

Figure 10.12 Chip-scale packaging, including electroplated CSP and single-mask CSP [55]. (Photo courtesy of Unitive Advanced Semiconductor Packaging, as published in *SMT* Magazine.)

advantages of conventional packages while reducing size. There may be as many as 100 different CSPs, each of which has strengths and weaknesses on the basis of their applications. In general, CSPs are simple and cost-effective. Figure 10.12 shows a CSP that includes electroplated CSP and single-mask CSP [52–54].

10.2.7 Wafer-Level Packaging

Major economic savings occur when packaging and testing are done before attaching dies to the wafer. In wafer-level packaging (WLP) technology, the die and package are manufactured and tested on the wafer prior to sigulation [56]. Wafer-level packages use solder balls to connect the package to the board since the size of the package and the size of the package mounted on the board equal the size of the die [57]. The cost of WLP is expected to be less than that of traditional packages. To be more competitive, WLP should have (1) all the interconnections under the die, (2) a very high I/O die with very small solder balls [down to 2 mil (50 μm)], and (3) a high-density board to interconnect the I/O. To implement these, microvias are necessary, and these are very expensive.

10.2.8 Three-Dimensional Packaging

Over the past few years there has been an emerging trend for stacking two, three, or more memory dies in a single package. The die on the bottom is the largest, the die on the top is the smallest, and they are all wire bonded to the package. Most Asian companies have adopted this three-dimensional method (Figure 10.13), and a collaborative effort in Europe has developed a method in which dies are embedded in a three-dimensional interconnect structure [58–60].

10.3 PACKAGE SEALING

Achieving high yield in hermetic sealing requires good sealing equipment with a proven and certified process and proper electrode material and designs

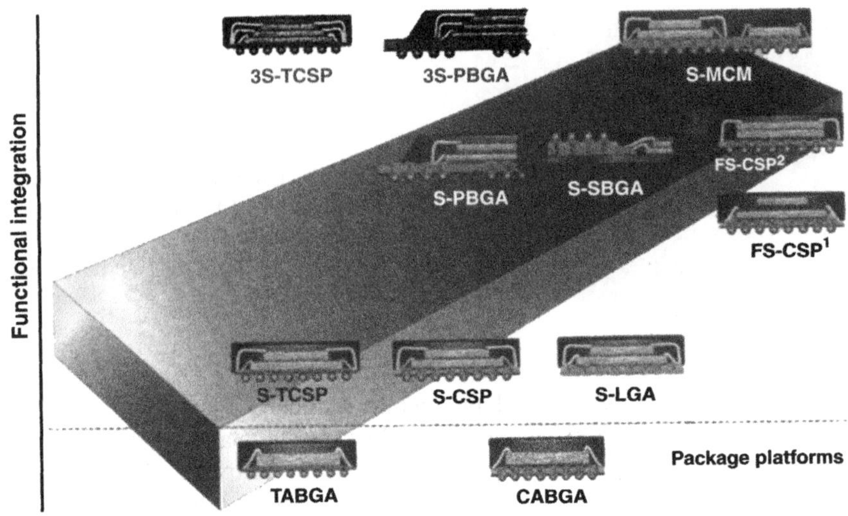

Figure 10.13 Three-dimensional packaging application road map [61]. (Courtesy of Amkor Technology, Westchester, Pennsylvania.)

[61,62]. The design and configuration of the electrodes are extremely important, especially when a sealing machine is used to seal a variety of package sizes and configurations on a daily basis. To decrease contamination, the most widely used technique is potting or encapsulating the components in a silicone rubber solution after the components are assembled into the package. By filling the voids with encapsulant, any chance of gross leakage and subsequent contamination are eliminated. Potting or encapsulating by this technique is good for most packages from 10^{-6} to 10^{-7} hermeticity. In addition, postassembly encapsulation increases reliability by minimizing vibrations caused by mechanical shock. Regarding design for hermeticity: One can almost always assume that leakage will follow the lead path. As a consequence, temperature changes affecting contraction and expansion of the lead and plastic can be a major cause for leakage. Hermeticity can be improved by selecting a lead material with a coefficient of expansion best matching that of the plastic.

Hermetic glass-to-metal *matched seal* devices involve an insulated metal lead, package shell or body, and a glass insulator. Usually, a kovar [(Fe–Ni–Co)-F-15 alloy] (Carpenter Technology) alloy is sealed with 7052 (Corning) glass under oxidizing conditions in this glass-to-metal seal junction [63,64]. Glass-to-metal seals for lead feed-through thus offer flexibility of lead configuration and true hermetic sealing capabilities for MIL Standard applications. The final sealing of a lid on such packages is often accomplished by resistance welding. Successful welding requires exceptional care in adjustment of the welding equipment as well as suitable design of lids and weld flanges of the packages. Careful operators can achieve hermeticity yields in excess of 98%. Various packaging systems characteristics are given in Table 10.3.

TABLE 10.3 Various Packaging System Characteristics[a]

Package Type	Package Material	Typical Lead Spacing mils	Typical Lead Spacing mm	Typical Maximum Lead Count
DIP	Plastic	100	2.54	64
PGA	Ceramic	100	2.54	300
SOIC (surface mounting)	Plastic	50	1.27	28
Flatpack (surface mounting)	Plastic	40	1.00	
LCC J or gullwing	Plastic or ceramic	50	1.27	
		40	1	
		30	0.75	

Source: From [1]; © reprinted by permission of Pearson Education, Inc.

[a] *Insertion packaging:* DIP, ZIP (zigzag-inline package), SIP (single-inline package), S-DIP (shrink dual-inline package); *SMT packaging:* SOP (small-outline package), QFP (quad flat package), SOJ (small-outline J-leaded package), PLCC (plastic-leaded chip carrier), QFJ (quad flat J-leaded package).

10.4 THERMAL EFFECTS ON ELECTRONIC PACKAGING

The electronic package system in hybrid circuits uses various metals [65–68], their alloys [67–69], plastics [70], ceramics [71,72], various epoxies [73], and some materials whose composition is not known precisely. The accuracy of an analysis for a cooling system depends greatly on how closely the thermal conductivities of various parts of the system are known. As a consequence, electronic packaging engineers have been forced to resort to experimentation and/or approximation when making thermal analyses of their designs. Typical temperature ranges for commercial devices are 0 to 70°C during storage and −55 to 125°C during operation. For military parts, operation and storage temperature ranges can be as much as −55 to 125°C and −65 to 150°C, respectively.

To meet the temperature requirements, heat must be removed rapidly from hybrid devices. The heat conductivity mechanism involves conduction, convection, or radiation. However, conduction is the only mechanism of practical significance in the immediate vicinity of hybrid circuit packages, the other mechanisms applying to larger assemblies. Thermal conduction in metals is due to free electrons in much the same manner as electric charge. Crystalline insulators can have thermal conduction by phonons (sound waves) [74], sometimes very effectively, as in the case of diamond and beryllia. In liquids it is difficult to distinguish between thermal energy transported by conduction or radiation. In gases, intermolecular bonds are negligible and thermal conductivity is low [75,76].

In solids thermal conductivity is affected by composition of the metal alloys and differs from metal to metal. To measure the conductivity of a solid, a relevant approach is to measure an actual device. The heat conduction equation can be represented by

$$J = -k\,\Delta T \tag{10.1}$$

where J is the heat flux, k the thermal conductivity of the material, and ΔT the thermal gradient. The generic heat conduction equation expressed in Eq. (10.1) can be used to define thermal conductivity as

$$\Theta = \frac{T_J - T_0}{P} \tag{10.2}$$

where Θ is the thermal resistance of the package, T_J and T_0 are the junction and ambient temperatures, and P is the total power dissipation.

10.4.1 Power Dissipation

The design engineer will concentrate on calculating the dissipating power of circuit elements individually once the size of the substrate, substrate material, number of layers, partitioning of thick and thin films, and number of discrete circuit elements are determined. The total power consumption by all these components is summed up to determine the ability of the package to dissipate the actual power. Recent demands for smaller and more numerous electronic circuits have been associated with greater complexity, more multifunctionality, and higher reliability. To meet these requirements, high-density multichip packaging (MCP) technology has been introduced [77,78]. The greater the scale of high-density MCP, the higher the heat dissipation per unit area. Therefore, the thermal analysis of individual devices and the assembled circuitry in designing high-density packaging, especially when high-power transistor chips are mounted on a multichip ceramic substrate with other ICs, should be prime considerations. As a rule of thumb, in designing MCP, power dissipation is estimated at about 6.45 W/in^2 (1 W/cm^2) [79]. Figure 10.14 shows a silicon transistor bonded eutectically to alumina substrate.

10.4.2 Thermal Design Calculations [80,81]

Temperature Drop of a Single Emitter and Substrate Pad The flow of heat in watts [81] from an emitter junction (T_1) to a pad (T_2) per unit time is

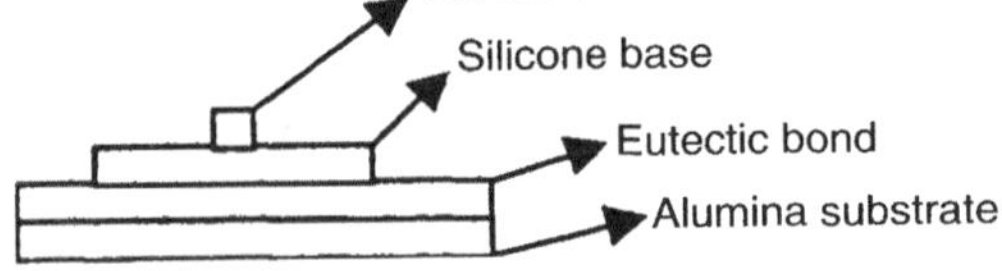

Figure 10.14 Silicon chip bonded to alumina substrate.

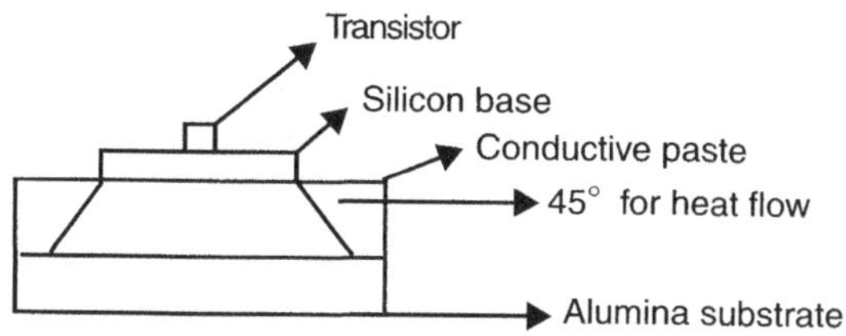

Figure 10.15 Heat flow through a thermally conductive path.

$$Q = \frac{\Delta T}{\Theta} \tag{10.3}$$

where $\Delta T = T_1 - T_2$ and Θ is the thermal resistance [and is equal to L/KA, where A is the area ($w \times d$, where w is the width and d is the thickness)]. The thermal resistance between the emitter junction and the pad is

$$\Theta_{\text{jp}} = \frac{1}{4K\sqrt{lw}} \tag{10.4}$$

where K is the thermal conductivity of a silicon transistor and l is the thermal path. Assuming that $l = 12$ μm, $w = 3$ μm, K for silicon at room temperature = 1.45 W/cm·°C, the value of $\Theta_{\text{jp}} \approx 287$°C/W. Assuming a heat dissipation of about 25 mW, the temperature rise = $\Theta_{\text{jp}} \times$ dissipation factor = 287°C/W × 0.025 W = 7.2°C.

Generally, discrete devices are mounted on the substrate using either conducting paste or eutectic bonding. If the conducting pad is thick, the calculation for thermal resistance will be different. Figure 10.15 shows the mechanism of heat flow from a transistor to an alumina substrate through a conductive (heat) path.

For a thermally conductive path (l) thicker than the heat source, a good approximations of the heat flow from the source is taken to be at about 45°. The thermal resistance is much lower in this case than in the preceding one, where heat flow was calculated following straight path calculation [82]. The thermal resistance in this case will be

$$\Theta = \frac{1}{K}\int_0^l \frac{1}{A}\,dl \tag{10.5}$$

where A is the surface area at any cross section of heat flow and dl is the incremental path length. Figures 10.16 and 10.17 show the heat conduction mechanism through square and rectangular devices.

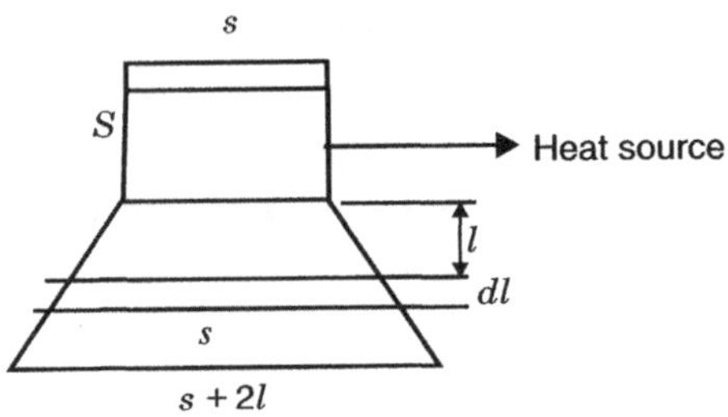

Figure 10.16 Square heat source.

Square Heat Source

$$\Theta = \frac{1}{K}\int_0^l \frac{1}{A}\,dl = \frac{1}{K}\int_0^l \frac{dl}{(S+2l)^2}$$

$$= \frac{1}{2K}\left(\frac{-1}{S+2l} + \frac{1}{S}\right) \tag{10.6}$$

$$= \frac{1}{KS(S+2l)} \tag{10.7}$$

Rectangular Heat Source

$$\Theta = \frac{1}{K}\int_0^l \frac{dl}{A} = \frac{1}{K}\int_0^l \frac{dl}{(S_1+2l)(S_2+2l)} \tag{10.8}$$

$$= \frac{1}{(2K)(S_1 - S_2)}\ \ln\left[\frac{S_1}{S_2}\,\frac{S_2+2l}{S_1+2l}\right] \tag{10.9}$$

Circular Heat Source

$$\Theta = \frac{1}{K}\int_0^l \frac{dl}{A} = \frac{1}{K}\int_0^l \frac{4l}{\pi(D+2l)^2} \tag{10.10}$$

$$= \frac{1}{K\pi[-2/(D+2l)]} = \int_0^l \frac{4l}{K\pi(D^2+2lD)} \tag{10.11}$$

where D is the diameter of the circular source.

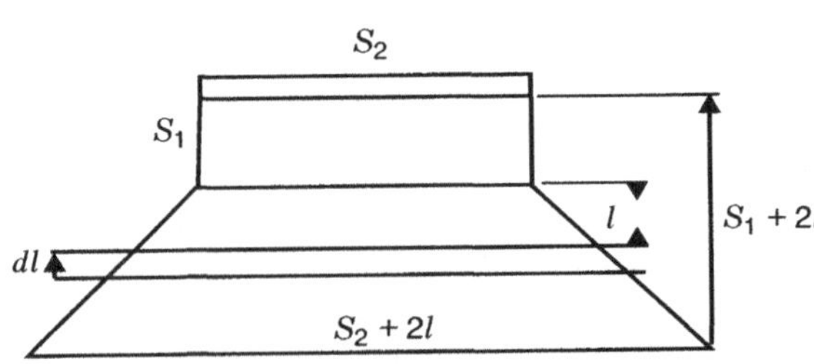

Figure 10.17 Rectangular heat source.

TABLE 10.4 Thermal Properties of Materials at Room Temperature

Material	Density (g/cm^3)	Specific Heat (J/g·°C)	Thermal Conductivity, K (W/cm·°C)	Thermal Diffusivity (cm^2/s)	Melting Point (°C)
Air	1.2×10^{-3}	1.0	2.6×10^{-4}	0.22	
Alumina (Al_2O_3)	3.68	1.0	0.35	0.95	2035
Beryllia (BeO)	2.9	1.0	2.5	0.86	2450
Copper (Cu)	8.85	0.38	4.0	1.19	1083
Gold (Au)	19	0.13	2.96	1.21	1063
Aluminum (Al)	2.6	0.90	2.3	0.94	660
Nichrome (Ni–Cr)	8.25	0.45	0.13	0.035	1350
Tungsten (W)			1.7		
Tantalum (Ta)	16.6	0.14	0.55	0.24	3000
Molybdenum (Mo)	10.2	0.25	1.4	0.51	2610
Silicon (Si)	2.33	0.65	1.5	0.96	1420
Solder	8.7	0.17	0.38	0.26	361
Resistor ink	2.7	0.93	0.01	0.004	1350
Phenolic	1.62	1.03	1.6×10^{-3}	1.0×10^{-3}	1635
Epoxy	2.2	0.84	4×10^{-3}	2×10^{-3}	—
Glass	2.5	0.75	0.014	7.5×10^{-3}	—

Source: From [83].

The formulas above will give approximate values for the thermal resistance. For complex situations, computer simulation programs are available. Table 10.4 presents thermal properties of the materials used in hybrid packaging systems.

10.5 NON-STEADY-STATE HEAT TRANSFER MODEL

Heat transport from the bottom and top surfaces of the multilayer substrate (MLS) inside MCPs is in the form of thermal diffusion by thermal radiation and natural convection, whereas in the inside of MLS, heat is transported only by conduction. Inside a kovar shell area, heat transport is due to heat conduction of the nitrogen gas and thermal diffusion by radiation, whereas from the topside surface area, heat transport is due to thermal diffusions by radiation and natural convection [84]. Figure 10.18 shows the heat transfer mechanism through a multichip package.

10.5.1 Thermal Resistance inside Substrate

Inside the ceramic substrate, heat flow is due only to conduction. The thermal resistance due to the flow of heat per unit time Q is

$$\Theta_c = \frac{\Delta T}{Q} \tag{10.12}$$

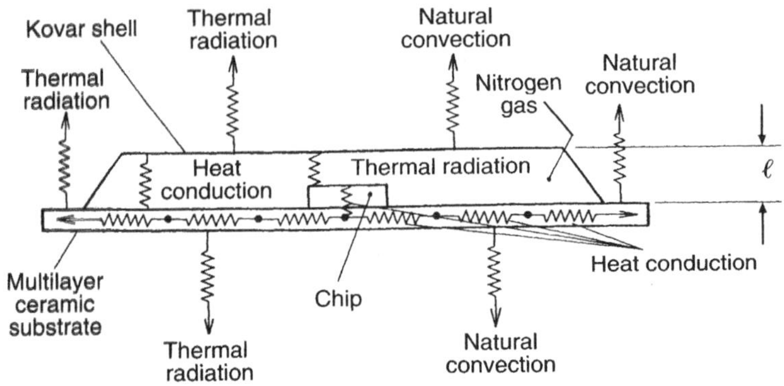

Figure 10.18 Heat transfer mechanism through a multichip package [84].

where $Q = (KA/l)\,\Delta T$ = flow of heat per unit time (K is the thermal conductivity, A the area, and l the thermal path).

10.5.2 Natural Convection

The thermal resistance due to natural convection can be written as

$$\Theta_{\text{conv}} = \frac{1}{hA} \tag{10.13}$$

where h is the heat transfer coefficient and A is the MLS area. If the natural value of convection is laminar flow, h can be written as

$$h = 1.9 \times 10^{-4} p \left(\frac{\Delta T}{l}\right)^{0.25} \tag{10.14}$$

where the value of p is determined from the placement of the flat plate ($p = 0.56$ when the plate is perpendicular, $p = 0.52$ when the plate is horizontal and the top portion of the plate is hot, and $p = 0.26$ when the plate is horizontal but the bottom surface of the plate is hot) and l is the floating path of the cooling air on the hot surface of the material.

10.5.3 Thermal Radiation

The thermal resistance for radiation is given by

$$\Theta_r = \frac{1}{h_r} \tag{10.15}$$

where h_r is the thermal radiation heat transfer coefficient and is equal to

$$h_r = 5.66 f \varepsilon \frac{[(T_1 + 273)/1000]^4 - [(T_2 + 273)/1000]^4}{T_1 - T_2} \tag{10.16}$$

where f is a geometric factor and ε is the emissivity of the material.

10.5.4 Thermal Resistance of Nitrogen Gas

The thermal resistance of nitrogen gas inside the kover shell is given by

$$\Theta_N = \frac{l}{KA} \tag{10.17}$$

where l is the length of the Kovar shell, K the thermal conductivity, and A the heat conduction area.

10.5.5 Heat Conduction inside the Kovar Shell (Area 3)

$$\Theta_3 = \frac{1}{h_{r3} + h_{d3}} \tag{10.18}$$

where

$$h_{r3} = 95.66 f \varepsilon_2 \frac{[(T_l + 273)/1000]^4 - [(T_s + 273)/1000]^4}{T_l - T_s}$$

$h_{d3} = k/l$ [k is the conductivity of kovar and l is the height of the kovar shell (ca. 0.3 cm)], $\varepsilon_2 = 0.4$, T_l is the MLS surface temperature, and $T_s = T_a + 0.5(T_l - T_a)$ (T_a is the ambient temperature).

The important package parameters are the junction-to-case thermal resistance (Θ_{jc}) and the junction-to-ambient thermal resistance, where the ambient is free air (Θ_{ja}). The most commonly used package unit and values of Θ_{jc} and Θ_{ja} are given in Table 10.5.

Most thick-film circuits are printed on 96% alumina substrate. The heat dissipation factor for 15- to 25-mil (0.380- to 0.635-mm)-thick alumina (96%)

TABLE 10.5 Thermal Dissipation Rates of the Package Unit

	Dissipation Rate (C/W)	
Package	Θ_{jc}	Θ_{ja}
8-lead TO-99 package	40	250
14-lead 3/8 by 3/8 in. flatpack	45	125
30-lead 1 by 1 in. flatpack	20	60

substrate is about 10 W/in^2 (1.55 W/cm^2). The heat generated on the top surface of the substrate should conduct through the thickness of the substrate and then into the package. If the power to be dissipated is too high for alumina, beryllia (BeO), with its high thermal conductivity, is recommended as a substrate. The thermal conductivity of BeO is approximately eight times that of alumina. However, BeO is somewhat weaker than alumina, toxic if pulverized, and more costly to produce in substrate form. High-density circuits demand substrates that are more heat conductive and are able to dissipate more heat. Aluminum nitride, silicon carbide, and metal-cored ceramic-coated substrates have been highlighted in the ongoing effort to establish the compatibility of the system with current processing technologies for hybrid microcircuits.

10.6 FLIP-CHIP TECHNOLOGY

Heat dissipation in a flip-chip configuration is through the solder joints to the substrate, in contrast to the case where the back side of the chip is bonded to the substrate eutectically. The amount of heat dissipation depends on the thermal cross section, which is determined by the number and size of solder joints [85–87]. If we consider one-dimensional heat flow through N solder joints of area A, the flux F_{th} of thermal energy (heat) transmitted per unit time per unit area in a temperature gradient dT/dx through a total area A_T is

$$F_{\text{th}} A_T = -K_{\text{th}} N A \frac{dT}{dx} \tag{10.19}$$

where the minus sign indicates a positive flow of heat toward lower temperatures, K_{th} is the thermal conductivity in W/cm·K, and N is the number of solder joints. Table 10.6 presents thermal conductivities and thermal expasion coefficients of materials used in a hybrid packaging system, and Table 10.7 provides some useful data for core materials.

TABLE 10.6 Thermal Conductivities and Thermal Expansion Coefficients of Materials

Material	Conductivity		Thermal Expansion Coefficient (K^{-1})
	W/cm·K	W/in.-K	
Alumina (90%)	0.17	0.43	6.5
Solder, Pb–5% Sn	0.63	1.6	29
Silicon (Si)	1.5	3.8	2.5
Gold (Au)	2.96	7.4	14.2
Au–3% Si	0.27	0.7	12.3
Epoxy (Ag-loaded)	0.008	0.2	53

Source: S. M. Sze, ed., *VLSI Technology*, McGraw-Hill, New York, 1985; reproduced with permission of The McGraw-Hill Companies.

TABLE 10.7 Physical Properties of Core Materials

Material	Young's Modulus (dyn/cm^2)	Poisson's Ratio (dimensionless)	CTE (TCE) (ppm/°C)	Thermal Conductivity (W/m·°C)	Density (g/cm^3)
Molybdenum	32.41	0.3	5.0	1.4	10.2
Invar	13.8	0.3	1.7	0.11	8.45
Copper	12.5	0.3	17.0	4.0	8.85
Graphite	1.28	0.25	4.5	1.4	2.04
Be-40	29.2	0.3	7.5	2.3	2.33
Be-60	33.3	0.3	6.2	2.4	2.62
MCX-693	20.7	0.24	6.8	1.7	3.20

10.7 PACKAGING MATERIAL RELIABILITY

The thermal and structural stability of the core materials used for packaging are of paramount important in surface-mounting technology [51,88–90]. The thermal and vibration fatigue reliability of solder joints and package leads are strong functions of the coefficient of thermal expansion [90–94]. The junction temperatures of the semiconductor devices and their expected service lives are strongly depend on the thermal resistance of the hybrid circuit substrate/core assembly. The weight of airborne electronic modules for commercial and military applications becomes an important factor in the selection of core materials [95]. The commercially available core materials, which are considered to be potential candidates for SMT are copper–invar–copper (CIC), copper–molybdenum–copper (CMC), molybdenum–graphite–molybdenum (MGM), aluminum particulate–silicon carbide (AlSiCp), and beryllium particulate–beryllium oxide (BeBeOp).

CIC and CMC are rolled all-metal products that contain invar and molybdenum (Mo) center sections sandwiched between copper (Cu) layers [95]. The commonly used Cu and Mo and Cu-to-invar thickness ratios are 13/17/13 and 20/60/20, respectively. MGM consists of graphite sandwiched between molybdenum sheets and bonded by a high-temperature vacuum brazing process that utilizes titanium (Ti) as the filter material [96,97]. Stiff metallic layers protect the graphite layer against shock and impact loads.

The structure of an AlSiCp metal matrix composite consists of silicon carbide particles in an aluminum matrix. AlSiCp is commonly fabricated with liquid-metal infiltration techniques. The precursor of the composite is a ceramic preform, which contains up to 70% vol % SiC in a matrix of polymer-based sacrificial binder. The organic binder is replaced by liquid aluminum during the infiltration process. The SiC content of the composite determines its CTE, modulus, toughness, and merchantability. SiC particulate may have bimodal and trimodal size distributions to achieve high packaging densities. The maximum particulate sizes of commonly used bimodal and trimodal distributions are approximately 0.6 mil (15 μm) and 2.9 mils (70 μm), respectively [97]. Be/BeO MMC consists of BeO whiskers and/or particulates in a Be metal matrix. Powder metallurgy and sintering techniques are commonly used in the fabrication of Be/BeO MMCs.

Many of the silicone and epoxy materials used in packaging are filled with electrically or thermally conductive particultes for higher conduction and to modify thermal expansion properties [98,99]. As high-density packaging becomes more prominent, manufacturability becomes more of a concern because of the physical properties of these filler materials. Spherical filler particles have minimal contact area with each other, the die, and the substrate, which could reduce the thermal conductivity of the material [100]. This is not necessarily a concern for underfill applications, since the primary reason for filling an underfill material is to control its CTE [100]. To date, silver flake has performed well in via fill materials, but with decreasing via size and aspect ratio, there may be a need to examine the behavior of other materials, such as boron nitride and other advanced materials.

REFERENCES

1. H. B. Bakoglu, *Circuits, Interconnections, and Packaging for VLSI*, Addison-Wesley, Reading, MA, 1990.
2. J. K. Lumpp, Hybrid assemblies, in *Electronic Packaging Handbook*, G. R. Blackwell, ed., CRC Press, Boca Raton, FL, 2000, Chap. 7.
3. J. Baliga, High-density packaging: the next interconnect challenge, *Semicond. Int.*, Vol. 23, No. 2, 2000, p. 91.
4. B. S. Matisoff, *Handbook of Electronic Packaging Design and Engineering*, Van Nostrand Reinhold, New York, 1982.
5. 1999 ITRS Road Map, *Semiconductor International*, July 2000.
6. R. C. Musa, Hybrid packaging, *Proc. IEEE Workshop Hybrid IC Technol.*, Mar. 1968, pp. 6-1 to 6-4.
7. J. Lau, C. P. Wang, J. L. Princes, and W. Nakayama, *Electronic Packaging Design, Materials Processing, and Reliability*, McGraw-Hill, New York, 1988.
8. G. K. Fehr, Microcircuit packaging and assembly state of the art, *Solid State Technol.*, Vol. 13, No. 8, Aug. 1970, pp. 41–47.
9. R. R. Tummala, E. Rameszewski, and A. Koperstein, *Handbook of Electronic Packaging*, Chapman & Hall, London, 1997.
10. J. Fjelstad and T. Distefano, Electronic packaging and production, *Semicond. Int.*, Dec. 1999, pp. 16–20.
11. R. Tummala, Georgia Tech opens low-cost electronics packaging research center, *Adv. Microelectron.*, Jan./Feb. 1995, p. 30.
12. R. T. Bilson, M. R. Herper, and J. P. McCarthy, The impact of surface mount chip carrier packaging on thermal management in a hybrid microcircuit, in *Thermal Management Concepts in Electronic Packaging*, ISHM Technical Monograph, Sr. 6984-003, ISHM, Reston, VA, 1984.

13. W. D. Brown, Introduction and overview of microelectronics packaging, in *Advanced Electrical Packaging*, W. D. Brown, ed., IEEE Press, Piscataway, NJ, 1999.
14. M. G. Petit, R. Agarwal, P. McClusky, T. Dishough, S. Javadpour, and R. Mahajan, *Electronic Packaging Materials and Their Properties*, CRC Press, Boca Raton, FL, 1999.
15. A. B. Timberlake, Ceramic materials, in *Electronic Packaging and Interconnection Handbook*, 3rd ed., C. A. Harper, ed., McGraw-Hill, New York, 2000, p. 158.
16. A. Krum and J. E. Sergent, Heat generation in an electronic circuit, in *Thermal Management Handbook*, J. E. Sergent and A. Krum, eds., McGraw-Hill, New York, 1998.
17. Advanced Technology Group Advertisement, *Microwave J.*, Vol. 37, No. 7, July 1994.
18. T. C. Evans, Thermal properties of electronic materials, in *Thermal Management Handbook*, J. E. Sergent and A. Krum, eds., McGraw-Hill, New York, 1998.
19. N. R. Keltner, Heat flow management theory and applications, in *Thermal Management in Electronic Cooling*, K. Azar, ed., CRC Press, Boca Raton, FL, 1997, p. 273.
20. K. Gilleo, IC transition to copper flip chip technology, in *Electronic Packaging and Interconnection Handbook*, 3rd ed., C. A. Harper, ed., McGraw-Hill, New York, 2000.
21. H. Domingos, Transient temperature rise in microcomponents, in *Thermal Management Concepts in Electronic Packaging*, ISHM Technical Monograph, Sr. 6984-003, ISHM, Reston, VA, 1984, p. 67.
22. A. W. Postlethwaite, Hermetic and non-hermetic packaging, *Solid State Technol.*, Vol. 13, No. 8, Aug. 1970, pp. 67–70.
23. C. Cohn and M. T. Shih, Hermetic package lead sealing, in *Electronic Packaging and Interconnection Handbook*, 3rd ed., C. A. Harper, ed., McGraw-Hill, New York, 2000, p. 7-89.
24. A. Okuno, K. Nagai, Y. Tsukasaki, N. Oyama, K. Nakahira, and T. Hashimoto, Unique epoxy resin and printing encapsulation system (PES) for advanced multichip modules PLCC, BGA, PGA, TAB, COB, and flip chip, *Int. J. Microelectron. Electron. Packag.*, Vol. 17, No. 2, 1994, p. 143.
25. C. Cohn and M. T. Shih, Packaging and interconnection of integrated circuits, in *Electronic Packaging and Interconnection Handbook*, 3rd ed., C. A. Harper, ed., McGraw-Hill, New York, 2000, Chap. 7; H. Tsunetsugu, K. Katsura, T. Hayashe, F. Ishitsuka, and S. Hata, A new packaging technology using microsolder bumps for high-speed photoreceivers, *IEEE Trans. Comp. Hybrid Manuf. Technol.*, Vol. 15, No. 4, Aug. 1992, pp. 578–582.
26. P. Garrou, Thin-film packaging and interconnections, in *Thin-Film Technology Handbook*, A. E. Riad and R. D. Barlow III, eds., McGraw-Hill, New York, 1998.
27. R. A. Rikoski, Packaging configuration, in *Hybrid Microcircuits: Thick Film*, Wiley, New York, 1980, p. 173.
28. *Integrated Circuit Engineering Corporation Monograph*, 5th ed., ICE, Phoenix, AZ, 1966, p. 10-4.
29. R. Bowlby, The DIP may take its final bows, *IEEE Spectrum*, Vol. 22, No. 3, June 1985, pp. 37–42.

30. R. R. Tummala, Ceramic Packaging, in *Microelectronic Packaging Handbook*, R. R. Tummala and E. J. Rymaszewski, eds., Van Nostrand Reinhold, New York, 1989.
31. R. C. Jaeger, Dual-line packages, in *Introduction to Microelectronic Fabrication*, Vol. V, Addison-Wesley, Reading, MA, 1990, Sec. 8.5.2, p. 158.
32. C. Cohen, Japan's packaging goes world class, *Electronics*, Nov. 11, 1985, pp. 26–31; J. R. Howel, Reliability study of plastic encapsulated copper lead frame epoxy die attach packaging system, *Proc. Int. Reliab. Phys. Symp.*, 1981, pp. 104–110.
33. E. C. Blackburn, VLSI packaging reliability, *Solid State Technol.*, Jan. 1984, pp. 113–116.
34. J. Lau, A brief introduction to flip chip technologies, in *Flip Chip Technologies*, McGraw-Hill, New York, 1996, Chap. 1.
35. J. C. Blakenhorn, *BGA, COB, and Flip Chip PCB Design*, IMAPS, Washington, DC, 1990.
36. J. S. Hwang, *Ball Grid Array and Fine-Pitch Peripheral Interconnection*, IMAPS, Washington, DC, 1990.
37. K. Gilleo, Flip chip technology: materials and processes for the next generation of high performance, in *Electronic Packaging and Interconnection Handbook*, 3rd ed., C. A. Harper, ed., McGraw-Hill, New York, 2000, Sec. 1-2-1.54.
38. M. S. Cole and T. Caufield, Ceramic ball grid array packaging, *Adv. Microelectron.*, Jan./Feb. 1995, p. 18.
39. J. J. Liu, H. Berg, Y. Wen, S. Mulaonker, and R. Bowlby, Pin ball grid array, *Adv. Microelectron.*, Jan./Feb. 1995, p. 17.
40. J. W. Kartz, Surface mount technology, in *Electronic Packaging and Interconnection Handbook*, 3rd ed., C. A. Harper, ed., McGraw-Hill, New York, 2000, Chap. 8.
41. D. Walshak and H. Hashemi, Ball grid array technology, *Adv. Microelectron.*, Jan./Feb. 1995, p. 38.
42. C. L. Chen, L. J. Mahoney, and D. J. Tsang, Bond wireless multichip packaging technology for high-speed circuits, *IEEE Trans. Comp. Hybrid Manuf. Technol.*, Vol. 15, No. 4, Aug. 1992, pp. 451–456.
43. A. W. Lin and C. P. Wong, Encapsulant for nonhermetic multichip packaging applications, *IEEE Trans. Comp. Hybrid Manuf. Technol.*, Vol. 15, No. 4, Aug. 1992, pp. 510–518.
44. E. Davidson, Large chip vs. MCM for a high-performance system, *1998 IEEE Micro*, July/Aug. 1998, pp. 33–41.
45. J. Baliga, Micropallet packaging system leverages silicon technology, *Semicond. Int.*, May 1997, p. 52.
46. S. Kamaran, Flash memory, packaging options optimize designers, *Adv. Packag.*, Nov./Dec. 1995, p. 18.
47. J. Baliga, Packaging Research Center, Georgia Institute of Technology, *Semicond. Int.*, July 2000, p. 181.
48. J. Baliga, Packaging provides viable alternatives to SOC, *Semicond. Int.*, July 2000, p. 169.

49. R. Dejule, Advances in SOC fuel growing market, *Semicond. Int.*, Apr. 2000, p. 66.
50. R. R. Tammala, SOP vs. SOC: needs, challenge, and status, 2000 European VLSI Packaging Techniques and Manufacturing Technologies Workshop, Cork, Ireland, May 8–9, 2000.
51. L. Lau, C. P. Wong, J. L. Prince, and W. Nakayama, *CSP in Electronic Packaging: Design, Materials, Processes, and Reliability*, McGraw-Hill, New York, 1998.
52. J. H. Lau, A brief introduction to flip chip technologies for MCM, in *Flip Chip Technologies*, J. Lau, ed., McGraw-Hill, New York, 1998, Chap. 1.
53. S. Tsui, C. J. Vath, and C. W. Cheng, CSBGA: new challenge for ball-placement process and equipment, *Adv. Packag.*, Sept. 2000, p. 55.
54. M. Amagai, H. Sano, and T. Maeda, Development of CSP for center pad devices, *Electron. Comput. Technol. Conf. Proc.*, 1997, pp. 343–352.
55. B. Lanzone, Emerging flip chip and WL-CSP technologies, *Surface Mount Technol.*, Aug. 2001.
56. J. H. Lau, Solder bumped flip chip attach on SLC board and MCM, in *Chip-on-Board Technology for MCM*, J. H. Lau, Van Nostrand Reinhold, New York, 1994; N. Kelker, R. Mathew, H. Takiar, and L. Nguyen, Micro SMD: a wafer-level chip-scale package, *IEEE Trans. Adv. Packag.*, Vol. 23, 2000, p. 227.
57. P. Garrou, Wafer level chip scale packaging: an overview, *IEEE Trans. Adv. Packag.*, Vol. 23, 2000, p. 198; P. Garrou, Wafer-level packaging has arrived, *Semicond. Int.*, Oct. 2000.
58. P. Garrou, Wafer-level chip-scale packaging: an overview, *IEEE Trans. Adv. Packag.*, Vol. 23, 2000, p. 198.
59. J. Baliga, Packaging provides viable alternatives to SOC, *Semicond. Int.*, July 2000, p. 174.
60. J. Wolf, P. Gerlach, E. Beyne, M. Topper, L. Dietrich, K. H. Becks, N. Wermes, O. Ehrmann, and H. Reichl, High-density pixel detector module using flip chip and thin-film technology, *Proc. Int. Syst. Packag. Symp.*, San Diego, CA, Jan. 1999.
61. J. Baliga, 3-D IC packaging, *Semicond. Int.*, May 1998, p. 20.
62. H. B. Bakoglu, Packaging technology, in *Circuits, Interconnections, and Packaging for VLSI*, Addison-Wesley, Reading, MA, 1990, Chap. 3.
63. D. P. Seraphim, R. Lasky, and C. Y. Li, *Principles of Electronic Packaging*, McGraw-Hill, New York, 1989.
64. J. A. Sartell, ed., *Electronic Packaging Materials and Processes*, American Society for Metals, Metals Park, OH, 1986.
65. J. E. Ahearn, *A*. Duplex glass enhancement for improving hybrid microelectronic packages, *Proc. ISHM*, Minneapolis, MN, 1987, p. 275.
66. M. Pecht, R. Agarwal, P. McCluskey, T. Dishough, S. Javadpour, and R. Mahajan, Properties of electronic packaging materials, in *Electronic Packaging Materials and Their Properties*, CRC Press, Boca Raton, FL, 1999, Chap. 1.
67. J. J. Licari, Thin and thick film, in *Electronic Materials and Processes Handbook*, C. A. Harper and R. N. Sampson, eds., McGraw-Hill, New York, 1994.
68. G. W. A. Dummer, Contact materials applications and properties, in *Materials for Conductive and Resistive Functions*, Hayden Book Co., New York, 1970, Chap. 12.

69. L. H. Van Vlack, Alloys of metals, in *Materials Engineering*, Addison-Wesley, Reading, MA, 1990, Chap. 5.
70. L. T. Manzione, *Plastic Packaging of Microelectronic Devices*, Van Nostrand Reinhold, New York, 1990.
71. D. M. Mattox, Ceramic glass and diamond, in *Electronic Materials and Processes Handbook*, C. A. Harper and R. N. Sampson, eds., McGraw-Hill, New York, 1994.
72. A. B. Timberlake, Ceramic materials, in *Electronic Packaging and Interconnection Handbook*, 3rd ed., C. A. Harper, ed., McGraw-Hill, New York, 2000.
73. J. H. Lee and K. Neville, *Handbook of Epoxy Resins*, McGraw-Hill, New York, 1967.
74. N. F. Mott and E. A. Davis, Phonons and polarons, in *Electronic Process in Non-crystalline Materials*, Clarendon Press, London, 1979, Chap. 3.
75. D. Q. Kern and A. D. Kraus, *Extended Surface Heat Transfer*, McGraw-Hill, New York, 1972.
76. H. S. Carslaw and J. C. Jaeger, *Conduction of Heat in Solids*, 2nd ed., Oxford University Press, Oxford, 1959.
77. D. A. Doane and P. D. Franzon, Package selection, in *Multichip Module Technologies and Alternatives: The Basics*, Van Nostrand Reinhold, New York, 1993, Chap. II.
78. C. L. Chen, L. J. Mahoney, D. Z. Tsang, and K. M. Molvar, Bond wireless multichip packaging technology for high-speed circuits, *IEEE Trans. Comput. Hybrid Manuf. Technol.*, Vol. 15, No. 4, Aug. 1992, p. 451.
79. T. T. Taraone, Thermal analysis of semiconductor devices assembled by various packaging technologies, *Proc. Int. Symp. Hybrid Microelectron.*, 1969, pp. 207–215.
80. A. Bar-Cohen and A. D. Kraus, *Advances in Thermal Modeling of Electronic Components and Systems*, Vol. I, Hemisphere Publishing, Philadelphia, 1988.
81. R. D. Jones, Thermal considerations, in *Hybrid Circuit Design and Manufacture*, Marcel Dekker, New York, 1982, Chap. 10.
82. R. H. Winkler, Thermal properties of high-power transistors, *IEEE Trans. Electron. Devices*, Vol. 14, No. 5, May 1967.
83. H. Domingos, Transient temperature rise in microelectronic components in thermal management, in *Thermal Management Concepts in Microelectron Packaging*, R. T. Howard, S. S. Furkay, R. F. Kilburn, and G. Monti, eds., Technical Monograph 6984-003, ISHM, Reston, VA, 1984.
84. Y. Fukuoka and M. Ishizuka, Transient temperature rise for multichip packages, in *Thermal Management Concepts in Microelectron Packaging*, R. T. Howard, S. S. Furkay, R. F. Kilburn, and G. Monti, eds., Technical Monograph 6984-003, ISHM, Reston, VA, 1984.
85. J. W. Mayer and S. S. Lau, eds., *Electronic Materials Science for Integrated Circuits in Si and GaAs*, Macmillan, New York, 1990.
86. C. Lea, *A Scientific Guide to Surface Mount Technology*, IMAPS, Washington, DC, 1990; L. J. Fried, J. Evans, J. S. Lechaton, J. S. Logan, G. Pall, and P. A. Totta, A VLSI bipolar metallization design with three-level wiring and area array solder connections, *IBM J. Res. Dev.*, Vol. 26, 1982, p. 362.

87. R. J. Wassink and M. M. F. Verguld, *Manufacturing Techniques for Surface-Mounted Assemblies*, IMAPS, Washington, DC, 1991.

88. M. Ohring, *Reliability and Failure of Electronic Materials and Devices*, Academic Press, San Diego, CA, 1998.

89. W. Engelmier, Functional cycles and surface mounting attachment reliability, in *Thermal Stress and Low Cycle Fatigue*, S. S. Mazumdar, ed., McGraw-Hill, New York, 1966, Chap. 4.

90. A. J. Wager and H. G. Cook, Modeling the temperature dependence of integrated circuit failures, in *Thermal Management Concepts in Microelectronic Packaging*, G. S. Furker, R. F. Kilburn, and G. Mouli, eds., ISHM, Reston, VA, 1984, p. 21.

91. H. K. Charles, Thermal mechanical stress behavior in electronic packaging, in *Electronic Packaging and Interconnection Handbook*, 3rd ed., C. A. Harper, ed., McGraw-Hill, New York, 2000, Chap. 3.

92. S. S. Manson, *Thermal Stress and Low-Cycle Fatigue*, McGraw-Hill, New York, 1966.

93. N. Coopman and S. Nangalia, Fluxless flip chip solder joint, in *Flip Chip Technologies*, J. Lau, ed., McGraw-Hill, New York, 1998, Chap. 2.

94. B. Ozmat, Engineering properties of materials for SMT applications, *Int. J. Microcircuits Electron. Packag.*, Vol. 17, No. 2, 1984, p. 196.

95. B. S. Matisoff, Heat transfer and thermal control, and materials and process, in *Handbook of Electronic Packaging Design and Engineering*, Van Nostrand, New York, 1982, Chaps. 6 and 14.

96. S. G. Konsowsku, Composite in electronic packaging, in *Electronic Materials and Processes Handbook*, C. A. Harper and R. N. Sampson, eds., McGraw-Hill, New York, 1994, Chap. 11.

97. J. Baliga, Assembly and packaging, *Semicond. Int.*, Vol. 23, No. 5, May 2000.

98. R. H. Estates and F. W. Kulesza, Conductive adhesive polymer materials in flip chip applications, in *Flip Chip Technologies*, J. H. Lau, ed., McGraw-Hill, New York, 1996, Chap. 6.

99. R. P. Prasad, *Surface Mount Technology*, Van Nostrand Reinhold, New York, 1989.

100. T. K. Gupta, Effect of temperature on electrical conduction of carbon-black filled polyimide, *IEEE Trans. Comp. Hybrid Manuf. Technol.*, Vol. 12, No. 4, Dec. 1989.

RECOMMENDED READING

Alvino, W. M., *Plastics for Electronics: Materials, Properties, and Design Applications*, McGraw-Hill, New York, 1995.

Bar-Cohen, A., and A. D. Kraus, *Advances in Thermal Modeling of Electronic Components and Systems*, Vol. 1, Hemisphere Publishing, Philadelphia, 1988.

Dean, D. J., *Thermal Design of Electronic Circuit Boards and Packages*, Electrochemical Society, Ayr, Scotland, 1985.

Garrou, P. E., *Multichip Technology Handbook*, McGraw-Hill, New York, 1998.

Harper, C., *Electronic Packaging and Interconnection Handbook*, 3rd ed., McGraw-Hill, New York, 2000.

King, J. A., *Materials Handbook for Hybrid Microelectronics*, Artech House, Norwood, MA, 1988.

Lau, J. M., *Ball Grid Array Technology*, McGraw-Hill, New York, 1995.

Manzione, L. T., *Plastic Packaging of Microelectronic Devices*, Van Nostrand Reinhold, New York, 1990.

Matisoff, B. S., *Handbook of Electronic Packaging Design and Engineering*, Van Nostrand Reinhold, New York, 1989.

Pecht, M., *Handbook of Electronic Package Design*, Marcel Dekker, New York, 1991.

Seely, J. H., and R. C. Chu, *Heat Transfer in Microelectronic Equipment: A Practical Guide*, Marcel Dekker, New York, 1972.

Sergent, J. E., and A. Krum, *Thermal Management Handbook*, McGraw-Hill, New York, 1998.

Tummula, R., ed., *Microelectronics Packaging Handbook*, Van Nostrand Reinhold, New York, 1989.

11

MULTICHIP MODULE AND MICROWAVE HYBRID CIRCUITS

11.1 MULTICHIP MODULE CIRCUITS

Simplicity, cost, and quality are three ideas that have long plagued the production of multichip modules (MCMs), manifesting themselves in never-ending questions concerning how to balance cost and quality, enhance simplicity, and improve performance. Benefits of MCMs include footprint optimization, circuit density, high performance, and improved time to market [1–4]. Some observers claim that the density/cost ratio (performance versus cost) is the primary impetus. The rationale is that if the IC chips (performance) are removed, the density/cost influence becomes the chief packaging concern [5]. However, design, piece-part supply, chip source(s), assembly, test, and so on, have made it difficult to produce low-cost, high-volume MCMs [6,7].

In an effort to balance cost and quality and to enhance performance, a group of MCM manufacturers met together with supporting groups in February 1994. Three months later, Microelectronics and Computer Technology Corp. in Austin, Texas extended its help to support the program and build a prototype super SPARC chip set featuring a CPU, cache controller, and eight SRAMS using a proprietary process [8,9]. Inspiration and enthusiasm continued to grow to enhance performance, and in 1995, Advanced Research Project Agency (ARPA) granted $12.1 million for a two-year research program to support assembly processes and general design requirements for low-cost MCMs [10]. The efficient design and substrate fabrication of MCMs requires thermal management of the circuit [11]. Various substrates, including dia-

Handbook of Thick- and Thin-Film Hybrid Microelectronics, By Tapan K. Gupta
ISBN 0-471-27229-9

mond, have been used to fabricate an efficient, cost-effective MCM circuit. Diamond, with its excellent thermal conductivity ($K = 200$ W/m·K), low thermal expansion (ca. 1.5×10^{-6}/°C), high electrical resistivity (10^{12} Ω·cm), and high mechanical strength is an ideal choice for both two- and three-dimensional MCM packaging. Unfortunately, the cost of diamond substrates makes it difficult to produce low-cost, high-volume MCMs. However, recent developments in chemical vapor deposition techniques have reduced the deposition cost for large-area diamond substrates and ultimately, the manufacturing cost of MCMs [12–15]. According to computer simulation results, diamond with 20 to 30% less thickness could be used to achieve similar thermal management performance, such as for ceramic or low-temperature cofired ceramic (LTCC) substrates [16,17].

The advantage of an MCM, which is a hybrid for wafer-scale integration (WSI), includes the ability to mix chip technologies [18,19]. In addition to that in MCM circuit design, good dies can be selected from wafers and bulky IC packaging can be avoided. It is expected that in the future, MCM assembly will be less costly, smaller, and lighter than SMDs, and circuit repair will be easy [13]. Mixed-signal applications pose a variety of challenges to engineers responsible for electronic system packaging. The emergence of advanced MCM packaging systems has reduced interference (EMI), crosstalk, feedback, and other causes of unwanted signal distortion [20,21]. MCM is now considered a technology that will improve final product cost and performance [15]. Multichip packaging technology offers significant advantages in miniaturization, electrical performance, thermal performance, lower cost, and improved reliability in avionics industries [22]. A variety of multichip module functions have been demonstrated successfully in recent years in various divisions of Rockwell International Corporation. As a whole, benefits of MCMs include footprint optimization, circuit density, high performance, and improved time to market. To be competitive in the microelectronics market, MCMs should be able to show that the manufacturing cost of a module satisfies the performance of the module [23].

The MCM is a hybrid for wafer-scale integration (H-WSI) and has a role similar to that of thick-film hybrid, which is substituting for less advanced IC technology [24]. H-WSI can be defined as an electronic circuit composed of multiple discrete unpacked integrated circuits and passive devices interconnected by means of an organic or inorganic thin film. The deposited interconnect structure has at least 10 times the complexity of the device technology available. However, MCM uses a variety of substrates according to the family to which they belong [24,25]. According to process and assembly requirements, the MCM family is generally divided into three categories: MCM-L (multichip module laminated), MCM-D (multichip module deposited), and MCM-C (multichip module ceramic) [26,27].

The issue for supplying and procuring known good dies (KGDs) is always problematic for MCM circuits. The cost of KGDs and packaged devices used in MCMs influences the fabrication cost of the modules. The key cost drivers are the number of KGD carriers, the complex ICs, the cost of the carrier, the number of burning hours, and KGD assembly/removal time. An ideal KGD will have a high pin count, a low burning time, and a long product life [28]. Each level of packaging reduces the number of connections required to the next level of packaging by a factor approximated by Rent's rule [29]. Using a smaller Rent's exponent, an MCM packaged system will have a better performance than that of a single-chip module system, given identical semiconductor technology [30].

As MCM technologies have grown out of hybrid microcircuit technology, one can expect that these technologies have common processing techniques. MCM technologies also use a wider range of substrate materials and metallization processes that provide much higher circuit density. The substrate materials on which MCM circuits are nested should have the following properties: (1) high electrical resistivity, (2) high thermal conductivity, (3) able to withstand high processing temperature, and (4) availability in the market for a reasonable price.

Substrates such as alumina (Al_2O_3), aluminum nitride (AlN), boron nitride (BN), silicon carbide (SiC), beryllia (BeO), and diamond are in very common use [31]. Alumina is used extensively in hybrid microelectronics circuits as substrate material and as multilayer structures for MCMs. Both high- and low-temperature cofired ceramic substrates are used. One of the most exciting developments in thin-film technology is the use of diamond films as substrate materials [32]. Diamond materials have been synthesized following high-temperature high- or low-pressure techniques. All low-pressure diamond deposition techniques are based on the same deceptively simple reaction chemistry. Hydrocarbons are broken down by plasma, hot filament, or high-temperature flame to produce gaseous hydrogen atoms, hydrocarbon radicals, and solid carbon [33,34]. Diamond substrates have improved the packaging density of chips and modules over that of alumina and silicon substrates [35]. MCM circuits on diamond substrates have been fabricated successfully for static and dynamic random access memory (SRAM and DRAM). Table 11.1 presents some of the material properties of MCM substrates.

11.1.1 Conductor Materials

Almost all the metals used in thick- and thin-film hybrid technologies are used in MCM circuits. Thick-film conductor materials are available in the form of a paste called *ink* and are deposited on the substrate through a screen. Thin-film conductors are deposited either by vacuum evaporation or sputtering. Gold, silver, palladium, and copper are the main conducting materials used. Sometimes, some binary metal compounds are also used to promote adhesion and to minimize electromigration at the cost of the conductivity of the more conductive metal. Copper and copper-based metals are developed as a low-cost sub-

TABLE 11.1 Substrate Material Characteristics

Property	MCM-L	MCM-C, LTCC	MCM-C, HTCC	MCM-D, Ceramic	MCM-D, Silicon
Dielectric constant	3.0–5.0	9.6	4.8–8.0	2.4–4.0	3.5–4.0
Maximum single layers	15	—	—	5–10	5
Line width range (μm)	15–150	200–300	100–150	100–150	10–75
Spacing range (μm)	40–75	100–150	50–75	50–75	5–30
Metal layer range	1–50	1–75		1–9	1–8
Thermal conductivity range (W/m·K)	<15	—	20–300	—	50–150
Typical maximum area (cm^2)	500	225	225	100	<100
Substrate I/O style	Peripheral and array	Peripheral and array	Peripheral and array	Peripheral and array	Peripherial
Tooling and set-up cost	Low	Low	High	Medium	High
Potential applications	Broad	Medium	Limited	Limited	Limited
Technology maturity	Good	Good	Good	Limited	Limited
Relative cost (low level)	Low	Medium	Medium/ high	High	High
Relative cost (high volume)	Low	Medium/ high	Medium	Medium	High

Source: From [36]. Reprinted from the September 1993 edition of *Solid State Technology*. Copyright 1993 by PennWell Publishing.

stitute for more expensive metals. These copper-based metals are being used extensively because of their solderability, leach resistance, low resistivity [37], and cost.

MCM-L (Laminated) The precise definition of MCM-L (*L* stands for *laminated*) is difficult, because of the diversified way of processing and the materials requirement for the fabrication of MCM-L. The substrate material is similar to PCBs that interconnect unpackaged, bare chips regardless of the number of the chips or the substrate density [i.e., to include all chip-on-board (COB) applications]. However, compared to COB, packaging efficiency is more than 30% in MCM-L circuits [38–40]. To establish some arbitrary demarcation line between "normal" COBs and *true* MCM-Ls, some use the number of the chips on a substrate and some use module termination methods. Figure 11.1 shows some MCM-L multilayer circuits on laminated substrate.

MCM-Ls vary from 1 by 1 in. (2.54 by 2.54 cm) to 6 by 6 in. (15.2 by 15.2 cm) although 4 by 4 in. (10 by 10 cm) is a favorable choice for the design engineers. The substrate is a thin laminate material with a high glass transition temperature (T_g). However, circuits are now fabricated on polytetra-

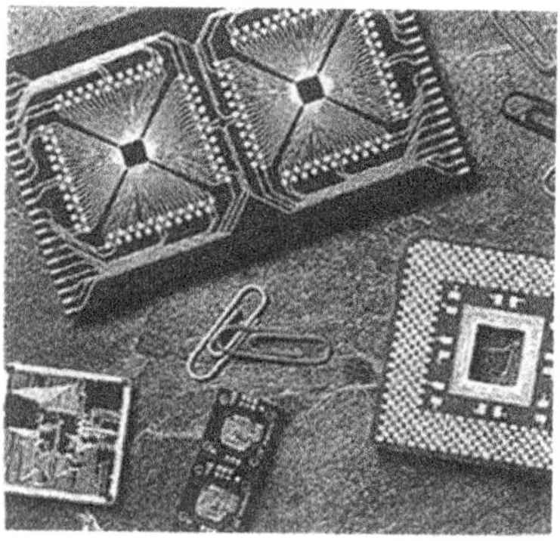

Figure 11.1 MCM-L multilayer circuits on laminated substrates [41]. (Reproduced with permission from the July/August 1994 issue of *Advanced Packaging*™ magazine. Copyright PennWell Corporation. All rights reserved.)

fluoroethylene laminates and glass-reinforced laminates for high-speed and high-frequency applications and the substrate thickness does not exceed 62 mils (1.58 mm) even with 12 layers of laminations. Hole diameters are made between 8 and 12 mils (0.2 to 0.3 mm) in pads of 12 to 18 mils (0.3 to 0.46 mm). Surface pad pitches are kept well below 10 mils. Line widths are between 3 and 5 mils (75 to 125 μm) with 3-mil (75-μm) spaces in between. The design grid is generally 50 mils (1.27 mm), although 40 mils (1.0 mm) is also recommended with a 3-in. (7.6-cm) space in between the conductors [41].

As the pattern dimensions shrink and the line widths decrease, great emphasis is placed on the search for more stable, flatter, and thinner laminates [42]. At the same time, use of low-K materials are essential to meet the higher-frequency requirements of the circuits. Although organic materials have been deemed superior in many areas, the new inorganic dielectrics, such as aluminum nitride, IBM's glass ceramic, and Johnson & Matthey's Sol Gel Borosilicate are proving themselves equal to the task by featuring lower dielectric constants, improved TCE, and better signal transmission than those of conventional inorganic materials [18,43–46]. Table 11.2 presents some of the material properties of MCM-L substrates.

Since it is difficult to test the MCM-L substrates electrically on a bed-of-nails tester, most manufacturers now rely on automated optical instruments to identify defects. The most favored method of chip termination is wire bonding, especially thermosonic bonding, which does not expose polymeric boards to extreme temperature [47]. Tape automated bonding (TAB) termination is also

TABLE 11.2 Properties of MCM-L Substrates

Substrate Material	X/Y Construction	TCE (ppm/°C)	T_g (°C)	Cost ($)
FR-4	4.8	16	125	1
BT epoxy	4.2	14	180	1.8
Polyimide	4.5	13	250	3.5
Cyanate ester	4.5	15	230	4.5

used. TAB has the ability to pretest chips and soldering [48]. It has been found that plastic ball grid arrays offer a significant board mounting advantages over conventional leaded plastic packages with MCM-L [49]. The plastic package can be expected to survive long past the prospective product lifetime except for military and invasive medical applications.

A new flip-chip technology [50] combines gold balls 2.4 mils (60 μm) wide at the base and 2 mils (50 μm) high with solder plating (95% Pb/5% Sn) capped with indium (In) layer offers a high packaging density and higher reliability. Indium provides both a wettable surface to gold and forms a very low contact resistance (<10 mΩ). The alloying temperature is above the melting point of In (m.p. 160°C), which is less than the reflow temperature of 95% Pb/5% Sn solder (m.p. 325°C). The contacts are capable of accepting a thermal shock of 1000 cycles and 1500 h of thermal aging at 150°C.

Processing time and yields are affected significantly in designing and processing MCM-L circuits. First, design requires sequential lamination, which consumes more processing time. Second, special attention must be paid when design incorporates cavity or multiple bond shelves, which typically lowers the yield. Finally, to increase the wire bondability, internal layers are sometimes oxidized prior to gold or nickel plating [51,52]. As a result, the processing time and yields are both affected. The most significant drawback of a MCM-L circuit is that the supportive passive components take up as much or more of the real estate as does the bare device. As an alternative, planar resistive elements are finding more use in MCM-L circuits. Unfortunately, planar resistors require three photo-imaging and developing steps, which require more manufacturing cycle time and ultimately the production cost. Despite all this, some manufacturers think that as the need for high-speed interconnects reduces packaging and lower cost continues, MCM-L will remain at the forefront [40]. MCM-L technology is used mostly in the miniframe computer and workstation industries. It is also used for test equipment, measurement equipment, optical displays, calculators, and advanced memory cards.

Following are some of the basic parameters of MCM-Ls.

- Up to 12 layers can be accommodated (most have six to eight layers).
- The design grid is 0.05 or 0.04 in. (0.13 to 0.1 cm) with 3 in. (7.6 cm) of space between conductors.
- Line widths and spaces are between 0.003 and 0.005 in. (0.0076 to 0.127 cm).
- Hole diameters are between 0.008 to 0.012 in. (0.02 to 0.03 cm) in pads of 0.012 to 0.018 in. (0.03 to 0.045 cm).
- Surface pad pitches can go down to 0.010 or 0.008 in. (0.025 to 0.03 cm).
- Soft gold is generally used for wire bonding and solder for tape automated bonding.
- MCM-Ls are thin; even with 12 multilayers, thickness is less than 0.062 in. (0.15 cm).

Figure 11.2 Typical MCM-C circuit on a ceramic board [36]. (From the September 1993 edition of *Solid State Technology*. Copyright 1993 by PennWell Publishing.)

- Sizes vary from 1 by 1 in. (2.54 by 2.54 cm) to 6 by 6 in. (15.2 by 15.2 cm), but the majority of MCM-L substrates are 4 by 4 in. (10 by 10 cm).
- Laminate material is generally high-T_g epoxy or polyimide.

The features of MCM-L Assembly are as follows:

$$\text{Signal propagation delay:}\quad t_{\text{pd}} = (L/C)(K)^{1/2}$$

$$\text{Microstrip geometry:}\quad t_{\text{pd}} = 1.017(0.475K + 0.67)^{1/2}$$

$$\text{Strip line geometry:}\quad t_{\text{pd}} = 1.017(K)^{1/2}$$

where C is the speed of light, L the distance the signal travels, and K the dielectric constant of the insulator.

MCM-C (Ceramic) MCM-C basically means that multichip modules are built on ceramic substrates [53]. Increasingly, MCM-C substrates (Figure 11.2) are being used in miniframe applications, implantable medical electronics, and various defense applications. The technology is very similar to thick- and thin-film hybrid microcircuit technologies. In general, this technology uses two types of substrates: high-temperature cofired ceramic (HTCC) and low-temperature cofired ceramic (LTCC). The processing of these two substrates differs primarily on the firing temperatures. Thus the selection of thick and thin materials, which are used in MCM-C circuits, is different [54]. HTCC substrates are designed to be fired finally at about 1600°C, which is very high for common conductor materials used in hybrid circuits. To cope with high temperature, refractory metals such as tungsten, molybdenum, and manganese alloys are common choice since at that high temperature these metals and metal alloys are susceptible to form oxides. To avoid oxidation, firing processes are accomplished in an inert ambient. Thus this process is not a cost-effective way for making MCM-C modules. Another challenge for HTCC MCM-C circuit

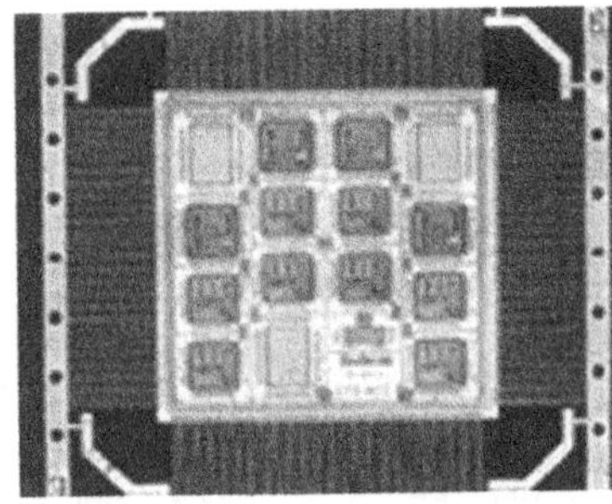

Figure 11.3 MCM-C circuit build on LTCC substrate [36]. (From the September 1993 edition of *Solid State Technology*. Copyright 1993 by PennWell Publishing.)

processing is the selection of thick-film slurry (ink). During firing, the volatile materials should not leave any microcracks or produce shrinkage, which might introduce stress. During via filling, this might cause gaps, creating problems in the interconnections.

On the other hand, the LTCC substrates generally contain glass/ceramics and are designed to fire at much lower temperatures (850 to 1050°C). Thus standard thick-film materials such as palladium/silver and gold, which have much lower sheet resistance, can be used. Thus LTCC MCM-C processing is cost-effective (Figure 11.3). For LTCC, substrate resistors and capacitors compatible with a green tape process may also be fabricated [3]. Sometimes in the LTCC MCM-C process, a sandwiched resistor layer between two layers of green tape material is formed and fired. These resistors have shown a high degree of stability and the temperature coefficient of resistance is within 100 ppm/°C. For accuracy and exact values of the resistors, laser trimming prior to lamination of these resistors is carried out. Figure 11.4 shows a typical MCM-C circuit used in the engine controller of a Saab automobile.

MCM-D (Deposited) MCM-D has recently gained increased attention among the MCM family because of its ability to support high-density packaging capability despite its smaller size than the other two members of the family [55]. The MCM-D (D means "deposition") process is very similar to the thin-film deposition process, and the deposition is done inside a vacuum

Figure 11.4 MCM-C in a ceramic substrate showing the circuit partitioning. The circuit is used in the engine controller of SAAB [10]. (From the May/June 1991 issue of *Advanced Packaging*™ magazine. Copyright 1994 PennWell Corporation.)

chamber [56]. The conductors are selected from such metals as gold, aluminum, and copper and the substrates from alumina, aluminum nitride, and silicon [57]. Polymers like benzocyclobutanes (BCBs) can be used as substrate material. BCBs offer distinctive advantages such as low permittivity, excellent planarization, low moisture absorption, good compatibility with copper, and the ability to be cured by rapid thermal techniques [58]. The dielectrics are used primarily in liquid state and are generally spin coated. A selective etching method adopting photolithography creates fine structures on the metals and oxides of metals and vias in the dielectric films. Line widths are generally 0.4 mil (10 μm) with a via diameter of 0.6 mil (15 μm). This MCM-D technology is predicted to provide excellent interconnection environment for high-clock-rate digital integrated circuits [59–61].

MCM-D is quite a mature technology and the cost of processing has been greatly reduced by continuous research and development in thin-film and IC production. Despite its maturity and the down trend in cost, commercial houses are looking for more from this family, especially MCM-D [62]. To be in the forefront of high-density packaging technology and in high-frequency digital circuits, some of the requirements that remain to be fulfilled are the ability to accommodate more metal layers without heavy overlayers to promote adhesion; decoupling capacitors that must be integrable during fabrication time and not exhibit unwanted series inductance; integral transmission line termination resistors fabricated directly onto the substrate and exhibiting low inductive parasites; and processing that can perform very short flip-TAB attachment, including solder flip-chip solder ball attachment [63,64].

MCM-D based on multilayer thin-film substrate technology is appearing in breakthrough products in the personal computer (PC) industry, which is a high-volume, cost-sensitive market. This technology is being utilized in all three (type I, II, and III) PCMC1A (personal computer memory card international association) cards. Type I cards have a 3.3 Z-axis and are used primarily for flash and other memory applications, Type II cards, which feature a 0.2-in. (5.0-mm) Z-axis, are used primarily for I/O peripheral functions such as local area networks and in modulator–demodulators (modems). Type III PCMCIA cards, which are the largest of the standard sizes, with a 0.41-in. (10.5-mm) Z-axis, have some *comb cards*, which combine I/O peripherals, and RF cards are also appearing in PCMICA type III cards. MCM-L and MCM-C are unusable in these cases, because of allowable substrate thickness, via density, and Z-axis requirements [65]. Most of the performance degradation on MCM-D circuits is due to wirebonded chips, which added inductance and increased module area. The performance of an MCM-D module fabricated with μBGA (micro-ball grid array) packages should be comparable to that of μBGA/laminated substrate modules. In this case, factors such as thickness, cost, and reliability could be introduced to extend the comparison [66].

Figure 11.5 shows Gigabit Logic's 10G023 GaAs chip wire-bonded to a MCM-D circuit. The wire-bond signal and power/ground attachments are fabricated with long wires. However, shorter wire or ribbon bonding with very

Figure 11.5 Photomicrograph of Gigabit Logic's 10G023 GaAs chip attached and wire bonded to MCM-D [61]. (Reproduced from *Electronic Packaging*, with permission from ISHM/IMAPS, Washington, DC.)

short TAB structures is recommended for better performance of the circuit [61]. Figure 11.6 a copper polyimide MCM-D circuit, designed at Mayo and fabricated at 3-M Corporation.

Advantages and Disadvantages of MCM-L MCM-L can be produced in large quantities with high yields. The established infrastructure and processes for MCM-L also ensure a quick turnaround of prototype orders. On the other hand, to produce a large quantity of MCM-Ds requires a large wafer fabrication facility. For large-volume applications, the conductor density requirements

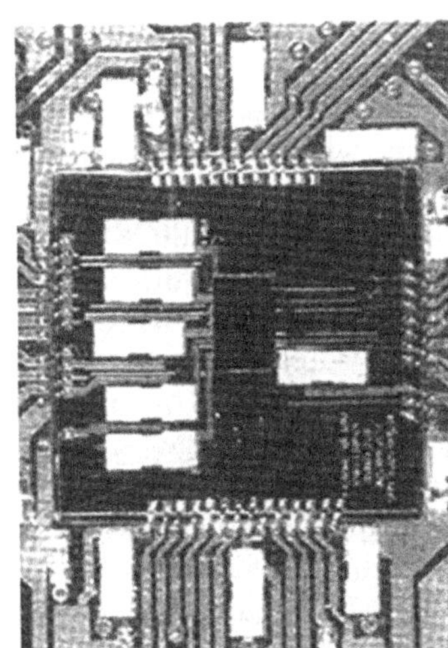

Figure 11.6 Photomicrograph of 3M's copper polyimide MCM-D. The GaAs chip is a flip adhesive-mounted at the center [61]. (Reproduced from *Electronic Packaging*, with permission from ISHM/IMAPS, Washington, DC.)

TABLE 11.3 Properties of Organic Dielectrics

Material	Application	Dielectric Constant	TCE (ppm/°C)	Tg (°C)	Water Absorption
MS Resin (Hitachi)	MCM-L	3.1	68	>300	0.50
FR4	MCM-L	4.8	13–18	125	0.10
BT-epoxy	MCM-L	4.3	15–16	180	0.05
Polyimide	MCM-L	4.2	10–14	250	0.35
Cyanate ester	MCM-L	3.6	8–10	230	0.08
RT/Duroid 5880DS fluorpolymer composite (Rogers Corp.)	MCM-L	2.2	16	327	0.05
Epoxy/nonwoven aramid	MCM-L	3.9	6–9	180	0.85
Teflon (DuPont)		2.0			
MD4212 (Amoco)	MCM-D	2.9	50	300	<0.4
PROB-400 (Ciba–Geigy)	MCM-D	3.3	40		<0.4
BCB (Dow Chemical)	MCM-D	2.7	52	>350	<0.4
P1-2555	MCM-D	3.3	40	>320	<0.35
P1-2722 (DuPont)	MCM-D	3.3	40	310	<0.35
P1-2741	MCM-D	2.9	10	365	<0.35
P1Q-13 (Hitachi)	MCM-D	3.4	45–58	>350	<0.4
EL-5010 (National Starch)	MCM-D	3.2	38	214	<0.4
Si			3.0		
Alumina			7.0		

for MCM-Ls are not extreme. MCM-L can be used at frequencies in the range of 100 MHz. Procurements of known good dies are a problem with MCM-L [67]. Another difficulty with MCM-L is the mismatch between the high TCE of PCB substrates (about 15 to 17 ppm/°C) and the low TCE (3.5 ppm/°C) of silicon chips. This mismatch presents no problems for the packaging of smaller chips, but for the mounting of larger chips, substrate mobilization is necessary to prevent cracking of chips or terminals during heat cycling of the system. Finally, one of the greatest impediments to the rapid proliferation of MCM technology in general, and of MCM-L in particular, is lack of knowledge on the part of the design engineers about MCMs, since trade-off analyses about their usefulness are scarce. Some typical organic dielectric materials used in MCM circuits are tabulated in Table 11.3. The most obvious advantage of MCM-L over the other MCM types is cost. Switching to MCM results in at least a fivefold cost reduction, with only a moderate area increase, as shown in the Table 11.4.

11.1.2 Summary [4,5]

The present and future market for MCM-Ls is difficult to assess because of the lack of agreement on the precise definition of MCM-Ls. All agree that the

TABLE 11.4 Cost Comparison: MCM-D versus MCM-L

		Size		Cost	
Type	Layers	mil^2	mm^2	$	%
MCM-D	4–5	61×10^3	38	112	100
MCM-L	5–6	80×10^3	50	15	13

interconnections of bare chips on a substrate should be similar to those of PCB substrate. But the materials and processes used in MCM-L are completely different than PCB. The materials and process divergences are so high that it is difficult to define MCM-L very distinctively. One approach to chip attachment is to include any type of PCB substrate that interconnects unpacked, bare chips regardless of the number of chips or the substrate density (i.e., to include all chip-on-board applications). The other approach is to establish some arbitrary demarcation lines between "normal" COBs and "true" MCM-Ls. For such delineation, some use the number of chips on a substrate; some use module termination methods. Achieving packaging efficiency over 30% can be a useful criterion for distinguishing MCM-Ls from COBs. Plastic ball grid arrays offer significant board-mounting advantages over conventional leaded plastic packages with MCM-L. A plastic package can be expected to survive long past the prospective product lifetime for all but a few applications, such as some military and invasive medical uses.

Although widespread use of MCM technologies had been predicted by consultants and other industrial pundits for more than a decade, the rate of implementation and growth of this segment of the hybrid microelectronics business largely failed to materialize as expected [68]. IBM's recent developments in the MCM-C and MCM-D arenas have produced IBM power PC microprocessors, Cadence Design System Allegro-E, Mentor Graphics, and Advanced Data Transfer Module (ADTM), used by the U.S. Air Force. ADTM includes 10 Atmel AT 28c0100 EEPROM memory devices, and a $1 \times$ xilinx Xc 4005A field-programmable gate array [69].

With electronic system designers demanding ever-increasing packaging densities, PCB manufacturers are pushing existing technology to the limit to produce finer features for MCM-Ls [70]. The complexity of microelectronics circuit is increasing rapidly. Figure 11.7 shows some of the major product types that are driving printed working board and MCM-Ls (e.g., portable communications, notebook computers, PCMCIA cards, and digital multichip modules). MCM-L technology has been able to fabricate a 900-MHz transceiver, a PCMCIA card [71]. An MCM-L circuit with the same size, weight, and circuit density has proved to be a better choice than an MCM-D circuit. The ability to create a large number of via holes is a key technology to allow circuit densities to approach more than 200 in./in^2 (508 cm/cm^2) in a layer pair [72]. It is expected that MCM-L technology will permit arranging more chips per area and will provide higher wiring densities in high-power designs. As a whole,

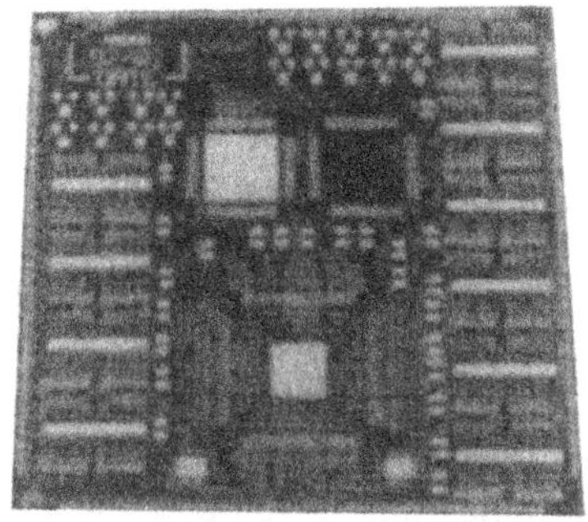

(*a*)

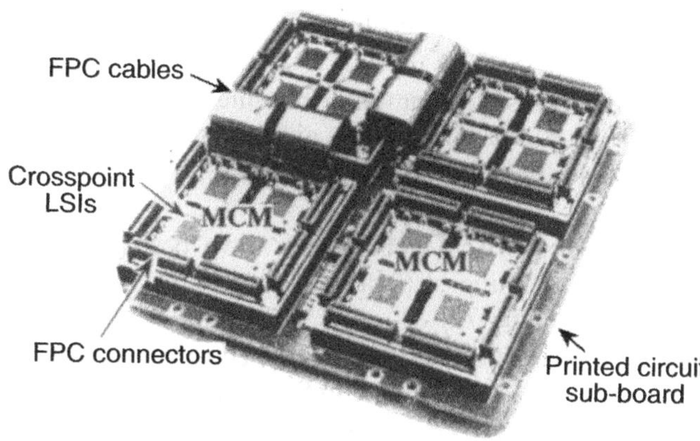

(*b*)

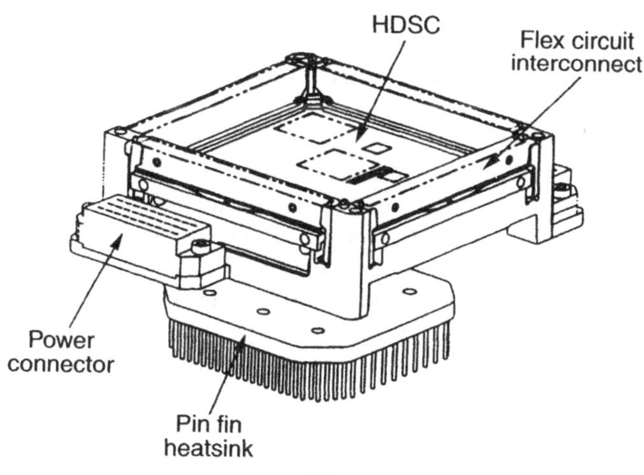

(*c*)

Figure 11.7 (*a*) Completed workstation MCM-L PCB; (*b*) MCMs mounted on the PCB subboard; (*c*) digital packaging view for VAX 9000 computer. (Reprinted with permission from ISHM/IMAPS, Washington, DC; from *Microcircuits and Electronic Packaging*, ISHM publication, Reston, VA: (*a*) Vol. 17, 1994; (*b*) Vol. 18, 1995; (*c*) Vol. 15, 1992.)

MCMs will offer a cost-effective solution for high-density hybrid microcircuits [73].

Applications

- *Miniframe computer.* Multichip module hybrid microcircuits are being used in miniframe computer systems such as the IBM 390/ES9000 (130), NEC SX3 (131), Hitachi M880 (132), and DEC Vax-9000 (17) [74].
- *Military.* Hybrid multichip modules are finding widespread applications for military use. Hughes has developed the HAC-32. MCM technology is also applied to build the F-22, which is an advanced technical fighter and supports naval electronic warfare systems [75]. Texas Instruments has developed a miniframe package for the military following MCM technologies [76].
- *Space.* The advanced space computer module (ASCM) system developed by Honeywell is an example of the application of MCM technologies. This computer module contains three CPUs and two floating-point processors [77].
- *Telecommunications.* MCM technologies have been used successfully in AT&T's 300-MHz timing generator [78] and in NTT's asynchronous transfer mode switching [79].
- *Consumer electronics.* In the consumer electronics sector, MCM technologies have been used extensively in Saab's engine controller [80] as well as in AT&T's telephone systems using silicon-based MCM technologies [64].

11.2 MICROWAVE HYBRID CIRCUITS

The term *microwave* relates to ac signals with frequencies ranging between 300 MHz (3×10^8 Hz) and 300 GHz (3×10^{11} Hz) [80]. James Clerk Maxwell laid down the field of microwave engineering as well as other areas of electromagnetics in 1873. Heinrich Hertz, a German professor and experimentalist, completely validated Maxwell's theory of electromagnetic spectrum that extends from dc to γ-rays [81,82]. The 1960s witnessed a large growth in hybrid microwave integrated circuits (HMICs), especially in the active microelectronics sector and monolithic microwave integrated circuits (MMICs) [83] based on silicon and gallium arsenide substrate. Soon, however, sophisticated functions such as multistage FET [84] amplifiers, 3- or 4-bit phase shifters, complete transmit/receive modules, and other circuits, were being fabricated as MMICs [85], following a decade of experience with HMICs. The 1980s brought complexity, higher-power, higher-density, higher-frequency circuits and it was felt that the well-established HMIC technology would be helpful in providing smaller devices at a much lower cost than for MMIC-based devices.

This has been possible because the discrete designs could usually be tuned and adjusted on a component or module level, after fabrication, to achieve optimum performance.

There are many methods for fabrication of microwave circuits working below 30 GHz. As the frequency of operation is raised to higher levels, more problems have arisen in the manufacturing areas. For example, small metallic waveguides require highly polished surfaces and less misalignment might increase losses. Selection of materials and process requirements for HMICs and MMICs are far more stringent than for low-frequency circuits. Conductor and substrate properties are of particular importance because they affect circuit design and performance substantially. However, thick- and thin-film hybrid microcircuits have introduced a new era in the history of microelectronics, adding sophistication and reliability at relatively low cost for manufacturing these microwave circuits. The ability of IC technology to fabricate 0.13-μm feature size and continuing improvement in packaging technology have strengthened the HMIC and MMIC technologies. Continuing efforts have been able to reduce the original 2-lb (ca. 1-kg) amplifier, with a volume of 140 in^3 (23.3 m^3) to 0.5 oz (ca. 15 g) and 0.1 in^3 (3 cm^3) [86]. However, above 60 GHz, screen-printed thick-film or selectively etched thin-film lines show a higher transmission line loss.

Figure 11.8 shows the picture of a bat, a sophisticated live "device" that communicates in the air through an ultrasonic (ca. 20-kHz frequency) sound system. A microwave circuit can also be described as a major communication application (Figure 11.9) [87,88] in the higher-frequency range, 300 MHz to 300 GHz. The major applications of microwave circuits are (1) point-to-point line-of-sight microwave radio [89]; (2) satellite relay systems [90]; (3) point-to-multipoint radio for cellular mobile systems [91]; (4) avionics systems; (5) navigation systems; (6) landing systems [92,93]; (7) weather radar; (8) defense systems (Figure 11.10*b*), including ground-based and airborne radar (Figure 11.10*a*) and space tracking radar [94]; (9) guidance systems; (10) industrial

Figure 11.8 Sophisticated live "antenna" that communicates in the frequency range around 20 kHz [87]. (Courtesy of EEV Microwave Products, Elmsford, New York.)

(a) (b)

Figure 11.9 Circuits used in (a) a communication system and (b) a missile system [88]. (Courtesy of ST Microwave Corporation, Chandler, Arizona.)

systems, including drying, alarm, and cooking equipment; (11) home satellite receivers [95–98] and radio astronomy and spectroscopy [99–101]; and (12) medical science [102].

11.2.1 Major Circuit Requirements

Substrates Ideally, substrates should offer not only mechanical stability for circuits but should have atomically smooth surfaces (especially when the circuit is being used in the high frequencies). They should show inertness to most chemicals and to most nesting materials (deposited thick- and thin-film conductors, resistors, and dielectrics) with zero porosity, good dimensional tolerances and planarity, and good adherence to the deposited materials. Low cost will be another criterion for a good substrate material. As these substrates are to be used in the very high frequency ranges (300 MHz to 300 GHz), surface roughness and dielectric and conductive losses should be minimal. Substrate roughness increases line lengths and ultimately, can attenuate signal. This is particularly important for high-frequency circuits. It can also cause discontinuities in thin-film structures and current crowding and can introduce noise in the resistors deposited. To keep the mechanical strength of the substrate

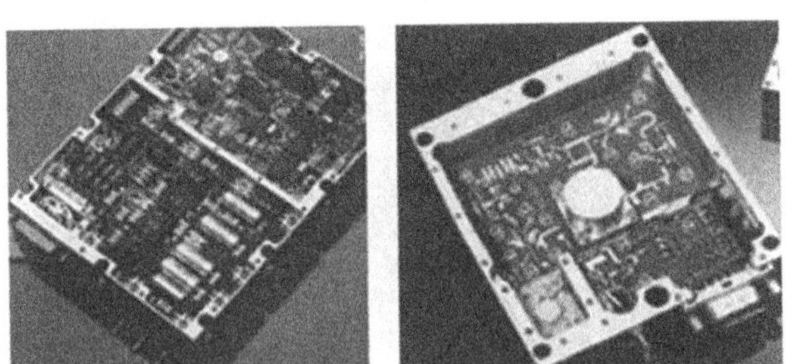

(a) (b)

Figure 11.10 Circuits used in (a) an X-band radar system and (b) a communication system [88]. (Courtesy of ST Microwave Corporation, Chandler, Arizona.)

high, there should be a minimum number of vias for interconnections. Vias are needed for various purposes, especially in high-frequency circuits, where shorter ground planes are important. Vias are also detrimental for Z-direction thermal expansions, especially when the aspect ratios are high. Typical substrate thickness is between 10 and 25 mils (ca. 250 to 625 μm). The substrate thickness for the onset of higher-order modes of propagation is inversely proportional to its relative permittivity. Except for iron, nickel, and cobalt, the permeability of most of metals is 1.

Gallium arsenide (GaAs) is another substrate material used in HMIC and MMIC circuits from 600 MHz to 10 GHz for broadband intermediate- and radio-frequency amplifiers operating in the L, C, and S bands. Indium phosphide (InP) is also used as a substrate material, for highly sophisticated microwave devices in the millimeter wave (e.g., imaging radiometers and passive interferometers). Potential commercial applications for these InP-based devices include automobile, collision avoidance radar, mobile telecommunication systems, and clear-air-turbulence radiometry. Use of inexpensive substrate material (e.g., stainless steel) is also being used in the high-frequency range to develop prototype cellular telephones based on thick-film superconducting materials (Y–Ba–Cu).

The choice of substrate material [103] is an important part of high-frequency hybrid microcircuit design, especially when speed is a concern. Alumina (90%) with dielectric constant of about 10 is acceptable, in the frequency range 800 MHz to 1 GHz. However, above 1 GHz, materials with a low dielectric constant ($K = 5.0$ to 6.0) will be a better choice [104]. Some core materials made of polytetrafluoroethylene [PTFE, $(CF_2–CF_2^-)_n$] with a glass transition temperature of about 130°C are being used with or without woven fiberglass pressed under conditions of high pressure and temperature. Sometimes silanes are added to these materials to prevent moisture absorption. Another impregnated woven glass mat substrate, prepared from glass fiber mixed with resin, flame retardant, and polymers (ability to be cross-linked), is used in hybrid microwave circuits. The formulation is proprietary; however, the hardness and processing temperatures are different for different additives, which are mixed for the promotion of adhesion, hardness, and stability. The primary product in this sector is FR-4. Most of these dielectric materials are have low dielectric constants ($K = 2$ to 3). Electro-deposited copper or pressed copper foil is used as a cladding material. To match the thermal expansion of copper-clad and core material, ceramic filler is used between the copper and the PTFE core. Heat resistance is frequently improved in amorphous organic materials such as resins, varnishes, and compounds when they are combined with filling agents. In a number of cases, when the filling increases, the heat resistance and mechanical strength of an impregnated system increases at the cost of surface smoothness. (Details of copper electroplating are given in Chapter 7.) An optional thermal barrier treatment with brass, zinc, or nickel allows clad laminate adhesion to be maintained despite the thermal processing conditions [51].

Smaller substrate sizes with buried integral passive components such as

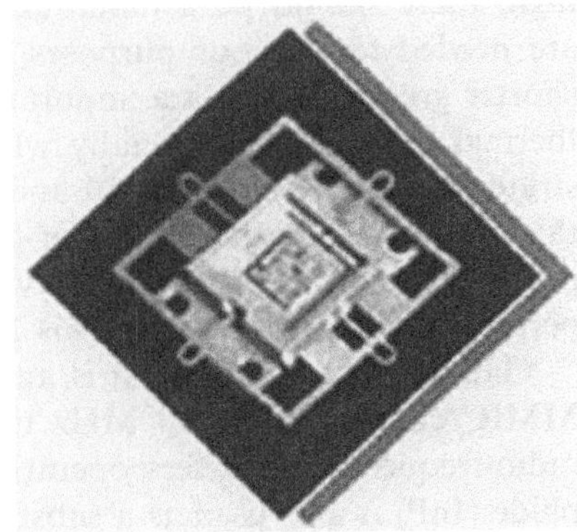

Figure 11.11 Commercial VCO (frequency range 1 to 18 GHz) [58]. (Courtesy of ST Microwave Corporation, Chandler, Arizona.)

spiral inductors (Figure 11.9*a*) are being designed and fabricated for circuits in the gigahertz range (especially for telecommunication systems). Capacitors are generally created with multilayer packages by layer lamination and manipulating ceramic tape size between metallized planes. This technology has reduced both the size and cost of high-frequency wireless systems such as digital mobile phones, duplexers, and varactor coax oscillators (VCOs; Figure 11.11).

Except for small IC components, microwave circuit components are considered to be distributed elements rather than lumped elements, as they are used in their counterpart low-frequency systems. Coupling, which enables microwave energy to be extracted from or added to a device or circuit, must provide a low-loss connection between a package and a source or load or between a device and its circuit. The power levels for input and output ports for a microwave package should be within 5 W for coaxial connectors, and for 60 GHz and higher frequencies, rectangular waveguides should be used [80]. Modern microwave circuits generally utilize planar waveguides because of good coupling, excellent reliability, and better performance at moderate power levels. As this industry is maturing, aluminum oxide, and sometimes beryllium oxide, are used instead of fiberglass and quartz, because these materials are produced at reasonable purity levels and reasonable cost. High-conductive thick-film pastes from DuPont, EMCA, ESL, and others are being used on alumina or beryllia substrates for microwave applications. A copper-clad material called Epsilam 10 from 3M offers more versatility than aluminum. On the other hand, a wide variety of substrate materials (e.g., low-temperature cofired ceramics) are opening new vistas for microwave applications [53]. Substrate properties are summarized in Table 11.5.

Dielectric Loss Most of the time, alumina is the substrate used in hybrid microcircuits. The alumina used in hybrid microcircuits is not a single crystal. It has been observed that matrix contributions in polycrystalline materials contribute to the anisotropic nature of the material because of its preferred orientation. Sometimes, the fillers in organic dielectric materials also contrib-

TABLE 11.5 Properties of Substrates Used in Hybrid Microwave Circuits

Material	ε_r (approx.)	Tan $\delta \times 10^4$ at 10 GHz	Surface Roughness (μm)	Thermal Conductivity (KW/cm·°C)	Remarks
RT-duroid 5880	2.16–2.24	5–15	0.75–1[a] 4.25–8.75[b]	0.0026	Cu-plating, flexible strip line, microstrip
RT-duroid 6010	10.2–10.7	10–60	0.75–1[a] 4.25–8.75[b]	0.0041	Cu-plating, flexible strip line at L-band microstrip
Epsilam-10	10–13	20	—	0.0037	Cu-plating, flexible strip line at L-band microstrip
Alumina, 99.5%	9.6–10.4	0.5–3	0.05–0.25	0.37	Cr–Au layer/microstrip, microstrip-like lines, slot line coplanar lines
Fused quartz, 99.9%	3.75–3.8	1	0.006–0.025	0.01	Cr–Au layer/optical finish microstrip-like lines at mm wave frequency
BeO	6.6	1	0.05–1.25	2.5	High-conductivity coupled substrate
Glass	5.0	20	0.025	0.01	Lossy/lumped element
Kapton	3–3.5	—	—	—	Flexible substrate/strip line
Cu–fleon	2.1	4.5	—	—	Flexible substrate/strip line
Rutile (TiO_2)	100	4	0.25–2.5	0.02	High dielectric constant/microstrip line, coplanar lines
Sapphire (single crystal)	$\varepsilon_{perp} = 9.4$ $\varepsilon_{parallel} = 11.4$	—	0.005–0.025	0.4	Well-defined and repeatable electrical properties

Source: From [167].

[a] ε_r = permittivity of the medium.

[b] δ is called the dielectric loss angle and tan δ is equal to the ratio between the active and reactive currents.

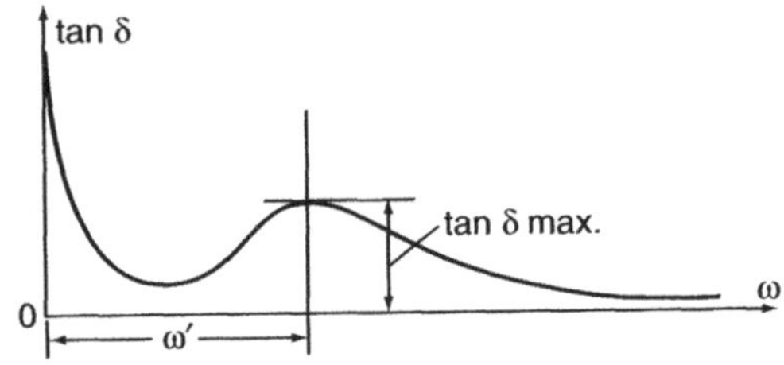

Figure 11.12 Behavior of the loss tangent on the applied voltage frequency for a polar dielectric [105].

ute to the anisotropic nature of the substrate. High-density polycrystalline alumina composed of many grains of sapphire, if perfectly oriented, would yield a fired body with a dielectric constant as high as 10.8 (which is not acceptable above 1 GHz). Anisotropy in the substrate materials introduces additional firing capacitance. This additional capacitance introduces additional impedance. Additional capacitance also changes the time constant, especially where narrow lines and structures are used in filters and couplers. In good dielectric materials the loss due to conduction should be minimum and should be able to withstand high electric potentials. Under the influence of electric fields, electric charges in the form of polarized dipoles align themselves along the direction of the applied field. As the frequency of the field is increased, the electric charges reorient themselves. This reorientation ability with time determines the amount of loss of the charge in the form of heat. Its loss tangent characterizes the loss of energy in the dielectric, which is equivalent to the ratio of lost energy to stored energy [105]. In a good dielectric the lost energy is generally recovered totally. High-density ceramics have losses because they respond elastically to electronic movement. In nonpolar materials such as polyethylene, the field response is limited to ionic and electronic displacement, so that within the microwave frequencies, the dielectric losses are very low. In polyimides, which have polar cross-linked materials, the loss is high. Figure 11.12 shows the variation of $\tan\delta$ values with applied voltage frequency. The losses are purely dipole mechanism losses. The value ω corresponds to such a ratio between the period of an external field and the relaxation time τ of dipoles as required to observe the greatest loss of energy to overcome the resistance of the viscous medium by the dipoles [105]. Figure 11.13 shows that the $\tan\delta$ losses for nonpolar dielectrics with applied voltage frequencies undergo several transitional stages.

Conductors For a good conductor, the decay of fields from the interface at $Z = 0$ will be very rapid, and right-hand end of the surface (S) can be made far enough away from $Z = 0$ [80]. The plane wave normally incident on a conductor is mostly reflected, and transmitted power is dissipated as heat within a

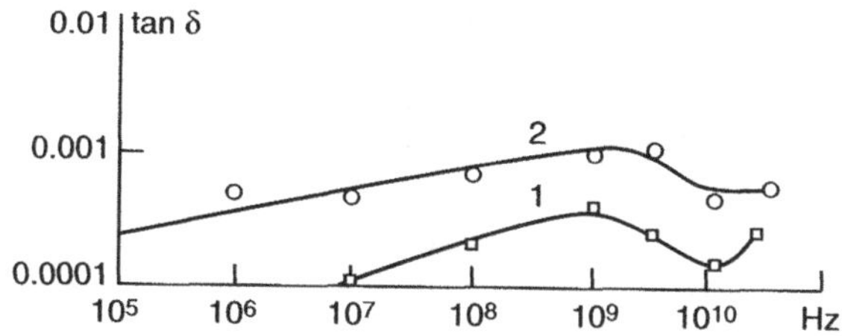

Figure 11.13 Tan δ values versus applied voltage frequency for nonpolar dielectrics. Curve 1 represents the behavior of polytetraflluoroethylene and curve 2, the behavior of tetrafluoroethylene and hexafluoroethylene [105].

very short distance from the surface. For a good conductor this distance is very small (within a few skin depths of the material), and the surface impedance (η) through the conductor is very small compared to the air impedance (η_0). Another important criterion for a good conductor is that the surface of the conductor be very smooth, to minimize losses at very high frequencies (i.e., current crowding, increased length, and even discontinuities in the metal lines). The surface roughness depends on several things in the films deposited. For thick-film inks, the suspended glass and binders and the mesh size of the screen may contribute to the surface roughness. It has been observed that the dc resistances of fired thick-film conductors are generally higher than those of evaporated thin-film conductors. In thick films the propagating electromagnetic wave penetrates the glass binder before it interacts with the metal fillers. However, thin films are free from this, so better surface smoothness is achievable in thin films. Thick-film conductor loss has been estimated to be approximately 1.5 times more than that of thin films. The surface attenuation can be 60% or more if the surface roughness becomes equal to the skin depth of the substrate.

For RF magnetic fields which are tangential to conductors, most of the currents lie on the dielectric sides of the conductor. Losses in the microwave circuits are divided into two parts: conductor loss, caused by the current flow through glossy conductors; and dielectric loss, due to propagation of the field through the dielectric media. Conductive loss depends on the nature of the conducting system (i.e., the current flow distribution, conductivity of the material, and surface smoothness of the conductor).

Copper and gold are the two conductive metals most used in microwave circuits. Copper is less expensive and the film adheres better than gold. Addition of a second or third metal prior to the deposition of conductor metal can increase metal adhesion. Chromium, molybdenum, and titanium are the metals used most frequently as adhesion promoters. Amorphous transition metal nitrides are sometimes used between the silicon substrate and the metal film to minimize the diffusion of silicon into the metal. But the addition of a second metal layer either to promote adhesion or as a barrier can reduce the conductivity of the interconnecting metal layer. An elaborate description of the barrier

TABLE 11.6 Characteristics of Microwave Conducting Materials

Conducting Material	Surface Resistivity R_s (m/s)	Skin Depth at 10 GHz (cm $\times 10^{-7}$)	Thermal Expansion (ppm/deg)	Typical Deposition[a]
Silver (Ag)	0.025	6.40	21	EB
Copper (Cu)	0.026	6.60	18	EB and EP
Gold (Au)	0.03	7.86	15	EB and EP
Aluminum (Al)	0.032	8.14	26	RVH
Chromium (Cr)	0.032	8.1	8.5	RVH and SD
Nickel (Ni)	0.51	1.35	18	
Palladium (Pd)	0.65	1.6	11	RVH, SD, and ELP
Molybdenum (Mo)	0.5	1.0	4.6	SD and EB
Titanium (Ti)	0.62	1.70	9.0	SD and EB

Source: From [80].

[a] EB, electron beam evaporation; EP, electroplating; RVH, resistor vacuum heating; ELP, electroless deposition; SD, sputter deposition.

layers is given in Chapter 7. Characteristics of microwave conducting materials are summarized in Table 11.6.

Conductor Loss When ac flows through a conductor, the magnetic field generated by the oscillating current creates an opposing current. The magnitude of this current opposing the magnetic field depends on the resistivity and permeability of the conducting material. The permeability of a conductor does not greatly influence the loss since the conductor permeability used in microwave circuits is generally 1. The surface resistivity of the conductor (R_s) which actually contributes to the loss is related to the skin depth [106] and can be expressed mathematically as

$$R_s = \frac{1}{\sigma \delta_s} \tag{11.1}$$

where δ_s (cm) is the skin depth and is equal to

$$\left(\frac{\rho}{\mu \pi f}\right)^{1/2} \tag{11.2}$$

where ρ is the resistivity of the metal, μ the permeability, and f the frequency in hertz. For a good conductor, the resistivity should be very small, and most of the current should flow through a very thin surface layer. For a good conductor, $\delta_s = (2/\omega\mu\sigma)^{1/2}$; for a lossy medium, $\delta_s = 1/\alpha$ [$\alpha = \mathrm{Re}(\gamma)$ and $\gamma = j\omega(\mu\varepsilon)^{1/2}$, where ε is the permittivity]; and for lossless media, $\delta_s = \infty$. The amplitude of the field in the conductor decays by an amount 1/e, or 36.8%. Gold and silver show low losses and are generally chosen for microwave components [80].

Resistors The resistor value R in ohms can be presented mathematically as

$$R = \rho \frac{l}{A} = \rho \frac{l}{wt} = \frac{\rho}{t} \frac{l}{w} = \rho_s \frac{l}{t} \tag{11.3}$$

where ρ is the resistivity, l the length, w the width, t the thickness of the resistor, and ρ_s the sheet resistance. At high frequencies, the dimensions of the resistor are reduced, due to incremental inductance. The true value of the resistor will be

$$R = 2\pi f[L(\text{nominal dimensions}) - L(\text{dimensions reduced by } \delta_s/2)] \tag{11.4}$$

Thus at high frequencies, the dimensions of the resistors are high. There are two possible ways to achieve this: decrease the width, which is not possible, because to avoid the discontinuity in lines, a minimum width has to be maintained, or increase the length of the resistor. At high frequencies, a longer resistor will offer more stray inductive and capacitive effects. Another way to increase resistor length is by decreasing the sheet resistance. If the resistance is increased, decreasing the sheet resistance will cause the resistor to behave like a lossy transmission line, which is not desirable. So in designing resistors for microwave circuits, these points must be taken into account.

The current-carrying capacity at high frequencies through the conductors available for resistor termination is decreased by the skin effect, and most of the current is seen to flow near the dielectric layer. Thus conductors used for termination should be larger. At the same time, the resistor should be close to the ground. Long wires and conductor lines should be avoided. This is accomplished either by via holes through the substrate or by wraparound. Too many via holes will decrease the mechanical stability of the substrate. Moreover, the use of laser beams (either Q-switched Nd:YA or pulsed CO_2 laser) should be avoided, because the laser beams are absorbed by most of the substrates used in microwave circuits and will increase the temperature of the surrounding material. Even the heat from the beam can melt the materials deposited and can change their structure and grain size. Increased grain size may cause an increase in surface roughness.

In power-divider circuits, resistors are used as a shunt to ground or as isolation resistors. At high frequencies, resistor conductor termination design is important, because of the return losses due to shunt capacitance and inductance. Grounding can be accomplished either by joining the resistor terminals straight to ground or by tapering the ground terminal. The return losses (in decibels) can be high at higher frequencies (above 5 GHz).

Capacitors The capacitors used in microwave circuits are generally very small and can be formed in three different structures: interdigitated, metal–insulator–metal, and shunt capacitance formed between conductor and ground. Interdigitated capacitors arise from the fringing fields both through air and the dielectric and by direct fields through the conductor. Thus the quality and value of these capacitors depend mainly on the isotropic nature of the dielectric medium, the uniformity of the dielectric layer, and the gap between lines. The quality factor of the capacitor will depend primarily on the nature of the conductor and the working frequency.

Sandwiching the dielectric layer between two metal layers forms the second capacitor (MIM). Thus this capacitor can be considered to be a parallel-plate capacitor, and the value of the capacitance can be calculated on that basis. At low frequencies the quality factor depends primarily on the dielectric film, but at high frequencies, conductor losses dominate (assuming a frequency-independent loss tangent). Thus the choice of good conducting plates and their thicknesses are important in the high-frequency range. In an MIM sandwich capacitor, the thickness of the bottom plate is kept at less than a skin depth, and the upper plate is made thicker (on the order of a few micrometers). The surface roughness of the metal plates is kept within 2 μm.

Inductors The inductors are deposited either by screen printing or by vacuum deposition as described in Chapters 5 and 7. To avoid interelectrode capacitance, the gap between the metal layers is kept small, and thus the overall diameter of the inductor is reduced. The total value of the inductor is proportional to the number of turns of the metal layers; the fewer the turns the smaller the inductor value. The central portion of the spiral inductor is made larger in order to have more open space and more flux lines to span through. The quality factor is seen to fall very rapidly after 6 GHz, which is supposed to be due to current crowding. At high frequency due to the skin effect, current will flow through into a limited depth of the outer layer of the outer surface of the conductor and of the inner layer of the inner part of the conducting layer. Thus the outer surface of the outer layer of the conductor should be very smooth.

Interconnections Interconnections in hybrid microwave circuits are done both by gold/aluminum wires and deposited metal lines. The parasitic inductance in free space is directly proportional to the length and inversely proportional to the logarithmic value of the diameter. Thus to minimize the parasitic inductance, the connecting wire should be as short as possible and the diameter of the wire should be greater. Generally, the bonded wire diameter in ICs is 0.7 mil (17.5 μm), so the only alternative left is to reduce the wire length. There should be no looping in the wire, as a larger looping diameter will increase the parasitic inductance. Vacuum-deposited metal lines photographically etched and bonded to the devices have been found to lower the parasitic inductance. When the width and the geometry are tightly controlled and the strip lines are insulated, transmission has always been better with less interference and crosstalk. Electroplated air-insulated crossover on a sacrificial metal layer has been used successfully in a number of microwave circuits. This technology has shown low parasitic capacitance and inductance in insulated metal lines and very low impedance in the metal layers.

11.2.2 Waveguides

Waveguides are considered as fundamental microwave components to distribute microwave power from one point to another. In the early 1950s, waveguides were developed, but the industry was struggling to develop small, complex circuits using waveguides, coaxial cables, and two-wire transmission lines using hybrid microcircuit technology. The higher costs of these circuits severely hindered the efforts of design engineers [107].

Parallel-Plate Waveguide The parallel-plate waveguide can be described as the simplest type of guide used in microwave circuits. The design of such a waveguide consists of two parallel-plate strips of width w separated by a distance d [80,108,109]. The space between the two plates is filled with a dielectric material of permittivity ε and permeability μ. The impedance Z_0 for such a device can be written as (TEM mode) [80]

$$Z_0 = \frac{V}{I} = \frac{\eta d}{w} \tag{11.5}$$

where, V is the voltage between the two plates; I the current in amperes; η, the wave impedance, $= \sqrt{\mu/\varepsilon}$; d the separation; and w the width of the strip. The phase velocity v_p is given by

$$v_p = \frac{\omega}{\beta} = \frac{1}{\sqrt{\mu\varepsilon}} \quad \text{and} \quad \text{wavelength } \lambda = \frac{v_p}{f} \tag{11.6}$$

where β is a propagation constant. For the TM mode,

$$Z_{\mathrm{TM}}(\text{impedance}) = \frac{\beta\eta}{k} \tag{11.7}$$

where β is equal to $(k^2 - kc^2)^{1/2}$ and $kc = n\pi/d$. For the TE mode,

$$Z_{\mathrm{TE}}(\text{impedance}) = \frac{k\eta}{\beta} \tag{11.8}$$

where β is equal to $[k^2 - (n\pi/d)^2]^{1/2}$.

Rectangular Waveguide This is one of the earliest waveguides [80] used in couplers, detectors, isolators, attenuators, and slotted lines [110,111]. Since this

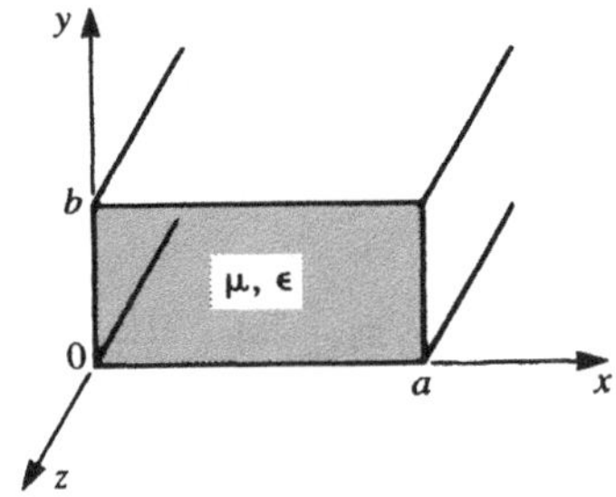

Figure 11.14 Outline of a rectangular waveguide [80].

waveguide has only one conductor, it can propagate TM and TE modes and not TEM waves. The standard waveguide bands are from 1 to over 220 GHz. Figure 11.14 shows the outline of a rectangular waveguide. The guide is filled with material of permittivity ε and permeability μ. Then for the TE mode, the lowest cutoff frequency, called the *dominant mode* ($a > b$), is given by [80]

$$f = \frac{1}{2a\sqrt{\mu\varepsilon}} \tag{11.9}$$

where $1/\sqrt{\mu\varepsilon}$ is the phase velocity (v_p).

Attenuation can occur in a rectangular waveguide due to dielectric or conductor loss. Dielectric loss can be treated by making the permittivity ε complex [$\varepsilon = \varepsilon_0\varepsilon_r(1 - j\tan\delta)$] and using a Taylor series approximation. α_d(dielectric loss) $= (k^2 \tan\delta)/2\beta$ (TM or TE mode) in nepers/m, where k is the wavenumber, β the propagation constant, and NP $= 8.686$ dB. One can treat conductor loss using the perturbation method (ε is the permittivity, $\varepsilon_0 = 8.85 \times 10^{-12}$ F/m, and ε_r is the relative permittivity; μ is the permeability $= \mu_0\mu_r$, $\mu_0 = 4 \times 10^{-7}$ H/m, and μ_r is the relative permeability of the material). For the TM mode,

$$fc_{11} = \left(\frac{1}{2\pi\sqrt{\mu\varepsilon}}\right)\left(\sqrt{\left(\frac{\pi}{a}\right)^2 + \left(\frac{\pi}{b}\right)^2}\right) \tag{11.10}$$

$$\text{phase velocity } v_p = \frac{1}{\sqrt{\mu\varepsilon}} \quad \text{and} \quad \text{dielectric loss} = \frac{k^2 \tan\delta}{2\beta} \quad \text{NP/m} \tag{11.11}$$

Circular Waveguide [80] A hollow metal tube of circular cross section constitutes a circular waveguide (Figure 11.15). The inner diameter of the tube is a. A circular waveguide can support both the TE and TM waveguide modes. For the TE mode,

$$\text{cutoff frequency} = \frac{k_c}{2\pi\sqrt{\mu\varepsilon}} = \frac{p'_{nm}}{2\pi a\sqrt{\mu\varepsilon}} \tag{11.12}$$

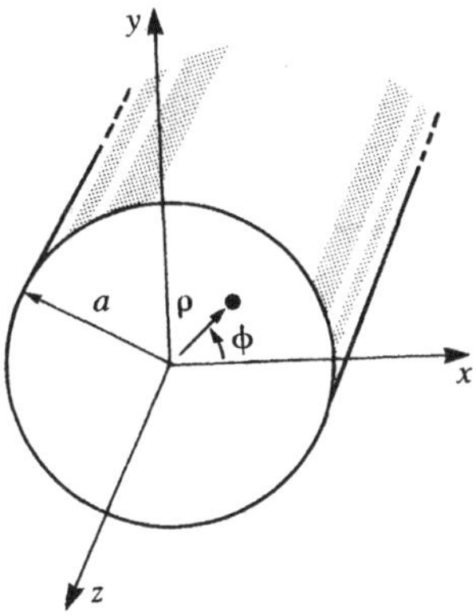

Figure 11.15 Outline of a circular waveguide [80].

Table 11.7 presents some of the TE mode P'_{mm} values. P'_{nm} values represent the mth root of $J'_n(x)$: that is, $J'_n(P'_{nm}) = 0$; $J_n(x)$ represents Bessel functions of the first kind.

$$\text{wave impedance} = Z_{\text{TE}} = \frac{\eta k}{\beta} \tag{11.13}$$

$$\text{attenuation constant} = \alpha_c = \frac{R_s}{ak\eta\beta}\left(k_c^2 + \frac{k^2}{p_{11}'^2 - 1}\right) \quad \text{NP/m} \tag{11.14}$$

$$\text{TE}_{11}\ fc = \frac{p'_{11}C}{2\pi a\sqrt{\varepsilon_r}} \tag{11.15}$$

$$\text{TM}_{01}\ fc = \frac{p_{01}C}{2\pi\sqrt{\varepsilon_r}} \tag{11.16}$$

Table 11.8 shows comparative values of TM_{nm} and TE_{nm} mode values for

TABLE 11.7 TE Mode p'_{nm} Value

n	P'_{n1}	P'_{n2}	P'_{n3}
0	3.832	7.016	10.174
1	1.841	5.331	8.536
2	3.054	6.706	9.970

Source: From [80].

TABLE 11.8 TM Mode Characteristics

TM_{nm}	TE_{nm}
$V_p = \omega/\beta$	$V_p = \omega/\beta$
$Z_{\text{TM}} = \eta\beta/k$	$Z_{\text{TE}} = \eta k/\beta$

Source: From [80].

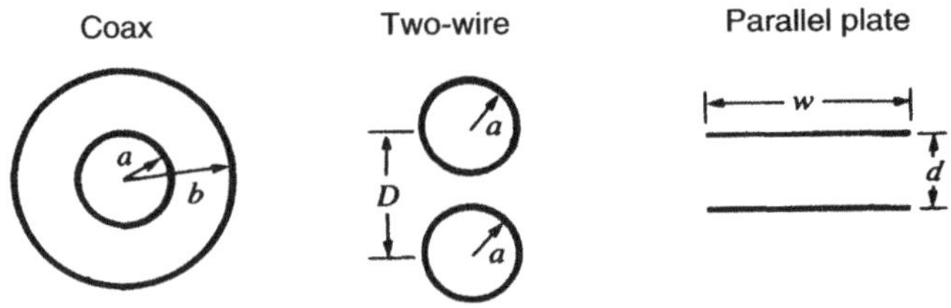

Figure 11.16 Constructional details of a coaxial cable [80].

phase velocity v_p and impedance Z. k_c is the cutoff wavenumber (k), β is the propagation constant, and R_s is the surface resistivity. Figure 11.16 shows the constructional details of a coaxial cable; a and b are the inner and outer diameters of the coaxial cable [112]. Table 11.9 presents a comparative analysis of transmission line parameters.

Coaxial Connectors [113]

- *Type N:* male and female thread together (outer diameter female = 0.625 in., f = 11 to 18 GHz, standing-wave ratio (SWR) = 1.07
- *SMA:* female, diameter = 0.210 in., f = 25 to 40 GHz
- *SSMA:* scaled SMA, female, diameter = 0.156 in., f = 38 GHz
- *APC-7:* SWR less than 1.04, f = 18 GHz
- *BNC:* baby *N*-connector

Figure 11.17 shows different varieties of precision connectors, which are used in microwave circuits, and Figure 11.18 shows various types of coaxial cable assemblies.

TABLE 11.9 Comparative Analysis of the Transmission Line Parameters

	Coaxial	Two-Wire	Parallel Plate
Series inductance L per unit length (H/m)	$(\mu/2\pi)\ln(b/a)$	$(\mu/\pi)\cosh^{-1}(D/2a)$	$(\mu/d)/w$
Shunt capacitance C per unit length (F/m)	$2\pi\varepsilon'/\ln(b/a)$	$\pi\varepsilon'/\cosh^{-1}(D/2a)$	$\varepsilon' w/d$
Series resistance R per unit length (Ω/m)	$(R_s/2\pi)\times(1/a+1/b)$	$R_s/\pi a$	$2R_s/w$
Shunt conductance G per unit length (S/m)	$2\pi\omega\varepsilon''/\ln(b/a)$ $\varepsilon=\varepsilon'-j\varepsilon''$	$\pi\omega\varepsilon''/\cosh^{-1}(D/2a)$ $\mu=\mu_0\mu_r$	$\omega\varepsilon'' w/d$ R_s = surface resistivity

Source: From [80].

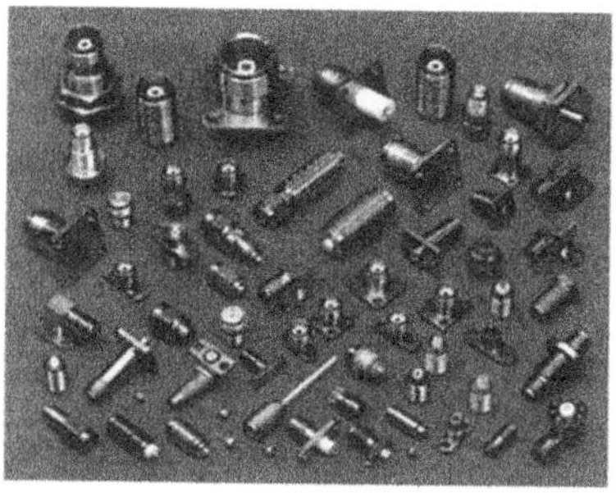

Figure 11.17 Varieties of precision coaxial connectors [114]. (Courtesy of Dynawave, Georgetown, Massachusetts.)

Ceramic Waveguide A ceramic waveguide (CWG) can be rectangular, circular, or coaxial with a ceramic dielectric [80]. The ceramic structure is lighter, compact, and provides low loss, with a high production rate. The integrated passive and active circuits and modules can be used at C through K_A bands [115]. For K_A bands, quartz is generally chosen because of its low loss tangent and low dielectric constant. However, the thermal conductivity is not as good as alumina. Beryllia is a good choice for the K_u and X bands but beryllia is toxic. However, alumina (Al_2O_3) is good for the X and C bands and environmentally friendly.

In the most common method of fabrication of ceramic waveguides, called a *green process*, ceramic is mixed with a binder and compressed to the shape desired and then *fired* to their final dimensions. The second method is to form the circuit by compression, sinter the form, machine to dimensions (allowing for proper shrinkage), and fire. This procedure permits greater latitude in developing more complicated configurations. An ultrasonic impact grinder forms the third method, which is called the *cookie cutter technique*. The first two methods provide pieces, which are less prone to breakage and cracking and have cleaner holes and edges than those of ground pieces. Sputter-coated chromium followed by gold is used for metallization. Later, gold is plated on the vacuum-deposited gold surface to minimize ohmic drop and power losses.

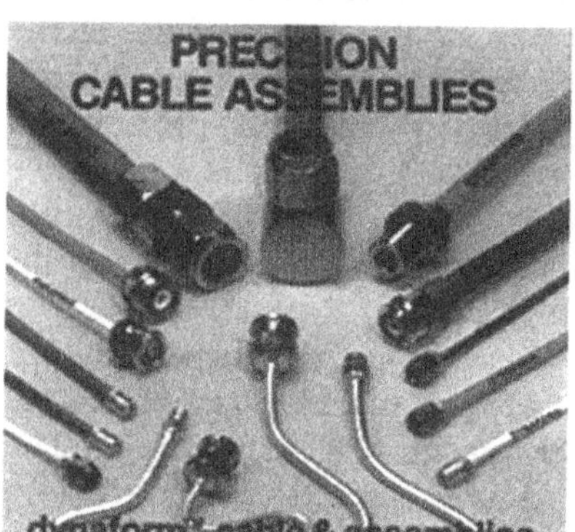

Figure 11.18 Types of coaxial cable assemblies [114]. (Courtesy of Dynawave, Georgetown, Massachusetts.)

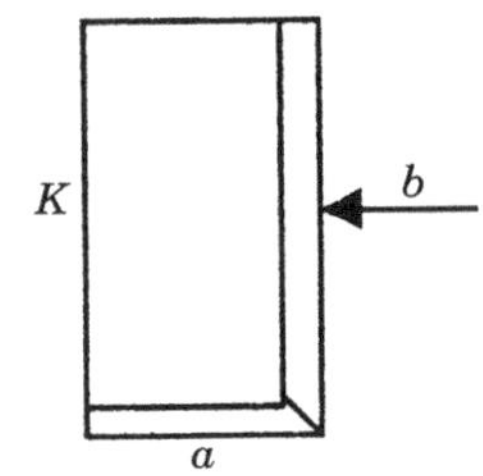

Figure 11.19 Outline of a rectangular ceramic waveguide.

The impedance of a ceramic (alumina) rectangular waveguide at the X-band, is about 150 Ω if sealed down in dimensions by $1/K$, where K is the dielectric constant. Theoretically, the loss in CWG is greater than the air waveguide (AWG), for two reasons: dielectric loss and due to an increase in current density at the metal dielectric sheath, which is considered to be a major loss in CWG. Increasing dimensions can minimize the conduction loss in CWG. The losses measured on an X-band alumina waveguide for $a = 0.29$ in. (0.74 cm) and $b = 0.128$ in. (0.325 cm) (Figure 11.19), with surface finishes of 10 to 25 in. (25 to 630 cm) and with sputtered chrome gold which are typically on unlapped ceramics, are 0.05 dB/in. (0.2 dB/cm). The power handling capacity of the waveguide is proportional to the cross-sectional area, divided by the square root of the dielectric constant and the dielectric breakdown values of the ceramics. Thus compared to AWG, CWG will be able to handle $l/\sqrt{K}$ times the peak power-handling capacity of AWG. However, this peak power-handling capacity in CWG can be reduced by surface roughness. On the other hand, CWG provides a reliable solution to the power-handling problem at high altitudes, since the dielectric constant of the transmission line remains the same at all altitudes. Figure 11.19 is the outline of a rectangular ceramic waveguide [80].

Circuit Design The first ceramic waveguide as a test circuit was developed at the X-band and included a slotted line, couplers, tapers, loads, transitions, and circulators. These circuits were used later in filter, oscillator, and amplifier circuits. The CWG circulator design is done following the design of the ferrite core waveguide, as the dielectric constants (K) for the ferrite and ceramic are similar. The preliminary design with a single $\frac{1}{4}\lambda$ step provides a 20-dB isolation bandwidth of 23% and 0.4-dB insertion losses. For X-band it is a two-pole 0.3-dB-loss 10% band filter, and for C-band it is a six-pole 4% bandwidth, with a 0.8-dB loss filter. These filters show three- to fourfold better results than those of microstrip.

Both Gunn and avalanche oscillators in ceramic cavities have provided the same power capability as in standard guides. Varactor coax oscillators (VCOs) using varactor tuning have given a minimum 10% tuning band. The problems of the CWG (made out of alumina, titania, silicates, and quartz) are (1) dimensional changes for most of CWGs with temperature and (2) frequency

TABLE 11.10 Comparative Studies between Microstrip and CWG

	Microstrip	CWG
Power handling	Tens of watts	Kilowatts
Two-port circuit analysis	No	Yes
Circuit-to-circuit isolation	40 dB	100 dB
Loss	0.2 dB/in. (0.008 dB/cm)	0.02 dB/in. (0.0008 dB/cm)
Q values	300	2500
Interconnection	Difficult	Standard techniques
Volume of active module	4 in^3 (11 cm^3)	$2\frac{1}{2}$ in^3 (6 cm^3)
Weight of active module	6 oz (170 g)	2 oz (57 g)
Metal adherence	Difficult	Simple and contiguous
Machining	Diamond operations	Performed
Breakage (size limitations)	1- to 2-in. lengths (3 to 6 cm)	4- to 5-in. lengths (11 to 13 cm)
Hermetic sealing of unpacked devices	Very difficult	Relatively simple
Heat sinking	Difficult and often adds weight penalty	Can be designed
Module costs in protection	Medium to high	Should be low

Source: From [117]. Reprinted with permission from ISHM/IMAPS, Washington, DC.

drifts of filter circuits with temperature. The temperature (T) dependence frequency (f) of a CWG filter is given by [116]

$$\frac{1}{f}\frac{df}{dT} = \frac{\alpha_T}{2} + \alpha_l \qquad (11.17)$$

where $\alpha_l = 1/l(dl/dT)$ is the linear expansion coefficient (where l is the length of the CWG filter) and $\alpha_T = 1/K(dK/dT)$ is the dielectric temperature coefficient (where K is the dielectric constant). For an alumina waveguide, α_l is less than 100 ppm/°C and α_T is 150 ppm/°C. Thus a filter must be designed to compensate ceramics which require that $\frac{1}{2}\alpha_T = -\alpha_L$. A comparative analysis of the characteristics of microstrip [117,118] and CWG are given in Table 11.10.

11.2.3 Transmission Lines

Transmission lines are distributed parameter networks where voltages and currents can vary in magnitude and phase over the entire length. They are used primarily to distribute microwave power from one point to another and thus be considered as fundamental microwave components [80]. The length of the entire line may be a considerable fraction of a wavelength or many wavelengths. It should always have at least two conductors, which are often represented schematically as a two-wire line. The R, L, G, and C elements of the conductors are defined as series resistance per unit length (Ω/m), inductance per

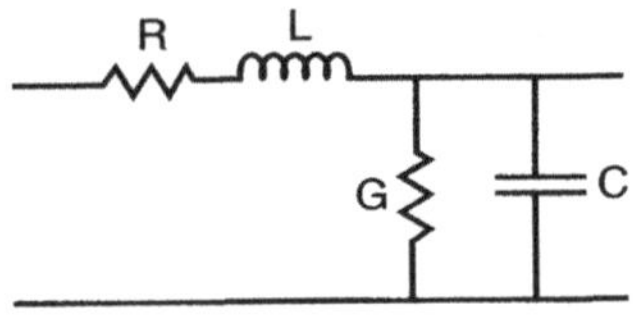

Figure 11.20 Electrical model of a uniform transmission line [124]. (Reprinted by permission of Pearson Education, Inc.)

unit length (H/m), shunt conductance per unit length (S/m), and shunt capacitance per unit length (F/m) (Figure 11.20). The characteristic impedance Z_0 can be defined as [120–123]

$$Z_0 = \frac{R + j\omega L}{\gamma} = \sqrt{\frac{R + J\omega L}{G + j\omega C}} \tag{11.18}$$

where the complex propagation constant $\gamma = \alpha + j\beta = \sqrt{(R + j\omega L)(G + j\omega C)}$ and β is the propagation constant.

$$\text{phase velocity } v_p = \frac{\omega}{\beta} \tag{11.19}$$

$$\text{wavelength } \lambda = \frac{2\pi}{\beta} \tag{11.20}$$

Currents into a long line cannot be increased indefinitely because of the surface loss and thus introduce a limit to the traveling voltage/current waveform. The current can be written as [124]

$$I = \frac{\Delta Q}{\Delta t} = \frac{\Delta CV}{\Delta t} \tag{11.21}$$

where Q is the charge, t the time in seconds, C the capacitance, and V the applied voltage. The propagation speed v is equal to

$$v = \frac{1}{(LC)^{1/2}} = \frac{c}{(\varepsilon_r \mu_r)^{1/2}} \tag{11.22}$$

where c is the speed of light and ε_r and μ_r are the permittivity and permeability of the propagation medium. Figure 11.21 shows the cross section of a typical transmission line structure.

If the total resistance of the line is greater or equal to its characteristic impedance, the wire will behave like a distributed RC line. As the operating frequency increases, the transmission line behavior becomes important, especially when the transmission line length is very long. Transmission line behav-

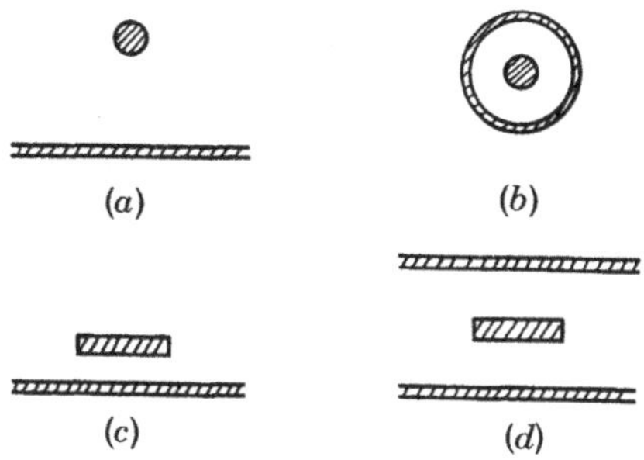

Figure 11.21 Cross section of a typical transmission line: (*a*) wire above ground; (*b*) coaxial line; (*c*) triplate strip line; (*d*) microstrip [124,125]. (Reprinted by permission of Pearson Education, Inc.)

ior is very visible when the time of flight delay along the line is longer than the rise time of the signal. The other factors that affect a transmission line are reflections, which are caused by discontinuities at the transmission line cross-talk, which are often caused due to mutual capacitance and inductance between the neighboring lines and simultaneous switching noise [123,124]. To minimize line impedance the wires should be shorter and closer to the ground plane. It has been observed that impedance offered by a dual inline package is highest (3 to 50 nH), followed by tape automated bonding (91 to 10 nH) and chip flip (0.25 to 1 nH) [123].

Strip Line Strip line [80,120] consists of two conductors with a homogeneous dielectric in between. It is a planar type of transmission line and is fabricated following photolithography and etching technology of the microwave ICs [126]. Higher-order TM and TE modes are usually avoided in practice with strip line circuits. The main difficulty with a strip line is that it does not tend itself to a simple analysis as did transmission lines and waveguides. In a strip line [127] a thin conduction strip, usually gold or copper of width w, is inserted in between two ground planes separated by a distance d, which is filled with a dielectric, either alumina (K of about 9) or other materials. The phase velocity of a TEM mode is given by

$$v_p = \frac{1}{\sqrt{\mu_0 \varepsilon_r \varepsilon_0}} = \frac{c}{\sqrt{\varepsilon_r}} \tag{11.23}$$

and the propagation constant of the strip line is

$$\beta = \frac{\omega}{v_p} = \sqrt{\varepsilon k_0} = \omega\sqrt{\mu_0 \varepsilon_0 \varepsilon_r} \tag{11.24}$$

where μ_0 is the permeability of free space ($= 4\pi \times 10^{-7}$ H/m), ε_r the permittivity of the strip line, ε_0 the permittivity of free space ($= 8.854 \times 10^{-12}$ F/m), c the velocity of light (3×10^8 m/s), and $\omega = 2\pi f$ (f is the frequency of the microwave).

Figure 11.22 Strip line circuit assembly with two hybrid rings [80].

Figure 11.22 shows a strip line circuit with two hybrid rings and a coupled line bandpass filter. One of the two hybrid rings is used as a power divider and the other is used in a balanced mixer [80]. The impedance Z_0 is given by [80]

$$Z_0 = \sqrt{\frac{L}{C}} = \sqrt{\frac{LC}{C}} = \frac{1}{v_p C} \tag{11.25}$$

where L and C are the inductance and capacitance per unit length and v_p is the phase velocity.

Microstrip Line Microstrip line [80,125,128] is a popular and commonly used type of transmission line. It can be fabricated by using the photolithography and etching technologies used for ICs and can easily be integrated with other passive and active components of microwave circuits [129–131]. A conductor, usually gold, is printed on a grounded dielectric substrate of thickness d and relative permittivity ε_r by standard thick-film screen-printing techniques or by selective etching of thin film deposited either by vacuum evaporation or sputtering. If the dielectric (K) is not present, we can take the microstrip line as a two-wire line consisting of two flat strip conductors separated by a distance $2d$. Figure 11.23 shows the geometry together with the electric and magnetic field lines through a microstrip transmission line.

The phase velocity v_p and the propagation constant β for microstrip lines can be expressed as $v_p = c$ and $\beta = k_0$ for simple TEM transmission when the dielectric is not present. In actuality, the exact fields of a microstrip line constitute a hybrid TM–TE wave; and the phase velocity is given by

$$v_p = \frac{c}{\sqrt{K_e}} \quad \text{and} \quad \beta = k_0\sqrt{K_e} \tag{11.26}$$

where K_e is the effective dielectric constant $(1 < K_e < K_r)$ and dependent on substrate thickness d and conductor width w. $k_0 = 2\pi f/c$, where c is the veloc-

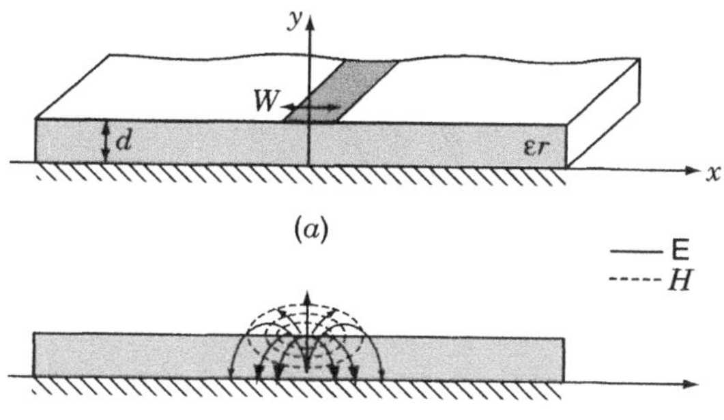

Figure 11.23 Microstrip transmission line: (*a*) geometry; (*b*) electric and magnetic flow lines [80].

ity of light. The impedance Z_0 of the microstrip line can be written as

$$Z_0 = \begin{cases} \dfrac{60}{\sqrt{K_e}} \ln\left(\dfrac{8d}{w} + \dfrac{w}{4d}\right) & \text{for } w/d \le 1 \quad (11.27) \\ \dfrac{120\pi}{\sqrt{K_e}\left[\dfrac{w}{d} + 1.393 + 0.667 \ln\left(\dfrac{w}{d} + 1.444\right)\right]} & \text{for } w/d \ge 1 \quad (11.28) \end{cases}$$

$$\frac{w}{d} = \begin{cases} \dfrac{8e^A}{e^{2A} - 2} & \text{for } w/d < 2 \quad (11.29) \\ \dfrac{2}{\pi}\left\{B - 1 - \ln(2B - 1) + \dfrac{K - 1}{2K} \right. \\ \qquad \left. \times \left[\ln(B - 1) + 0.39 - \dfrac{0.61}{K}\right]\right\} & \text{for } w/d > 2 \quad (11.30) \end{cases}$$

where d is grounded dielectric substrate thickness.

$$A = \left(\frac{Z_0}{60}\sqrt{\frac{K+1}{2}}\right) + \left[\frac{K-1}{K+1}\left(0.23 + \frac{0.11}{K}\right)\right] \qquad (11.31)$$

$$B = \frac{377\pi}{2Z_0\sqrt{K}} \qquad (11.32)$$

In hybrid microwave circuits small lumped R, L, and C elements are used up to a frequency of 60 GHz. Resistors for these microwave circuits are fabricated either from lossy material such as nichrome, tantalum nitride–doped semiconductor material, or thick-film counterpart. Small values of inductance on the order of nanohenries can be fabricated by either thick- or thin-film

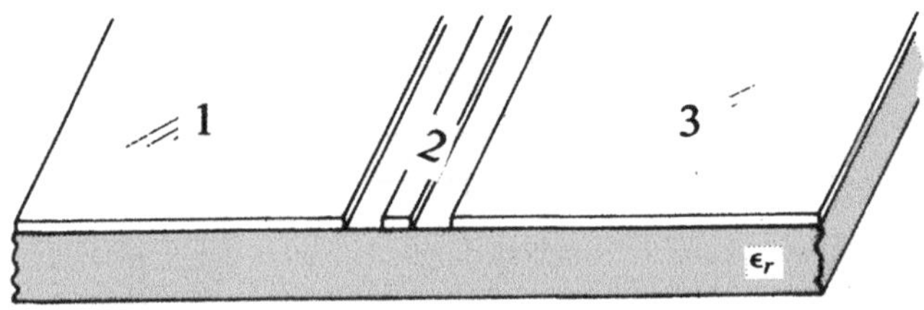

Figure 11.24 Construction of a coplanar waveguide.

technique. Larger inductors (more than 15 to 20 nH) can be realized with large spiral inductor, but the losses will be greater [132]. Moreover, it will add shunt capacitance to the circuit, which will limit the maximum operating frequency.

Hybrid microwave circuits on alumina substrate 25 mils (626 μm) thick has been fabricated with a 23.62-mil (600-μm) line width, 0.2 to 0.23 mil (5 to 6 μm) thick, to obtain 50-Ω impedance [133]. Passive resistors, capacitors, and active package components are added to the circuit by soldering. For metallization, thermal vacuum deposition of chromium followed by gold is performed to promote the adhesion [134]. Electroplated gold on vacuum-deposited gold brought the resistivity of the gold film down further. Integrated airgap crossovers, developed by Western Electric, provided a better design for the long coupler. To connect the background plane, holes are plated with electroless gold, and it is found to work better than evaporated gold [135].

Coplanar Waveguide Planar transmission lines such as microstrips, strip lines, and coplanar waveguides (CPWs; Figure 11.24) have become conventional structures to use in the design of microwave and millimeter-wave circuits due to flexibility provided in fabricating passive components. Although microstrips and strip lines have been utilized most in passive circuits, limitations in mounting devices have made coplanar waveguides more popular [80,136]. Coplanar waveguides can be thought of as a slotline with a third conductor centered in slot region [80]. The presence of this third conductor can support even or odd quasi-TEM modes, depending on whether the E-fields in the two slots lie in opposite or the same direction. This type of waveguide is particularly useful for fabricating active circuitry due to the presence of the center conductor and the close proximity of the ground planes.

The characteristic impedance Z_0 is given by [137]

$$Z_0 = \frac{1}{c} v_{\text{ph}} \tag{11.33}$$

where $c = (\varepsilon_r + 1)\varepsilon_0(2a/b)$, $\varepsilon_{\text{reff}} = [(\varepsilon_r + 1)/2]$, and $v_{\text{ph}} = [(2/\varepsilon r + 1)]^{1/2} v_a$ ($v_a = 3 \times 10^8$ m/s). Both microstrip and CPW have been used to feed log periodic arrays of dual-feed-patch radiators [80,138], which have a cardioid radiation pattern. Most important of all is the fact that the impedance remains almost constant from very low frequency up through the first resonance.

TABLE 11.11 Qualitative Comparison of Propagation Mode

Characteristic	Microstrip	Coplanar Waveguide	Coplanar Strips	Slot Line
Attenuation loss	Low	Medium	Medium	High
Dispersion	Low	Medium	Medium	High
Impedance range (Ω)	10–100	25–125[a]	40–250[a]	High
Connect shunt elements	Difficult	Easy	Easy	Easy
Connect series elements	Easy	Easy	Easy	Difficult

Source: From [126].

[a] Infinitely thick substrate.

Slot Line Slot line is almost as popular as microstrip line. The construction of a slot line is very similar to coplanar except for the third centered conductor [80,127,139]. Like microstrip, the two conductors of slot line lead to a quasi-TEM type of mode. Table 11.11 gives a qualitative comparison of propagation mode in microstrip, CPW, coplanar strip, and slot line. The attenuation loss is the lowest in microstrip and highest in slot line. Load impedance is complex in slot line.

A typical waveguide slotted transmission line is shown in Figure 11.25. The printed circuit slot line or notch antenna is an attractive choice for phased arrays and multibeam antennas because it is simple to fabricate [140,141]. Two types of slot lines, unilateral and bilateral, are very common [80,142]. The characteristic impedance of a slot line depends on the geometry of the slot line, such as height and width of substrate material and on the wavelength [143]. For frequency up to 20 GHz, a coaxial cable is preferable to a waveguide as a feeding transmission line, whereas above 20 GHz, waveguide has the advantage since it minimizes the attenuation. The most common methods of feeding end fine strip line antennas in microwave frequencies are coaxial cable, microstrip, or strip line. From the natural geometry, a unilateral slot line antenna uses a coaxial cable or microstrip feed, while bilateral slot line uses a strip line feed [143].

11.2.4 Lumped Circuit Elements [144]

Lumped microwave circuit elements can be used as high as 12 GHz. Smaller elements with high Q values are more difficult to fabricate. A lumped-element

Figure 11.25 X-band waveguide slotted line [80].

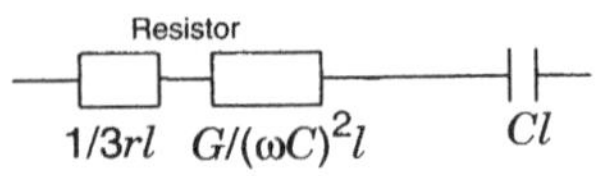

Figure 11.26 Equivalent-circuit elements of a lumped circuit.

resistor may be represented by assuming that $r \gg \omega L$, G tends to zero, $Z_L = 0$, and taking the limit l tends to zero. The equivalent-circuit element is shown in Figure 11.26 and the driving point impedance

$$Z_{\text{in}} = \frac{rl}{1 + j\omega rCl^2/3} \tag{11.34}$$

and Q for the lumped-element capacitor is given by

$$\frac{1}{Q} = \frac{1}{Q_c} + \frac{1}{Q_d} \tag{11.35}$$

where $Q_c = 3/\omega Clr^2$ and $Q_d = \omega C/G$.

11.2.5 Directional Couplers

Directional couplers are a four-port network and provide a simple, low-cost method of sampling and measuring power flow in an RF/microwave transmission path [145–147]. Figure 11.27 shows commonly used symbols for directional couplers and power flow conventions. Three quantities are generally used to characterize a directional coupler:

$$\text{Coupling:} \quad C = 10 \log \frac{P_1}{P_3} = -20(\log \beta) \quad \text{dB} \tag{11.36}$$

$$\text{Detectivity:} \quad D = 10 \log \frac{P_3}{P_4} = 20 \log \frac{\beta}{|S_{14}|} \quad \text{dB} \tag{11.37}$$

$$\text{Isolation:} \quad I = 10 \log \frac{P_1}{P_4} = -20 \log|S_{14}| \quad \text{dB} \tag{11.38}$$

The ideal coupler would have infinite directivity and isolation ($S_{14} = 0$). Then both α and β could be determined from the coupling factor C. S is the

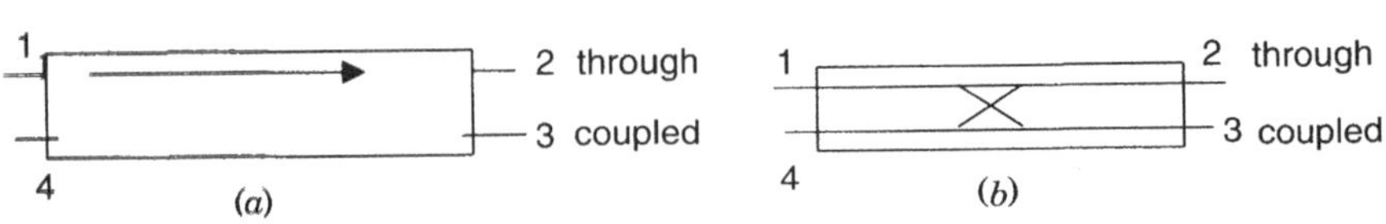

Figure 11.27 Commonly used symbols for directional couplers. 1, input; 2, through; 3, coupled; 4, isolated.

scattering matrix [80,148]. The directivity of a directional coupler is a measure of the coupler's ability to separate forward and backward wave components. So applications of directional couplers often require high (35 dB or greater) directivity. Poor detectivity will limit the accuracy of a reflectometer and will cause variations in the coupled power level from a coupler when there is a mismatch on the through line [110]. Thick-film technology has been used successfully to fabricate a 10-GHz four-port parallel topology directional coupler circuit. Line width with 5-mil (126-μm) spacing is accomplished with silver and palladium silver conductor ink. Insertion loss values as low as 0.5 dB have been observed for a 15-dB parallel topology coupler at 10 GHz.

11.2.6 Impedance Matching

The basic idea of impedance matching is to transfer maximum power to the load from the generating circuit and to minimize phase errors. Impedance matching also improves the signal-to-noise ratio of the system. The impedance-matching network should be ideally without loss, and reflections are to be eliminated from the feeding transmission line network, although there will be multiple reflections between the matching network and the load [149,150].

Designing a good and cost-effective impedance-matching network is an engineering problem. A much simpler, cheaper, more reliable, and less lossy network design is more desirable than a more complex one. Probably the simplest type of matching network is the L section, which uses two reactive elements to match arbitrary load impedance to a transmission line (Figure 11.28). To match an arbitrary complex load to a line of characteristic impedance Z_0, the real part of the input impedance to the matching network must be Z_0, while the imaginary part must be zero. This implies that a general matching network should have at least two degrees of freedom. For an ideal case, the load impedance Z_0, which is equal to the transmission line network impedance Z_0, can be written as [80]

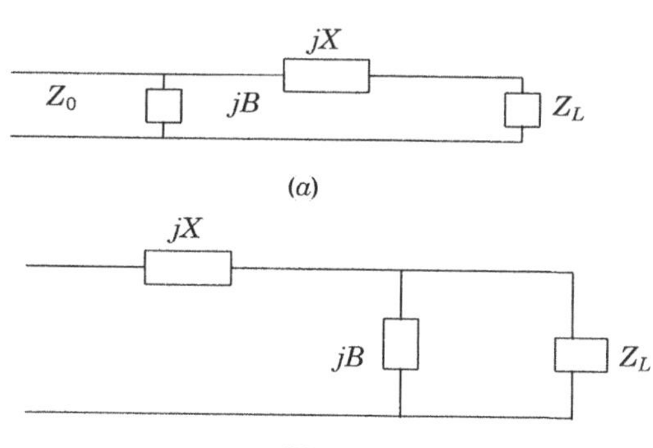

Figure 11.28 L-section matching network [80].

$$Z_0 = jX + \frac{1}{jB + (1/R_L + jX_L)} \tag{11.39}$$

where $R_L = Z_L - jX_L$. Rearranging and separating real and imaginary parts give two equations for two unknowns, X and B:

$$B(XR_L - X_L Z) = R_L - Z_0 \tag{11.40}$$

$$X(1 - BX_L) = BZ_0 R_L - X_L \tag{11.41}$$

Solving (11.40) for X and substituting into (11.41) gives a quadratic equation for B, and the solution for B is

$$B = \frac{X_L \pm \sqrt{R_L/Z_0}\sqrt{R_L^2 + X_L^2 - Z_0 R_L}}{R_L^2 + X_L^2} \tag{11.42}$$

Now $R_L > Z_0$, so the argument of the second square root is always positive. Then the series reactance X will be

$$X = \frac{1}{B} + \frac{X_L Z_0}{R_L} - \frac{Z_0}{BR_L} \tag{11.43}$$

From equation (11.42) it is clear that two solutions are possible for X and B.

Next let us consider the circuit (11.28b). The load impedance $Z_L = R_L + jX_L$ must be equal to

$$\frac{1}{Z_0} = jB + \frac{1}{RL + j(X + X_L)} \tag{11.44}$$

Rearranging and separating real and imaginary parts, we have

$$BZ_0(X + X_L) = Z_0 - R_L \tag{11.45}$$

$$X + X_L = BZ_0 R_L \tag{11.46}$$

Solving for X and B gives

$$X = \pm\sqrt{R_L(Z_0 - R_L)} - X_L \tag{11.47}$$

$$B = \pm\frac{\sqrt{(Z_0 - R_L)/R_L}}{Z_0} \tag{11.48}$$

Since $R_L < Z_0$, the arguments of the square roots will always be positive.

11.2.7 Microwave Integrated Circuits

For the last 20 years, microwave circuit technology has replaced bulky and expensive waveguide and coaxial components with small and inexpensive planar components. It is analogous to the digital circuitry [151]. This trend in hybrid microwave industry has brought revolution, and the result of this enormous effort has created the monolithic microwave integrated circuit (MMIC) [152]. This technology has reached the point where integrating a radar receiver or transmitter module requires a space of a few square millimeters [80].

In the late 1980s, the U.S. government's Defense Advanced Development Agency's (DARPA, now ARPA) massive research and development program called MMIC was very successful in allowing companies to develop cost-effective design and fabrication techniques to make commercial application of the technology possible. As a result, in the 1990s we have seen good progress toward commercial applications and use of MMIC technology, ranging from cell phones, wireless local area networks, direct broadcast satellite, and TV receivers to automotive collision avoidance radar. The conventional approach to microwave circuit design is through either monolithic or hybrid technology. The monolithic technology is called *monolithic microwave integrated circuits* (MMICs), and the hybrid approach is called *hybrid microwave integrated circuits* (MICs) or *discrete microwave integrated circuits*. The choice of MMIC versus the hybrid approach depends on several features (e.g., the types of substrates, passive components, active components, and interconnections used). In MMICs the substrate is either silicon or gallium arsenide, whereas in hybrid MICs the substrate is an insulator. The passive components in MMICs are deposited thin films patterned by photolithography. In MICs, however, the application engineer has a choice between discrete or deposited patterned film components. The key advantage of MMIC is its small size. The LNA, IF amplifiers, signal generation (VCO), signal conversion (mixer), and filtering can be accommodated on a 1 by 2 mm single chip. In contrast to MMIC, MIC lithography is inexpensive, and it is much easier to tune or repair a hybrid MIC circuit after fabrication than an MMIC circuit. As a matter of fact, MMIC is always packaged along with an MIC hybrid to make a practically useful system. As an example, Raytheon's radar module includes several MMICs and MICs interconnected by microstrip lines. The most significant cost-savings process of MMIC is the batch processing approach. Despite this fact, the cost of maintaining an MMIC manufacturing facility is extremely high. The yield in MMIC processes can be defined at several points: (1) the dc yield, which can exceed 85%; (2) the RF yoeld, which depends on the specifications, statistical process control, and the sensitivity of the circuit elements; and (3) the packaging yield, which can be around 55%, depending on the circuit.

MMIC is the latest development of hybrid and thin-film MIC [153,154]. Hybrid MICs have one layer of metallization for conductors and transmission lines, with discrete components (resistors, capacitors, inductors, transistors, and diodes) bonded to the substrate, while the thin-film counterpart is provided with some vacuum-deposited finer components. The selection of substrate material for hybrid MMIC is an important part of the circuit design. The design engineer has to take into considerations the physical, mechanical, and electrical (conductivity, dielectric constant, loss tangent) properties of the substrates, their behavior at elevated temperatures, and the cost of the material. In the computer industry, transistors are used in digital circuits as switches. The speed of a digital circuit with a silicon transistor is slow compared to that of a GaAs transistor because of certain material factors such as electron mobility. However, the speed advantage has not been considered by the computer industry to outweigh the significant economic advantages of much explored, low-cost silicon processing. Thus silicon-based microwave ICs are approaching higher-frequencies applications such as Ku-band direct DBS receivers. However, when higher resistivity (compared to alumina and sapphire) and higher speed are issues, GaAs definitely supersedes Si. Turning our attention to the passive components (except resistors; e.g., transmission lines, inductors, or capacitors, GaAs has much better properties than Si (low or almost no signal loss). GaAs high-electron-mobility transistors (HEMTs) are also very promising devices for ultrahigh speed (10- to 17-ps switching delays) in static RAM devices. Thus HEMT technology has opened the door to new possibilities for ultrahigh-speed large-scale integration and very large scale integration applications required for high-speed computers and signal processing systems. Alumina, with its high dielectric constant (9 to 10) and rigid ceramiclike material, is a good candidate for lower-frequency circuits (up to 10 GHz) [155,156]. Quartz, on the other hand, with its lower dielectric constant (ca. 4), is useful for a higher frequency range (up to 30 GHz). Teflon is soft and flexible, but its heat transfer capability is not as good as that of alumina and quartz. Copper and gold transmission conductor lines are typically used for hybrid MIC.

Two very important features for the consideration of high-frequency circuits are permittivity and permeability. Permittivity relates to electrical fields and indicates how a material becomes polarized under the influence of an electric field. Permeability is a measure of the magnetic field imposed on material. Mobility and saturated velocity also have a strong influence on the maximum frequency at which a microwave transistor can have useful gain. Typical substrate thickness varies from 0.010 to 0.025 in. (0.025 to 0.063 cm). At higher frequencies, substrate thickness must be decreased to prevent radiation loss and spurious effects. Substrate surface roughness also accounts for the loss in dB/cm and is a function of the root-mean-square value of the voltage applied.

A CAD system (either MICAD or CALMA) helps in laying out the circuit once the design has been completed. The metallization mask is first made of rubylith using a step-and-repeat camera and reduced to proper size. Then, using the master, the mask is made on a special glass. The metallized substrate is etched selectively using photolithography. Plated-through or via holes can be made by evaporating a layer of metal inside holes that have been drilled in the substrate. Experimental evidence shows that electroplating is a better way to metallize the vias. Finally, the discrete components are assembled on the substrate following surface-mounting technology [156]. A quasi two-dimensional simulation CAD system can be used for the design and modeling of submicrometer recessed-gate GaAs MESFETs. The model provides recessed-gate field-effect transistors (FETs) with nonuniform doping profiles. The model also accounts for surface depletion, a nonabrupt depletion region interface, substrate conduction, and gate conduction. Two-port MESFET are commercially available that obtain the s parameters of FETs.

MMIC Bandpass Filter Design For bandpass filters, the primary building block is a coupled line section. In general, the theories of coupled lines [157] and microstrip lines [158,159] have been covered previously. The coupled microstrip line section offers the designer a basic filter element that covers both a broad range of values and high Q values on common substrate. In the design of MMICs, all the devices are planar and all passive and active elements are diffused into the top surface of GaAs substrate, on which most MMIC devices are fabricated. Filters are fabricated on a variety of dielectric materials, including alumina [160].

Since the late 1970s, MMICs have been used in a variety of products, from military defense systems to domestic appliances. This progress is due primarily to the development of GaAs material. As reported in [161], the demand for GaAs MMIC devices for millimeter-wave applications is forecast to grow at a compound average annual growth rate of 29%. Transceiver shipments will grow from 1.09 million units in 2000 to over 5 million units in 2005. Automotive radar will be the fastest-growing sector. The promise of GaAs MMICs is so great that Signal Tech. Corp. of Denver, Massachusetts, together with Strategic Alliances, are translating their expertise in the design and manufacture of RF and microwave modules for defense and broadband wireless applications and for commercial 2.5G and 3G cell phones and wireless appliances into their production. GaN HEMT microwave devices and epi wafers are entering the MMIC microwave industry due to the joint efforts of Erec, Inc. in Durham, North Carolina and the Sensors Directorate of Air Force Research Laboratories [162]. GaAs-powered MMICs have enjoyed tremendous cost reductions and improved reliability compared to those of conventional hybrid microwave circuits in C-band phased-array radar. GaAs-powered MMICs can produce 3 W of output power at a 1-dB compression point with 7 dB gain and almost 26% power efficiency over C-band (4 to 6 GHz).

Some applications of common MMIC circuits based on GaAs technology now in heavy use are shown in Figures 11.29 to 11.32. Figure 11.29 shows a MMIC amplifier. Its cost is related to how many circuits can be built on a single chip and on the frequency range of the amplifier. At high frequencies, a transistor will have some intrinsic feedback from output to input. The design engineer must remember that the amplifier should not oscillate during this feedback. The Smith chart is another important aid in designing a microwave amplifier, considering constant-gain circles of constant noise and input/output stability. For better high-frequency performance, the device dimensions should

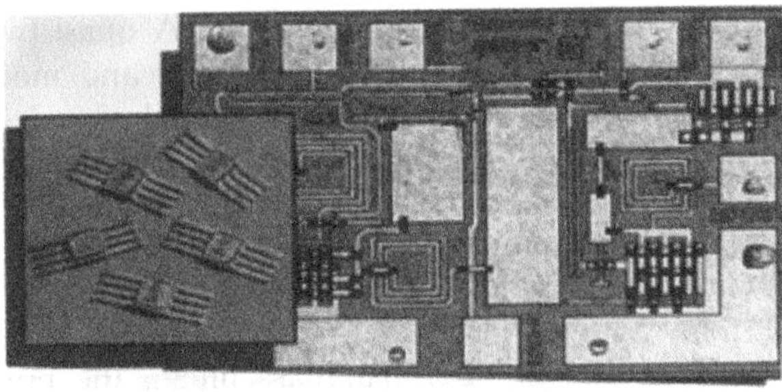

Figure 11.29 Monolithic microwave integrated circuit amplifier [163]. (Courtesy of M/A Com Microelectronics Division, Lowell, Massachusetts.)

be in the submicrometer level. The latest molecular beam epitaxy (MBE) has helped in growing high-quality epitaxial layers with controlled doping profiles in highly localized regions. MBE techniques have led to the development of HEMTs, which can operate at frequencies as high as 100 GHz. A HEMT is a variation of the metal electrode semiconductor field-effect transistor (MESFET) structure, which produces better performance (higher gain and a lower noise figure at a given frequency). Thus HEMT can be described as a recessed-gate FET. FETs are also used in mixers and oscillators with different configurations and biasing arrangements.

Figure 11.30 shows a microwave oscillator, which converts dc power to RF power. A solid-state oscillator circuit uses an active device such as IMPATT or Gun diode or a transistor. For high frequency (ca. 24 GHz), low-noise HEMT is used. Typically, common-source or common-gate FET configurations are used with positive feedback. One of the first design steps is characterization of the s parameter based on source feedback. The load impedance is chosen to match the input impedance. The load resistance R_L is chosen in such that $(R_L + R_i) < 0$, where R_L and R_i are the load resistance and input resistance, respectively.

Figure 11.30 Microwave oscillator (500 MHz to 18 GHz) [164]. (Used with permission of General Microwave, Amityville, New York.)

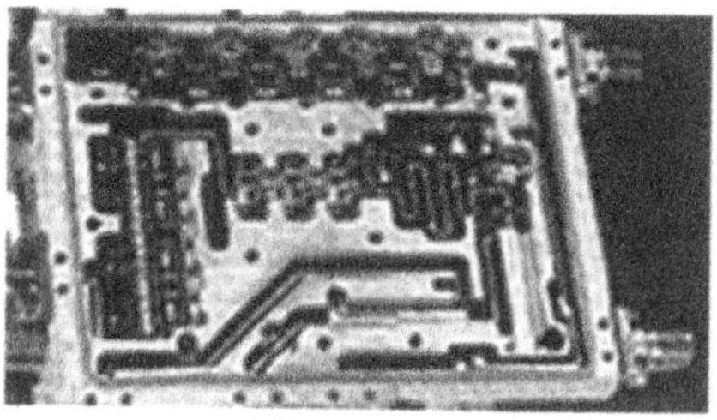

Figure 11.31 MICS custom design (100 MHz to 40 GHz) [164]. (Used with permission of General Microwave, Amityville, New York.)

Figure 11.31 shows the circuit diagram of a hermetically sealed MIC amplifiers from General Microwave. This hybrid microwave circuit is designed to work from 100 MHz to 40 GHz. Figure 11.32 shows the antennas on an Earth survey aircraft equipped with multifrequency synthetic aperture radar opening at wavelengths of 0.04, 0.23, 0.68, and 2.3 m.

11.2.8 Dielectric Resonators

A microwave dielectric resonator [166] is a small disk or cube of low loss made of high-dielectric-constant material, with a dielectric constant between 10 and 100. The high dielectric constant of the resonator ensures that most of the field is confined within the dielectric, but unlike metallic cavities, there is some field fringing or leakage from the sides and ends of the dielectric resonator. Conductor losses are absent, but dielectric loss increases with dielectric constant [80]. This resonator can achieve high Q values—up to several thousands. Dielectric resonators are particularly important in oscillator design. Typically, circuits realized using thick- and thin-film materials show low Q values, phase noise, and poor temperature stability [167]. Recent developments on high-Q materials have extended its commercial use above 20-GHz regions. Some of the properties of the substrates used in hybrid microwave circuits are tabulated in Table 11.12. To decrease size, the active elements are integrated inside the resonator, and low-loss, high-permittivity ceramic disks are preferred [168]. It has been observed that dielectric ring resonators offer better performance than that of a cylindrical resonator [169].

Figure 11.32 Antennas on Earth survey aircraft (Moscow air show) [165]. (Reprinted with permission from *Microwave Journal*, May 1994, p. 24.)

TABLE 11.12 Physical Properties of the Substrates Used in Hybrid Microwave Circuits

Material	Surface Roughness (μin)	Dielectric const.	Tan δ 10^{-4} (1 MHz)	Thermal Conductivity W/m·°K	Thermal Expansion ppm/°C	Resistivity Ω·cm
Sapphire	≪1	9.5–11.8	10–11	35–45	7.9–8.5	$>10^{14}$
Alumina 99.5%	<1	9–10	3.0–15	12.5–25.0	6.9–7.1	10^{13}–10^{15}
Beryllia 99.5%	1–4	6.7	3.0	146	8.0	10^{14}
Silicon (Si)	≪1	11.5–11.7		1.45–1.50	2.5–2.8	1.1×10^{15}
GaAs	<1	12.5–12.9		0.7–0.8	5.5–5.9	2.8×10^{7}
Fused Silica	≪1	3.5–3.8	<1	1	0.5	$>10^{16}$
PTFE (polytetrafluoroethylene)		2.2–2.6	1–7	0.1–2.5	15–88	$>10^{18}$

REFERENCES

1. D. A. Doane and P. D. Franzon, Introduction on overall properties for MCMs market, in *Multichip Module Technology and Alternatives: The Basics*, Van Nostrand Reinhold, New York, 1993, Chap. 1.
2. P. E. Garrou, Technical drivers, in *Multichip Module Technology Handbook*, McGraw-Hill, New York, 1998, Chap. 1.
3. A. J. Blodgett, A multichip ceramic multichip module, *IEEE Trans. Comp. Hybrid Manuf. Technol.*, Vol. 3, 1980, pp. 634–637.
4. C. P. Wong, J. M. Segelkon, and C. N. Robinson, Chip-on-board encapsulation, in *Chip-on-Board Technology for MCM*, J. H. Lau, ed., Van Nostrand, Reinhold, New York, 1994.
5. J. E. Sergent, The hybrid microelectronics and MCM technology, in *Electronic Packaging and Interconnection Handbook*, 3rd ed., C. A. Harper, ed., McGraw Hill, New York, 2000.
6. K. O'Brian and S. Patser, Establishing an MCM infrastructure, *Adv. Packag.*, May/June 1994, pp. 34–37.
7. M. Andrews, Building MCM infrastructure: a review of the issues and processes, *Proc. Int. Symp. Microelectron.*, Mar./Apr. 1993.
8. R. Miracky et al., MCM-D prototype of the super-SPARC chip set in two weeks, *ICEM Proc.*, 1994.
9. In the News, *Adv. Packag.*, Sept./Oct. 1995.
10. T. M. Mazzullo, Mixed signal MCMs: solving analog and digital circuit problems, *Adv. Packag.*, May/June 1994, p. 18.

11. P. A. Sandborn and H. Moreno, *Conceptual Design of Multichip Modules and Systems*, Kluwer Academic, New York, 1994.
12. J. E. Butler et al., Thin film diamond growth mechanism, *Philos. Trans. R. Soc. London*, Vol. A342, 1993, pp. 209–222.
13. A. P. Malshe, S. Jamil, M. H. Gordon, H. A. Naseem, W. D. Brown, and L. W. Schaper, Diamond for MCMs, *Adv. Packag.*, Sept./Oct. 1995, p. 50.
14. W. Kerry and V. Beanu, Chemical vapor deposition of inorganic films, in *Thin-Film Processes*, J. L. Vossen and W. Kern, eds., Academic Press, San Diego, CA, 1978, pp. 251–331.
15. K. E. Spear and D. P. Dismukes, eds., *Synthetic Diamond*, Wiley, New York, 1994; R. C. Eden, Application of diamond substrates for advanced high-density packaging, *Diamond Relat. Mater.*, Vol. 2, 1993, pp. 1051–1058.
16. R. W. Johnson, R. F. K. Teng, and J. W. Blade, Cofired ceramic technology, Part III, in *MCM System Advantages: Major Construction and Material Technology*, IEEE Press, Piscataway, NJ, 1991.
17. J. J. Licari, Thin and thick film technology, in *Electronic Processing Handbook*, C. A. Harper and R. N. Sampson, eds., McGraw-Hill, New York, 1994, p. 8.20.
18. H. B. Bokoglu, Packaging technologies, in *Circuits, Interconnections, and Packaging for VLSI*, Addison-Wesley, Reading, MA, 1990, Chap. 3.
19. R. A. Filton, A forecast on the future of hybrid wafer scale integration and technology, *IEEE Trans. Comp. Hybrid Manuf. Technol.*, Vol. 16, No. 7, Nov. 1993, pp. 615–625.
20. W. Stallings, Transmission impairments, in *Data Computer Communication*, Macmillan, New York, 1991, Chap. 2.
21. S. K. Mitra and J. F. Kaiser, eds., *Handbook of Digital Signal Processing*, Wiley, New York, 1993, Chaps. 1 and 14.
22. J. Nelson and R. Rhodes, The development of Unisys multichip modules, in *Multichip Module Technologies and Alternatives: The Basics*, Van Nostrand, New York, 1992.
23. F. W. Kear, Overview: design and development of hybrid assemblies, in *Hybrid Assemblies and Multichip Modules*, Marcel Dekker, New York, 1993.
24. G. Gengel, A process for manufacture of cost-competitive MCM substrates, *Int. J. Microcircuit Electron. Packag.*, Vol. 17, No. 4, 1994.
25. J. Blade, Crisis in technology: the questionable U.S. ability to manufacture thin-film multichip modules, *Proc. IEEE*, Vol. 80, No. 12, Dec. 1992, pp. 1995–2002.
26. G. Messner, I. Turlik, J. Blade, and P. Garrou, *Thin-Film Multichip Module*, ISHM, Reston, VA, 1992.
27. A. Blodgett, A multilayer ceramic multichip module, *IEEE Trans. Comp. Hybrid Manuf. Technol.*, Vol. 3, No. 4, 1980, pp. 634–637.
28. C. F. Murphy, Known good die selection tradeoffs: A cost model, *Int. J. Microcircuits Electron. Packag.*, Vol. 17, No. 4, 1994, p. 312.
29. B. S. Landman, On a pin vs. block relationship for partition of logic graphs, *IEEE Trans. Comp. Hybrid Manuf. Technol.*, Vol. 20, No. 12, Dec. 1971, pp. 1469–1479.
30. R. Amerson and P. Kuakes, The design of extremely large MCM-C, *Int. J. Microcircuit Electron. Packag.*, Vol. 17, No. 14, 1994, p. 377.

31. J. E. Sergent, The hybrid microelectronics and multichip module, in *Electronic Packaging and Interconnection Handbook*, 3rd ed., C. A. Harper, ed., McGraw-Hill, New York, 2000; T. J. Moravec, R. C. Eden, and D. A. Schaefer, The use of diamond substrates for implanting 3-D MCMs, *Proc. Int. Conf. MCMs*, Apr. 1992, pp. 85–91; W. L. Mineham, W. K. Weidner, R. J. Jensen, R. K. Spielberger, and W. F. Jacobson, Fabrication assembly, and characterization of stacked MCMs using hot pressed co-fired AlN, *Int. J. Microcircuits Electron. Packag.*, Vol. 17, No. 4, 1994, pp. 369–375.
32. J. V. Bushand and J. P. Dismukes, Diamond thin films, in *Synthetic Diamond: Emerging CVD Science & Technology*, K. E. Spear and J. P. Dismukes, eds., Wiley, New York, 1991, p. 581.
33. A. B. Harker, Diamond films: hard materials to beat, *Res. Dev.*, Mar. 1990, p. 84.
34. B. A. Fox, Diamond films, in *Thin Film Technology Handbook*, A. E. Riad and F. D. Barlow III, eds., McGraw-Hill, New York, 1998, Chap. 7.
35. W. Banholzer and C. L. Sipro, *Diamond Films Technol.*, Vol. 1, 1991, p. 115; P. W. May, Diamond thin films: a 21st century material, *Phil. Trans. R. Soc. Lond. A*, Vol. 358, pp. 473–495, 2000.
36. P. Lal and S. Bhagath, An overview of multichip modules, *Solid State Technol.*, Sept. 1993, p. 65.
37. S. P. Murarka, I. V. Verner, and R. J. Gitman, *Copper: The Fundamental Mechanisms for Microelectronic Applications*, Wiley, New York, 2000.
38. J. P. Herkin, Manufacturing MCM-L, *Adv. Packag.*, Sept./Oct. 1995, p. 62.
39. J. C. Blakenhorn, *BGA, COB, and Flip Chip PCB Design*, IMAPS, Washington, DC, 1994.
40. A. Okuno, K. Nagai, Y. Tsukasaki, N. Oyama, K. Nakahir, and T. Hashimoto, New packaging of a chip on a board by a unique printing method, *IEEE Trans. Comp. Hybrid Manuf. Technol.*, Vol. 15, No. 1, Feb. 1992, pp. 74–77.
41. J. P. Herkins, Manufacturing MCM-L, *Adv. Packag.*, Sept./Oct. 1995, p. 62.
42. G. Gengel, A process for manufacture of cost-competitive MCM substrates, *Microcircuits Electron. Packag.*, Vol. 17, No. 4, 1994, p. 330.
43. R. R. Tummala, Ceramics and glass-ceramic packaging, *J. Am. Ceram. Soc.*, Vol. 74, No. 5, May 1991, pp. 885–908.
44. Y. Shimada, K. Utsumi, M. Suzuki, H. Takamizowa, M. Nitta, and T. Watari, Low firing temperature glass ceramic multilayer substrate, *IEEE Trans. Comp. Hybrid Manuf. Technol.*, Vol. 6, No. 4, Dec. 1983, p. 382.
45. J. Jean and T. K. Gupta, Properties of a cristobalite-free low-K silica dielectric, *Int. J. Microcircuits Electron. Packag.*, Vol. 17, No. 2, 1944, p. 169.
46. L. L. Hench and D. B. Dove, eds., *Physics of Electronic Ceramics*, Part A, Marcel Dekker, New York, 1972.
47. J. F. Maguire, Metals joining of electronic circuits, in *Electronic Materials and Processes Handbook*, C. A. Harper and R. N. Sampson, eds., McGraw-Hill, New York, 1994.
48. J. H. Lau, *Handbook of Tape-Automated Bonding*, Van Nostrand Reinhold, New York, 1992.
49. P. Sandborn, H. Hashemi, and L. Bal, Design of MCMs for insertion into standard surface mount packages, *Proc. Nepcon West*, Feb. 1993, pp. 651–660.

50. J. Lau, A brief introduction to flip chip technologies, in *Flip Chip Technologies*, J. Lau, ed., McGraw-Hill, New York, 1996, Chap. 1.

51. F. D. Barlow III, A. E. Riad, and R. Brown, Film deposition techniques and processes, in *Thin-Film Technology Handbook*, A. E. Riad and F. D. Barlow III, eds., McGraw-Hill, New York, 1998, p. 1-27.

52. N. F. Mott, The theory of the formation of protective oxide films on metals, *Trans. Faraday Soc.*, Vol. 35, 1939, pp. 175–179.

53. J. J. Licari and L. R. Enlow, Fabrication processes (HTCC and LTCC, Secs. 1.6 and 1.7) in *Hybrid Microcircuit Technology Handbook*, Noyes Publications, Park Ridge, NJ, 1998, pp. 116–122.

54. W. K. Jones and G. Harsanyi, MCM-C, in *MCMs with Integrated Sensors*, Kluwer Academic, Boston, 1997, Chap. 2.

55. P. Garrou and I. Turlik, MCM-D thin-film processes and applications, in *Multichip Module Technology Handbook*, McGraw-Hill, New York, 1998, Sec. 3.1.

56. M. I. Maissel and R. Glang, eds., *Handbook of Thin-Film Technology*, McGraw-Hill, New York, 1983.

57. J. J. Licari, Thin and thick films, in *Electronic Materials and Processes Handbook*, C. A. Harper and R. N. Sampson, eds., McGraw-Hill, New York, 1994, Chap. 8.

58. B. Rogers, M. J. Berry, I. Turlik, P. Garrou, and D. Castillow, Soft mask for via patterning in benzocyclobutane, *Int. J. Microcircuits Electron. Packag.*, Vol. 15, No. 3, 1994, p. 210.

59. J. Tajadod, Processing and microwave characterization of multilevel interconnects using bezocyclobutane dielectric, *IEEE Trans. Comp. Hybrid Manuf. Technol.*, Vol. 16, No. 7, 1993, pp. 714–719.

60. M. Martinez, D. Gibson, L. Matthew, T. Distetno, and J. Cofield, Performance comparison of a novel laminated technology and MCM-D, *Int. J. Microcircuits Electron. Packag.*, Vol. 17, No. 4, 1994, p. 345.

61. B. K. Gilbert, W. I. McNeff, P. J. Zabinski, W. P. Kornrumpf, S. F. Ted, R. L. Maki, K. E. Carlson, and J. A. Gerber, Development of deposited multichip modules with unique features for application in GaAs processors operating above 1-GHz clock rates, *Int. J. Microcircuits Electron. Packag.*, Vol. 15, No. 3, 1992, pp. 144–159.

62. J. C. Demmin, An industry perspective on MCM-D, in *Advance Electronic Packaging with Emphasis on MCM*, W. D. Brown, ed., IEEE Press, Piscataway, NJ, 1999, p. 566.

63. M. Pecht, *Integrated Circuits, Hybrids, and Multichip Modules: Packaging and Design Guidelines*, Wiley, New York, 1994.

64. T. D. Dudderar, Y. Degani, J. G. Spadafora, K. L. Tai, and R. C. Frye, AT&T surface μ-mount assembly: a new technology for the large-volume fabrication of cost-effective flip chip MCMs, *Int. J. Microcircuit Electron. Packag.*, Vol. 17, No. 4, 1994, p. 361.

65. B. Lux and R. Haubner, in *Diamond Films and Coatings*, R. F. Davis, ed., Noyes Publications, Park Ridge, NJ, 1990; T. Moravec and A. Partha, Diamond takes the heat, *Adv. Packag.*, Special issue, Oct. 1993.

66. H. Konda, R. C. Mason, C. Okabe, J. D. Smith, and R. Velasquez, Buried resistors and capacitors for multilayer hybrids, *Proc. Int. Symp. Microelectron.*, 1994.

67. D. Keezer, Yield, testing and reliability, in *Thin-Film Technology Handbook*, A. E. Riad and F. D. Barlow III, eds., McGraw-Hill, New York, 1998, Chap. 11.
68. P. B. Matthews, *Adv. Microelectron.*, Vol. 22, No. 2, 1995, p. 16.
69. G. Lievan and P. Ludlow, The MCM consortium, *Adv. Electron.*, Vol. 22, No. 2, 1995, p. 20.
70. L. Lemke, Recent developments in the production and applications of advanced laminated-base MCM and chip carrier, *Adv. Electron.*, Vol. 22, No. 2, 1995, p. 28; J. M. Segelken, L. Wu, M. Y. Lau, K. L. Tai, R. R. Shively, and T. G. Garau, Ultra dense: an MCM-based 3-D digital signal processor, *IEEE Trans. Comp. Hybrid Manuf. Technol.*, Vol. 15, No. 4, Aug. 1992, pp. 438–443.
71. S. Roberts, The role of flexible materials in MCM substrate, *Proc. IPC Natl. Conf. MCM-L*, Minneapolis, MN, 1967.
72. D. Carey, H. Hashemi, and P. Hunter, MCM-L technologies and myths and experiences in multiple production scenarios, *Int. J. Microcircuits Electron. Packag.*, Vol. 17, No. 4, 1994, p. 336.
73. J. K. Hagge, State-of-the-art multichip modules for avionics, *IEEE Trans. Comp. Hybrid Manuf. Technol.*, Vol. 15, No. 1, 1992, pp. 29–35.
74. P. Garrou, Thin-film packaging and interconnections, in *Thin-Film Technology Handbook*, A. E. Riad and F. D. Barlow III, eds., McGraw-Hill, New York, 1991, pp. 9-46 to 9-51.
75. R. Brown and R. Boone, A comparison of MCM packaging teaching for 32-bit high performance processor, *Proc. ISHM*, 1993, p. 562.
76. R. E. Terril et al., Alladdin: lesson learned, *Proc. MCM Conf.*, Denver, 1995, p. 7.
77. W. Jackobsen et al., Application of MCM-C and MCM-D for space-born data processors, *Proc. Int. Conf. MCMS*, Denver, 1993, p. 525.
78. D. Tuckerman et al., A high-performance second-generation space MCM, *Proc. Int. Conf. MCM*, Denver, 1994, p. 314.
79. Y. Doi et al., An ATM switch hardware technology using multichip packaging, *IEEE Trans. Comp. Hybrid Manuf. Technol.*, Vol. 16, 1993, p. 60.
80. D. M. Pozar, *Microwave Engineering*, Addison-Wesley, Reading, MA, 1990.
81. J. C. Maxwell, *A Treatise on Electricity and Magnetism*, Dover Publications, Mineola, NY, 1954.
82. H. Hertz and D. E. Jones, trans., Macmillan, New York, 1893, and Dover Publications, Mineola, NY, 1962.
83. R. Goel, ed., *Monolithic Microwave Integrated Circuit Technology*, Artech House, Norwood, MA, 1989.
84. C. A. Mead, Schottkey barrier gate field transistors, *Proc. IEEE*, Vol. 54, 1966, pp. 307–308.
85. D. N. McQuiddy, Jr., J. W. Wassel, J. B. Lagrange, and W. R. Wisseman, Monolithic microwave integrated circuits: a historical perspective, *IEEE Trans. Microwave Theory Tech.*, Vol. 32, Sept. 1984, pp. 997–1008.
86. R. Brown, Thin-film microwave hybrids, in *Thin-Film Technology Handbook*, A. E. Riad and F. D. Barlow III, eds., McGraw-Hill, New York, 1997, Chap. 10.
87. EEV Microwave Products Advertisement, *Microwave J.*, Vol. 36, No. 7, July 1993.

88. ST Microwave Corporation Advertisement, *Microwave J.*, Vol. 36, No. 7, July 1993.

89. A. Wilson, A digital radio transmitter tester, *Microwave J.*, Vol. 37, No. 3, Mar. 1994, p. 118.

90. F. G. Stremler, *Introduction to Communication Systems*, Addison-Wesley, Reading, MA, 1977.

91. W. C. Brown, The history of power transmission by radio waves, *IEEE Trans. Microwave Theory Tech.*, Vol. 32, Sept. 1984, pp. 1230–1242.

92. C. Mahle, High-end microwave technologies in global communications, *Microwave J.*, Vol. 36, No. 12, Dec. 1993, pp. 22–30.

93. D. M. Pozar, Introduction to microwave systems, in *Microwave Engineering*, Addison-Wesley, Reading, MA, 1990, Chap. 12.

94. K. Miyouchi, Millimeter-wave communication, in *Infrared and Millimeter Waves*, K. J. Button, ed., Academic Press, San Diego, CA, 1983.

95. M. I. Skolnik, ed., *Radar Handbook*, McGraw-Hill, New York, 1990.

96. J. E. Cunningham, *Cable Television*, Howard W. Sams, Indianapolis, IN, 1980.

97. A. K. Jain, *Fundamentals of Digital Image Processing*, McGraw-Hill, New York, 1990.

98. F. C. Okress, *Microwave Power Engineering*, Academic Press, San Diego, CA, 1968.

99. W. F. Richardson, Microstrip antennas, in *Antenna Handbook*, Y. T. Lee and S. W. Lee, eds., Van Nostrand Reinhold, New York, 1988, Chap. 10.

100. F. T. Ulaby, R. K. Moore, and A. K. Fung, *Microwave Remote Sensing*, Addison-Wesley, Reading, MA, 1981.

101. S. G. Konsowski, Packaging of high-speed and microwave electronic systems, in *Electronic Packaging and Interconnection Handbook*, 3rd ed., C. Harper, ed., McGraw-Hill, New York, 2000, Chap. 12.

102. K. M. Tomera and D. K. Hellerstein, Transurethral microwave thermotherapy, *Microwave J.*, Vol. 35, No. 11, Nov. 1992, p. 24.

103. C. Makihara, K. Ikeda, H. Wada, Y. Sawa, and S. Tanahasi, Multifunctional ceramic substrate and packages for telecommunication application, *Proc. ISHM*, 1994, pp. 243–247.

104. R. E. Sigliano and A. Issacson, The need for speed, *Adv. Packag.*, Nov./Dec. 1995, p. 22.

105. B. Tareev, Dielectric losses, in *Physics of Dielectric Materials*, MiR Publishing, Moscow, 1979, Chap. 3.

106. H. A. Wheeler, Formulas for skin effect, *IRE Proc.*, Vol. 30, Sept 1942, pp. 412–424.

107. L. V. Blake, *Transmission Lines and Waveguides*, Wiley, New York, 1969.

108. H. A. Wheeler, Transmission line properties of parallel strips separated by a dielectric sheet, *IEEE Trans. Microwave Theory Tech.*, Vol. 10, Nov. 1965, pp. 777–781.

109. H. Guckel, P. A. Brennan, and I. A. Palocz, A parallel-plate waveguide approach to microminiaturized planar transmission lines for ICs, *IEEE Trans. Microwave Theory Tech.*, Vol. 15, Aug. 1967, pp. 468–476.

110. J. H. Bryant, Co-axial transmission lines, related two-conductor transmission lines, connectors, and components: a U.S. historical perspective, *IEEE Trans. Microwave Theory Tech.*, Vol. 32, Sept. 1984, pp. 970–983.

111. Y. H. Choung and W. C. Choung, Microwave- and millimeter-wave slotline transmission line design, *Microwave J.*, Vol. 37, No. 3, 1994, p. 71.

112. J. H. Bryant, Coaxial transmission lines, related two-conductor transmission lines, connectors, and components: a U.S. historical perspective, *IEEE Trans. Microwave Theory Tech.*, Vol. 32, Sept. 1984, pp. 970–983.

113. C. P. Hartwig, M. P. Lepie, D. Masse, A. Paladino, and R. A. Parcel, *Proc. Natl. Electron. Conf.*, Vol. 24, 1968, p. 314.

114. Dynawave Advertisement, *Microwave J.*, Vol. 35, No. 11, Nov. 1992.

115. MIT, *Radiation Laboratory Series*, McGraw-Hill, New York, 1948–1950.

116. D. D. Grieg and H. F. Engelmann, Microstrip: a new transmission technique for kilomegacycle range, *Proc. IRE*, Vol. 40, Dec. 1952, pp. 1644–1650.

117. W. H. Form, Ceramic wave guide integrated circuit, *Proc. Int. Symp. Microelectron.*, 1972, p. 3B-5-1.

118. I. J. Bahl and D. K. Trivedi, A designer's guide to microstrip line, *Microwaves*, May 1977, pp. 174–182.

119. R. Mitra and T. Itoh, Analysis of microstrip transmission lines, in *Advances in Microwaves*, Vol. 8, L. Young and H. Sobol, eds., Academic Press, San Diego, CA, 1974.

120. S. Cohn, Problems in strip line transmission lines, *IEEE Trans. Microwave Theory Tech.*, Vol. 3, Mar. 1955, pp. 119–126.

121. W. C. Johnson, *Transmission Lines and Networks*, Wiley, New York, 1950.

122. S. R. Seshadri, *Fundamentals of Transmission Lines and Electromagnetic Fields*, Addison-Wesley, Reading, MA, 1971.

123. R. E. Mattick, *Transmission Lines for Digital and Communication Networks*, McGraw-Hill, New York, 1969.

124. H. B. Bokoglu, Cross talk and power distribution noise, in *Circuits, Interconnections, and Packaging for VLSI*, Addison-Wesley, Reading, MA, 1990, Chap. 7.

125. R. Grag and I. J. Bhal, Charactersics of coupled microstrip lines, *IEEE Trans. Microwave Theory Tech.*, Vol. 27, No. 7, July 1979, pp. 700–705.

126. H. Howe, Jr., I. J. Bahl, and R. Grag, eds., *Strip Line Circuit Design*, Artech House, Norwood, MA, 1974.

127. K. C. Gupta, R. Grag, and I. J. Bahl, eds., *Microstrip Lines and Solutions*, Artech House, Norwood, MA, 1979.

128. G. I. Zysman and D. Varon, Wave propagation in microstrip transmission lines, *Dig. IEEE Symp. MTT*, 1969.

129. T. Cagnin, G. P. Ferraris, M. Pesetti, A. Tersalvi, and M. Ziboni, Further integration in hybrid thin film microwave circuits (HMICs), *Proc. Int. Symp. Microelectron.*, 1987, p. 79.

130. M. V. Schneider, Microstrip lines for microwave integrated circuits, *Bell Syst. Tech. J.*, May/June 1969, pp. 1421–1444.

131. T. C. Edwards, *Foundations of Microstrip Circuit Design*, Wiley, New York, 1984.

132. D. J. Masse and C. P. Hartwig, Losses in microstrip, *IEEE Trans. Microwave Theory Tech.*, Vol. 16, June 1968, pp. 342–350.

133. C. Bassani, G. P. Ferraris, C. Villa, and M. Ziboni, Improved performance of hybrid microwave film circuits using through-hole connections, *Proc. 6th Eur. Microelectron. Conf.*, Bournemouth, Hampshire, England, June 1987.

134. T. K. Gupta, Comparative studies of contact angle as a measure of adherence for photoresist, *Thin Solid Films*, Vol. 75, 1981, pp. 319–329.

135. R. F. Dayton, N. I. Dib, and L. P. B. Katchi, Design of micromachined high-frequency circuit components, *Proc. ISHM*, Vol. 18, No. 1, 1995, p. 19.

136. C. P. Wen, Coplanar wave guide: a surface strip transmission line suitable for nonreciprocal gyromagnetic device applications, *IEEE Trans. Microwave Theory Tech.*, Vol. 17, Dec. 1969.

137. E. A. Mariani, C. Heinzman, J. Agrios, and S. B. Cohn, Slot line characteristics, *IEEE Trans. Microwave Theory Tech.*, Vol. 17, Dec. 1969.

138. H. K. Smith and P. I. Mays, Log periodic array of dual-feed microstrip patch antennas, *IEEE Trans. Antennas Propag.*, Vol. 39, 1991, pp. 1659–1664.

139. S. G. Konsowski, Packaging of high-speed and microwave electronic systems, in *Electronic Packaging and Interconnection Handbook*, 3rd ed., C. A. Harper, ed., McGraw-Hill, New York, 2000, Chap. 12, p. 12.33.

140. Y. Choung and C. Chen, 44-GHz slot line phased-array antenna, *IEEE/APS Symp. Antennas Propag. Dig.*, Vol. 3, San Jose, CA, June 1989, pp. 1730–1733.

141. M. Povinelli and J. Johnson, Design and performance of wideband dual-polarized strip line notch arrays, *IEEE/APS Symp. Antennas Propag. Dig.*, Vol. 3, San Jose, CA, June 1988, pp. 30–203.

142. R. Ramaswamy and D. Schaubert, Characteristics impedance of a wide slot line on low-permittivity substrates, *IEEE Trans. Microwave Theory Tech.*, Aug. 1986, pp. 900–902.

143. K. Gupta, Expression for wavelength and impedance of slot line, *IEEE Trans. Microwave Theory Tech.*, Vol. 24, 1976, p. 532.

144. M. Sucher and J. Fox, eds., *Handbook of Microwave Measurements*, 3rd ed., Vol. II, Polytechnic Press, New York, 1963.

145. P. Dexter, S. Bogina, M. Hainswirth, P. Koerner, R. B. Beard, and K. Scoles, 8- and 10-GHz directional couplers using thick-film technology, *Proc. Int. Symp. Microelectron.*, 1987, p. 66.

146. A. E. Bailey, ed., *Microwave Measurement*, Peter Peregrinus, London, 1985.

147. R. E. Collin, *Foundations for Microwave Engineering*, McGraw-Hill, New York, 1966.

148. H. Sobol and M. Caulton, The technology of microwave integrated circuits, in *Advances in Microwaves*, Vol. 8, L. Young and H. Sobol, eds., Academic Press, San Diego, CA, 1974.

149. M. Caulton et al., Measurement of the properties of microwave transmission lines for microwave integrated circuits, *RCA Review*, Vol. 27, 1966, p. 377.

150. M. A. R. Gunston, *Microwave Transmission Line Impedance*, Van Nostrand Reinhold, New York, 1972.

151. S. B. Cohn, Microwave band pass filters containing high *Q* dielectric resonators, *IEEE Trans. Microwave Theory Tech.*, Vol. 16, Apr. 1968, pp. 218–227.

152. D. N. McQuiddy, Jr., J. W. Wassel, J. B. Lagrange, and W. R. Wisseman, Monolithic microwave integrated circuits: an historical perspective, *IEEE Trans. Microwave Theory Tech.*, Vol. 32, Sept. 1984, pp. 997–1008.
153. R. F. Hoffman, *Handbook of Microwave Integrated Circuits*, Artech House, Norwood, MA, 1987.
154. D. M. Pozar, Microwave integrated circuits, in *Microwave Engineering*, Addison-Wesley, Reading, MA, 1990, Chap. 11, (11.6).
155. A. Sweet, *MIC and MMIC Amplifier and Oscillator Design*, Artech House, Norwood, MA, 1990.
156. G. D. Vendelin, *Design of Amplifiers and Oscillators by S-Parameter Methods*, Wiley, New York, 1982.
157. G. L. Matthaei, L. Young, and E. M. T. Jones, *Microwave Filters, Impedance Matching Networks and Coupling Structures*, Artech House, Norwood, MA, 1980, Chap. 5.
158. T. Edwards, *Foundations for Microstrip Circuit Design*, Wiley, Chichester, West Sussex, England, 1992.
159. B. Bhat and S. K. Koul, *Strip Line–Like Transmission Lines for Microwave Integrated Circuits*, Wiley, New Delhi, India, 1989.
160. J. Sheinwald, MMIC compatible band pass filter design: a survey of applicable techniques, *Microwave J.*, Vol. 37, No. 3, Mar. 1994, p. 26.
161. News Update, From strategic analytics GaAs and high speed circuits service continuous information, *Adv. Semicond.*, III–V Review, Vol. 24, No. 27, Sept. 2001, p. 4.
162. III–V Review, *Adv. Semicond.*, Vol. 14, No. 8, Oct. 2001.
163. M/A Com Microelectronics Division Advertisement, *Microwave J.*, Vol. 35, No. 2, Dec. 1992.
164. General Microwave Advertisement, NY, *Microwave J.*, Vol. 36, No. 2, Dec. 1993.
165. D. K. Burton, The 1993 Moscow airshow, *Microwave J.*, Vol. 37, No. 5, 1994, p. 28.
166. S. J. Fiedziuszks and S. C. Holme, The dielectric resonator and its impact on miniaturization of microwave circuits, *Proc. Int. Symp. Microelectron.*, 1985, p. 146.
167. B. Bhat and S. Koul, *Strip Line Transmission Lines for Hybrid Microwave Integrated Circuits*, Wiley, New York, 1989; C. A. Harper, ed., *Electronic Packaging and Interconnection Handbook*, 3rd ed., McGraw-Hill, New York, 2000, p. 12.16.
168. D. Kajfez and P. Guillon, *Dielectric Resonators*, Artech House, Norwood, MA, 1986.
169. R. K. Mongia and P. Bhartia, Resonant frequency computations for ring resonators, *Microwave J.*, Nov. 1992.

GLOSSARY*

Abrasive trimming: trimming a film resistor to its nominal value by notching the resistor with a finely adjusted stream of an abrasive material such as aluminum oxide, applied directly against the resistor surface.

Active area (of a package): internal area of a package bottom, usually a cavity that is used for substrate attachment. Preferably, the term is applied to package cases of all-metal construction (as opposed to glass or ceramic).

Active components: electronic components such as transistors, diodes, electron tubes, and thyristers, which can operate on an applied electric signal to change its basic characteristics (i.e., rectification amplification, switching, etc.).

Active devices: discrete devices such as diodes, transistors, and integrated circuits used in analog or digital circuits in monolithic or hybrid form.

Active element: element of a circuit in which an electrical input signal is converted into an output signal by the nonlinear voltage/current relationships of a semiconductor device.

Active network: this is a question of semantics. From a purist standpoint, such an expression is a contradiction in itself. A network, by definition, is passive; by the addition of one or more active elements (devices), it becomes a circuit.

Active substrate: substrate in which active and passive elements may be laid down or attached to provide discrete or integrated devices.

* From G. S. Szekely, *Glossary of Hybrid Circuits*, ISHM, Reston, VA, updated edition—Summer 1987. Reproduced with permission from ISHM/IMAPS, Washington, DC.

Handbook of Thick- and Thin-Film Hybrid Microelectronics, By Tapan K. Gupta
ISBN 0-471-27229-9

Active trimming: trimming of a circuit element (usually, resistors) in a circuit that is electrically activated and operated to obtain a specified functional output for the circuit.

Additive patterning: processing a hybrid circuit substrate by sequentially depositing conductive, resistive, and insulate materials, each through a mask or patterned screen, thus defining the contours and areas of traces, pads, and elements.

Add-on components: discrete or integrated prepackaged or chip components that are attached to a film network to complete the circuit functions.

Add-on devices: *see* Add-on components.

Adhesion: property of a material to remain attached to another; measure of the bonding strength of the interface between a film deposit and the surface receiving the deposit may be another film or substrate.

Adhesive: substance used for firm attachment by exerting force of attraction between molecules of mating surfaces in an adhesive bond. In hybrid assembly, adhesive are basically organic polymers such as epoxies, together with their additives. When hardened after curing, adhesives will bond substrates to packages or dies and chip carriers to substrate lands, and so on.

All-metal package: hybrid circuit package made solely of metal, exclusive of glass or ceramic. Its main application is with microwave modules and large plug-ins.

Alloy: solid-state solution of two or more metals; to melt or make an alloy.

Alumina: aluminum oxide used as a substrate in hybrid microelectronic circuits.

Ambient temperature: temperature of atmosphere in intimate contact with an electrical part or device.

Amphoteric: literally, partly one and partly the other; specifically, capable of reacting as either an acid or a base. Amphoteric metallic coating materials may be applied to the back side of silicon very large scale integrated wafers to promote adhesion when die bonding in hybrid assembly.

Analog circuits: circuits that provide a continuous relationship between the input and output.

Angle bond: bond impression in which the first and second bonds are not in a straight line.

Angle of attack: angle between the squeegee face of a thick-film printer and the plane of the screen.

Angstrom: unit of measurement used especially in thin-film circuits; 10^{-10} meter.

Annealing: heating of a film followed by slow cooling to relieve stresses and stabilize the material.

Anodization: electrochemical oxidation process used to change the value of thin-film resistors or to prepare capacitor dielectrics.

Antistatic: electrostatic discharge protective material resisting turboelectric charging; usually some form of (impregnated) plastic, identified as antistatic, with surface resistivity controlled to be less than 1014 Ω/sq and more than 109 Ω/sq.

Application-specific integrated circuit: any integrated circuit that can easily be tailored to meet the unique requirements of a specific hybrid application.

Area array tab: tape automated bonding where edge pads on the inner surface area of a chip are addressed in the bonding scheme; practiced with extremely complex dies, very large scale integration, and so on; for use with integrated circuits where peripheral pad pitch cannot be reduced further and all inputs and outputs must be accommodated.

Array: group of elements or circuits arranged in rows and columns on one substrate.

Artwork: accurately scaled configuration or pattern, usually produced at an enlarged ratio, to enable the product to be made by photographic reduction of the working pattern; layouts and photographic film created to produce the working thick-film screens and thin-film masks.

As fired: values of thick-film materials on ceramic substrates are subjected to firing in a specific ambient and temperature before being subjected to the next operation.

Aspect ratio: ratio between the length of a resistor film and its width; equal to the number of squares.

Assembly: hybrid circuit which includes discrete or integrated components that has been attached; assembly of more than one such circuit on individual substrates; microelectronic device prior to packaging or encapsulation.

Assembly drawing: drawing showing a circuit with its components and interconnections mounted or soldered to a film network in their proper position. It may also show the assembly of more than one circuit of a given device prior to packaging/sealing.

Attach: inclusive term in hybrid assembly to denote permanent joining by means of an intermediate material(s).

Axial leads: leads coming from the ends of a discrete component or device along the central axis rather than from the sides.

Backfilling: gaseous filling of an evacuated hybrid circuit package with a dry inert gas prior to sealing.

Bake out: to subject an unsealed hybrid circuit package to an elevated temperature to bake out moisture and unwanted gases prior to final sealing.

Ball bond: bond formed when a ball-shaped end interconnecting wire is deformed by thermocompression against a metallic pad; also designated as a nailhead bond, from the appearance of the flattened ball.

Bathtub package: box-like package in which a substrate is mounted.

Beam lead: long structural member not supported everywhere along its length and subject to the forces of flexure, one end of which is permanently attached to a chip device and the other end intended to be bonded to another material, providing an electrical interconnection or mechanical support or both.

Beam-lead device: active or passive cheap component possessing beam leads as its primary interconnection and mechanical attachment means to a substrate.

Beam-lead tape: type of tape used in tape automated bonding in which the inner leads are formed as beam lids into an etched conductor pattern. The tape may be single-layer copper with a central row of simultaneously etched lead patterns or may include tapes of two, three, or more layers for specialized applications in high-frequency circuits.

Beam-lead tape bonding: mass gang bonding interconnection technique used for sequential automated tape bonding.

Beryllia: barium oxide; used as a substrate material in a hybrid circuit where high thermal conductivity is required.

Binders: additives to promote adhesion of thick-film inks (slurry).

Bleeding: during curing or firing, lateral spreading or diffusion of thick-film ink into adjacent areas.

Bond: permanent electrical and/or mechanical interconnection.

Brazing: joining of metals with nonferrous filler metal at temperatures above 448°C.

Breakdown voltage: voltage beyond which electrical current increases at an infinite rate.

Burn-in: electrical stress applied to a device for an adequate time to cause failure of proper functioning.

Camber: overall warpage present in a substrate.

Centerline average: arithmetical average of measured deviations in a surface profile from an imaginary mean central line located between the peaks and valleys.

Ceramic: inorganic nonmetallic (generally, oxides of metals) material used in hybrid microcircuits.

Cermet: intimate contact of an admixture of metal and ceramic.

Chemical vapor deposition: chemically reduced vapor-phase deposition of material under vacuum.

Chip: uncased, usually leadless form of electronic component part, either passive or active, discrete or integrated.

Chip carrier: special type of enclosure or package to house microcircuits and devices.

Coefficient of thermal expansion: ratio of the change in dimensions to the change in temperature.

Cofiring: processing thick-film conductors and resistors through the firing cycle at the same time.

Conductive epoxy: generally, a polymer resin with conductive filler used in hybrid microcircuits.

Conductivity: property of a material by virtue of which electric current can easily pass through; reciprocal of resistivity.

Contact angle: angle made between bonding material and bonding pad.

Contact resistance: apparent resistance that exists between two electrical contact points.

Crazing: fine cracks that may extend on or through layers of glass or plastic.

Crossover: crossing of metal paths without mutual electrical contacts.

Cure time: time for hardening of premixed polymers.

Dicing: total separation of dies from a wafer.

Die: unpacked discrete or integrated devices.

Dielectric: material that does not conduct electricity and that is used for making capacitors.

Dielectric constant: ability of a material to store electric charge when used as a capacitor.

Dielectric loss: power dissipated by a dielectric as the friction of its molecules opposes the molecular motion produced by an alternating electric field.

Dielectric strength: maximum electric field that a dielectric can withstand without physical breakdown.

Dissipation factor: ratio of the resistive component of a capacitor to the capacitive reactance of the capacitor.

Endothermic: reaction that occurs with the absorption of heat; opposite of **exothermic**.

Etchant: chemical mixture or solution to etch glass or any metal or polymer and to create fine structure.

Eutectic: mixture of two or more substances which has the lowest melting point.

Eutectic alloy: alloy that has the same temperature for melting and solidus.

Eutectic die attach: attachment of a die in microelectronic assembly with a eutectic alloy.

Exothermic: reaction that occurs with the evolution of heat; opposite of **endothermic**.

Failure mode: cause for rejection of a failed device as defined in terms of the specific electrical or physical requirement that it failed to meet.

Fatigue: failure of any structure caused by repeated application of stress over a period of time.

Fatigue factor: factor causing the failure of a device under repeated stress.

Ferrite: powdered, compressed, and sintered magnetic material having high resistivity.

Ferroelectric: crystalline dielectric that exhibits dielectric hysteresis.

Ferromagnetic: material that has a relative permeability noticeably exceeding unity; generally exhibits hysteresis.

Flip chip: leadless, monolithic structure containing circuit elements, designed to interconnect to the hybrid circuits.

Flip-chip mounting: method to mount a flip chip on a hybrid circuit without wire bonding.

Flux: composite rosin-based material and solvent generally used during soldering.

Frit: grounded glass composition used in thick-film compositions.

Fusing: melting and cooling two or more materials together so that they bond together in a homogeneous mass.

Glass transition temperature: temperature at which the thermal expansion coefficient of a material becomes constant.

Glassivation: covering the active part of a circuit with a thin layer of pyrolytic insulator.

Hermetic: gas-tight sealing of a package.

Hermeticity: ability of a packaging device to prevent any exchange between the inner and outer environments.

Hot spot: any spot in a circuit that cannot dissipate heat properly and operates at an elevated temperature.

Hybrid microcircuit: assembly of a circuit on a single ceramic substrate with dissimilar circuit components.

Hybrid module: special carrier of hybrid microcircuits and other components interconnected as a unit, as a component of an electronic subsystem.

Kerf: slit channel cut in a resistor during trimming by laser beam or by abrasive jet.

Kiln: high-temperature furnace used in firing ceramics.

Kovar: alloy of 17% cobalt, 53% iron, and 29% nickel, plus trace elements.

Lapping: smoothing the surface by rubbing against a flat surface with a liquid abrasive.

Leaching: removing portions of material by dissolving its soluble constituents into another material; in soldering, the dissolving of the part or surface to be soldered into the molten solder.

Life test: test of a component or circuit under load over the rated life of the device.

Loss tangent: decimal ratio of the irrecoverable part of the electrical energy introduced into an insulating material by the establishment of an electric field in the material.

Low-loss substrate: substrate that absorbs low energy in a microwave and high frequencies.

Master layout: original layout of a circuit.

Mean time between failures: measure applied statistically to reliability studies.

Mean time to failure: arithmetic average of the lengths of time-to-failure registered for parts or devices of the same type, operated as a group under identical conditions.

Microcircuit: small circuit (hybrid or monolithic) with a relatively high equivalent-circuit density which is considered as a single part on (hybrid) or with (monolithic) a single substrate to perform an electronic circuit function.

Micrometer: absolute unit of length equal to 10^{-6} meter.

Microstrip: microwave component, usually on a ceramic substrate.

Mil: unit equal to 0.001 in. or 0.0254 cm.

Multilayer circuit: composite circuit consisting of alternate layers of conductive circuitry and insulating materials bonded together with the conductive layers interconnected as required.

Multilayer substrate: substrate that has buried conductors so that complex circuitry can be handled. Fabricated either as a conventional multilayer circuit or as a cofired multilayer ceramic hybrid structure, in high- and low-temperature versions.

Multilevel processing: stepwise processing required to produce a multilevel substrate using thick-film technology.

Multiphase ceramic: ceramic substrate that contains more than one phase of that particular ceramic; for example, alumina has three phases: alpha, beta, and gamma; in an alumina substrate, more than one phase may be present.

Multiple-cavity package: flat package unit having multiple cavities, each one of which carries a hybrid substrate. Pins through the separator walls serve to interconnect the substrate circuits.

Negative temperature coefficient: when the electrical resistance of a material decreases with an increase in temperature.

Nugget: region of recrystallized material at a bond interface which usually accompanies the melting of material at the interface.

Ohmic contact: electrical contact between two materials which gives a linear voltage–current relationship when electric current is passed through a circuit containing these two materials.

Outgas: release of gas from a material over a period of time.

Overglaze: glass coating that is grown, deposited, or secured over another element, normally for physical or electrical protection purposes.

Oxidizing atmosphere: oxygen-containing atmosphere to oxidize any material during baking or firing.

Passive components: electrical circuit elements that do not change their characteristics when an electrical signal applied.

Passive substrate: substrate that serves as a physical support and thermal sink for a film circuit and does not exhibit resistance.

Passivation: formation of an insulating layer to protect circuit elements from oxidation or moisture.

Paste: synonymous with **composition** and **ink** when relating to screen-printable thick-film materials.

Pattern: outline of a collection of circuit conductors and resistors that defines the area to be covered by the material on a film network or circuit substrate.

Peak firing temperature: maximum temperature attained during the firing cycle of paste materials used in thick-film hybrid circuits.

Peel bond: separation of lead from a bonding surface; proceeds along the interface of the metallization and substrate insulation.

Phase diagram: state of a metal alloy over a wide temperature range; used to identify eutectic solders and their solidus/liquidus point.

Photo glass: flat glass plate used as a photo mask plate.

Photolithography: generation of fine structures, via a photo mask or rubylith, using light-sensitive polymer (photoresist) exposed to ultraviolet light. The exposed area is developed using developer.

Photoresist: photosensitive polymer that changes its chemical properties when exposed to ultraviolet light.

Pin grid array: package or interconnect scheme featuring a multiplicity of plug-in electrical terminals arranged in a matrix format or array; used with high-input/output-count devices.

Pinhole: hole that penetrates entirely through film and creates electrical and mechanical problems.

Pitch: peripheral center-to-center spacing as applied to pads, rows of bumps, pins, posts, exit leads, and so on.

Pit: depression produced in metal or ceramic surfaces by nonuniform deposition.

Polyimides: class of resin compounds containing an NH group which are derived from ammonia and are "imidized" from polyamic acid at temperatures high enough to initiate and complete imide ring closure.

Porosity: ratio of solid matter to voids in a material.

Positive temperature coefficient: changing a value in the positive direction with increasing temperature.

Postcuring: heat aging of a film circuit after firing to stabilize the resistor values through stress relieving.

Postfiring: refiring a film circuit after having gone through a firing cycle.

Poststress electrical: application of an electrical load to a film circuit to stress the resisters and to evaluate the resulting change in values.

Pot life: time period of reliably useful life for an organic compound, such as an epoxy or any adhesives.

Power density: amount of power (W/in^2) dissipated from a film resistor through a substrate.

Power dissipation: dispersion of the heat generated from a film circuit when a current flows through it.

Power factor: ratio of the actual power of an alternating or pulsating current as measured by an ammeter or voltmeter.

Purple plague: formation of a complex when two or more materials in a single compound are left in an atmosphere containing oxygen and moisture. Gold–aluminum compounds form purple plague when activated by reexposure to moisture and high temperature. Potentially, the compound becomes very brittle.

Pyrolyzed (burned): material that has gained its final form by the action of heat.

***Q*:** inverse ratio of the frequency band between half-power points (bandwidth) to the resonant frequency of the oscillating system.

Reflow soldering: soldering involving application of solder prior to actual joining.

Refractory metal: metal having high melting temperature.

Registration: alignment of a circuit pattern on a substrate.

Resistivity (ρ): proportionality factor characteristics of different substances equal to the resistance that a centimeter cube of the substance offers to the passage of electricity.

Resistor drift: change in resistor value due to aging; usually rated as percent change per 1000 hours.

Rheology: science dealing with deformation and flow of matter.

Risers: conductive paths that run vertically from one level of conductors to another in a multilayer substrate.

Scavenging: *see* **Leaching**.

Scribe: to score a substrate along a prescribed path as a preliminary to full separation.

Serpentine cut: special wiggly pattern cut during trimming.

Sheet resistance: electrical resistance of a thin sheet of pattern material with uniform thickness.

Shelf life: maximum time span, from the date of manufacturing, during which a material can retain its properties.

Sintering: heating a metal powder under pressure, causing the particles to bond together in a mass.

Skin effect: increase in resistance at microwave frequencies because of the tendency of current to concentrate at the conductor surface.

Slurry: thick paste made from a mixture of solids, binder, and solvents for making screen-printable material (ink) for the fabrication of thick-film hybrid circuits.

Smeared bond: bond impression that has been distorted by lateral movement.

Snapback: return of a screen to normal after being deflected by a squeegee moving across the screen.

Snap-off distance: screen printer distance setting between the bottom of the screen and the top of the substrate.

Soak time: length of time a ceramic material with film is held at the peak temperature of the firing cycle.

Soft solder: low-melting solder made out of an alloy of tin and lead, which melts below 425°C.

Solderability: ability of a conductor to be wetted by solder and to form a strong bond with the solder.

Solidus: locus of points in a phase diagram representing the temperature, under equilibrium conditions, at which each composition in a system begins to melt during heating or to freeze completely during cooling.

Solubility: ability of a substance to dissolve into a solvent.

Solute: substance dissolved in a solvent.

Specific heat: quantity of heat required to raise the temperature of 1 g of a substance 1°C.

Steatite: ceramic consisting primarily of a silicate of magnesium used as an insulator or circuit substrate.

Stratification: separation of nonvolatile components of a thick film into horizontal layers during firing, due to the large differences in density of the components.

Stray capacitance: unwanted capacitance developed from adjacent conductors separated by an air dielectric or any other material that might act as a dielectric.

Stripline: microwave conductor on a substrate.

Stylus: sharp-pointed probe used in making contact on the pad of a leadless device or film circuit.

Surface finish: peaks and valleys in the surface of a substrate, rated as μin./in. deviation.

Surface-mounting technology: method of assembling hybrid circuits and printed wiring boards, where component parts are mounted onto, rather than into, the printed wiring board.

Tinning: coating a surface with solder (generally, an alloy of tin and lead) before final bonding with another surface.

Tracking: degree to which two similar elements on the same circuit change values with temperature in relation to each other.

Transducer: device actuated by one transmission system, supplying related energy to another transmission system.

Vehicle: organic system (material) used in making thick-film paste.

Very large scale integration: monolithic circuit using transistors, diodes, and passive circuit elements (the total number of devices may be 1 million/in^2 area).

Via: opening in a dielectric layer(s) through which a raiser passes or one whose walls are made conductive.

Void: unintentional gap left behind by a continuous film.

Voltage gradient: voltage drop (or change) per unit length along a resistor or other conductance path.

Volume resistivity: measurement of the electrical resistance of a cube material.

Warpage: distortion of a substrate from a flat plane.

Waviness: series of elevations or depressions or both (i.e., wavelike surface when measured with a profilometer and plotted on paper) which are noticeable on the surface of a ceramic substrate.

YAG (yttrium–aluminum–garnet): single crystal grown from these elements.

CONVERSIONS

1 mil = 10^{-3} in.	1 mm torr = 10^{-3} torr (essentially identical to mmHg)
1 in. ≈ 2.54 cm ≈ 25.4 mm	1 torr = 133.322 Pa = 0.00131 atm = 0.01933 psi
1 mm = 10^{-3} m	1 Pa = 1 N/m^2 = 0.000001 MPa
1 cm = 10^{-2} m	1 Å = 10^{-10} m
1 mm = 10^3 μm	1 μm = 10^{-6} m = 10^{-3} mm = 10^4 Å
1 mm = 40 mils	1 cm = 10^4 μm
1 mil ≈ 25.4 μm ≈ 0.025 mm	1 cm ≈ 4.0 mils
1 μin. ≈ 254 Å	39 μin. ≈ 1 μm
1 mil^2 ≈ 645.16 μm^2	1 in^2 ≈ 645.16 mm^2
1 lb = 0.4535 kg = 16 oz	
1 mile = 1.61 km	1 km = 0.62 mile

1 cal s^{-1} cm^{-1} (°k)$^{-1}$ = 418.4 W (m)$^{-1}$ (°k)$^{-1}$

$T_f = (T_c \times 1.8) + 32$, $T_c = (T_f - 32)/1.8 = T_k - 273.15$, and $T_k = [(T_f - 32)/1.8] + 273.15$

(T_f, T_c, and T_k are temperature in degrees Fahrenheit, degrees Celsius, and kelvin)

INDEX

Printed and bound by CPI Group (UK) Ltd, Croydon, CR0 4YY
18/09/2026
14965014-0001